CW00937818

A World of Gardens

For EMILY

Published by Reaktion Books Ltd
33 Great Sutton Street
London EC1V 0DX, UK
www.reaktionbooks.co.uk

First published 2012

Printed and bound in China by Toppan Printing Co. Ltd.

British Library Cataloguing in Publication Data
Hunt, John Dixon.
A world of gardens.
1. Gardens – History.
2. Gardens – Design – History.
3. Gardens – Styles.
4. Gardens – Social aspects.
I. Title
712′.09-dc22

ISBN 978 1 86189 880 7

A World of Gardens

John Dixon Hunt

REAKTION BOOKS

Contents

Introduction: *The Garden World and the World of Gardens*

'Gardening, after all, is one index of the history of men.'
GEOFFREY GRIGSON

Gardens of all sorts come in all sizes and guises. And our interest in them also takes many approaches. We study the process of their design, their built forms, their materials and plantings, their meanings, their use or how they are experienced on the ground and represented in word and image, their decay and maybe their recuperation. We are interested in who their designers were, who commissioned them, and the motives of both designers and patrons, along with the political and social contexts in which gardens came into being. But sometimes we also construct our own memories of these places.

A history of gardens is best undertaken as a cultural history, even if its primary focus is design, botany, hydraulics or sculpture. People engage in place-making because our choice of habitation is of supreme importance, as we find our identity and a sense of belonging in the process of colonizing and cultivation, which the word 'culture' (derived from Latin *colere*) implies. It is not enough to look at gardens for their style (endlessly and emptily touted as 'formal' or 'informal', 'baroque', 'picturesque', 'arts and crafts'), nor even enough to assess their visual appearance. We need to ask why they came into being, what advantages and pleasures (including the visual, to be sure) accrue from them, and how and why they have survived, changed or vanished.

The range of places that can be envisaged within the category of 'garden' is also enormous and various, and it changes from locality to locality, and from age to age. Yet this diversity does not wholly inhibit us from knowing what it is that we want to discuss when we think of gardens. Above all, it is useful to think of the garden as typically a place of paradox, being the work of men and women yet created from elements of nature, the two held in some precious and often precarious tension. And while a garden is often acknowledged to be a 'total environment', a place that may be physically separated from other zones, it also answers and displays connections with larger environments and concerns, not least agriculture and cities. Gardens, in short, are both entities within themselves and a focus of human speculations, propositions and negotiations, concerning what it is to live in the world.

Therefore this book will not be a conventional history, following the garden from its earliest to the latest manifestation in a series of waves, each gathering strength, then cresting and falling backwards, sometimes passing their energies from one to another; nor will it attempt to allude to every significant designer or every known garden. It seemed less interesting to chronicle design innovations, which are always premised on the elaboration or even rejection of previous models; but as Joseph Brodsky noted about Ezra Pound's famous declaration 'Make It New', 'the true reason for making it new was that "it" was fairly old'. Rather, this book focuses upon men's and women's continuing responses to rural and urban forms and their re-workings of both the natural world and the spaces of human habitation. Even a modest survey of gardens 'from China to Peru' will notice the recurrence of types and uses of gardens in many different times and places: sacred landscapes, landscapes of play and amusement, scientific gardens, gardens for urban life, gardens for private seclusion and meditation, prosaic and practical sites, poetical and symbolic places, creations that celebrate locality or establish nationhood. Certainly cultural assumptions and local geography have always shaped such sites and their meanings, and those particular forms and local associations will be explored. Yet equally instructive is the fashion in which gardens are hospitable to a cluster of archetypal human needs and behaviour. So, while the structure of this sequence of essays does move chronologically from early examples to the most recent (and even future) ones, it also chooses to explore a more lasting, because synchronic, aspect of many types and sites. Since narrative process per se is not its ambition, an agenda of alternative issues will be canvassed: the recurrence of garden themes, *topoi* or commonplaces; the cross-cultural reception and exchange of forms and usage; the planning and development of gardens created for generic human activities; and the fashion in which gardens have shaped as well as given expression to basic human conditions and concerns.

While the chapters are offered largely in chronological order, readers may readily explore them in a different order, which is why occasionally quotations or sites recur in different discussions.

Each essay aims to be of similar length, despite the temptation to vary them: a kind of Procrustean measure has been applied, as it has been to the illustrations. These are also limited in number, though the images have been made as varied as possible in order to provide the fullest range of perspectives on garden art – a mixture of plans, engravings, contemporary imagery and modern photographs. There will be a minimum of notes (or references), but there must be opportunities to suggest further readings as well as my own need to record debts to many other garden writings; since the study of gardens has exponentially increased in the last decade, readers must be made aware of these recent resources as well as of established authorities.

1 The fighting giants in the Parco di Mostro at Bomarzo, near Viterbo, Italy.

1 Sacred Landscapes from Delphi to Yosemite

Throughout the world of gardens there are occasions when something particularly special about a place focuses and shapes both its design and its reception. This may be what was once called the 'sacred', as many early cultures declared. But in a modern world, grown secular and global, this often seems an inappropriate term. Nonetheless, there are places both old and new that call out for some recognition of this special site and its meanings. It may be that a particular place is mysterious, uncanny, even unsettling, because it is somehow removed from the ordinary world from which one has entered. We might think of Bomarzo, the wooded valley near Viterbo, where the items carved in the living rock – giants, mythical beings, strange inscriptional injunctions, leaning houses – still give one a strange feeling of disquieting puzzlement (illus. 1 and 10). Another place would be the Désert de Retz, created towards the end of the eighteenth century and later celebrated by surrealists such as André Breton and Jacques Prévert, who, like Salvator Dalí at Bomarzo, re-discovered and were thrilled by its mysterious enclave (illus. 2 and 167). Even modern place-makers have found themselves drawn towards this special resonance in a place: Lawrence Halprin can identify one of the outdoor rooms in his plans for the Franklin Delano Roosevelt Memorial on the Mall in Washington as 'sacred space' (illus. 3). And another garden-maker, Ian Hamilton Finlay, disturbed by what he saw as a wholly 'secular' world, sought what he called 'piety' in nature, the recognition of something that has its immaterial existence beyond the phenomena of the quotidian world: so in his garden at Little Sparta a sculpture of Pan-pipes is inscribed with the injunction 'When the Winds Blow Venerate the Sound' (illus. 4), and the ancient god lends his name, from which we derive the term for 'panic', to a moment of sublimity.[1]

This range of place-making from ancient China and Greece to Halprin and Finlay also extends to the reception of places found or discovered in the natural world for which, at least in the first place, no interventions were made. Mountains and wilderness have called out for the same response in sublime places like Yosemite or the Himalayas. But beyond these special places, where we also need to recognize sublimity's relationship to something that might be called sacred even in this apparently secular world where many profess no specific faith, there are feelings that demand some acknowledgement of what Yves Bonnefoy calls 'hauts lieux'. These are not simply, if at all, what the French term a 'high' place; but sites of significant moment, where one can better attain here rather

2. The Ruined Column and a fragment of the theatre at the Désert de Retz, Chambourcy, Yvelines, France.

than elsewhere a connection with one's own self ('rapport à soi qu'on recherche') These, then, are some of the themes to be explored at the start of this book, themes that in fact play an important role in many garden worlds.

The earliest human dedication to place-making was done to recognize something particularly sacred, mysterious or numinous that was discovered on a site or attributed to it. William Blake wrote about this in *The Marriage of Heaven and Hell* (*c.* 1793), where he also sounded what he saw as its threat from organized religion and rationalism:

> The ancient Poets animated all sensible objects with Gods or Geniuses, calling them by the names and adorning them with the properties of woods, rivers, mountains, lakes, cities, nations, and whatever their enlarged & numerous senses could perceive . . . Till a system was formed, which some took advantage of, & enslav'd the vulgar by attempting to realize or abstract the mental deities from their objects: thus began Priesthood . . .

When a place was thus identified or deemed to have such significance, some effort was made to draw attention to it, to mark its 'properties' of woods, rivers or springs in some way. Leon Battista Alberti wrote that the ancients would add 'dignity to places and groves', enhancing a specific site with often huge reworkings of the land, drawing 'some admirable and unusual property' from its local conditions.[2] Today we no longer share this notion of natural places instinct with 'unusual properties', for the zones now appropriated for the sacred are generally confined to specific religious institutions, or to burial sites like cemeteries. Nevertheless some people do subscribe to this appeal to the sacredness of place – Native Americans, for instance, are particularly alert to their need to respect landscapes which their people hold sacred. Yet while the sacred – as both a term and a specific feeling for place – seems frequently inappropriate for modern cultures, it is equally undeniable that people can feel that a place holds significance for them.

Early peoples found this consecration of place particularly important. It usually depended upon some physical reworking of a site. But it is worth recalling how Australian aborigines, as recounted

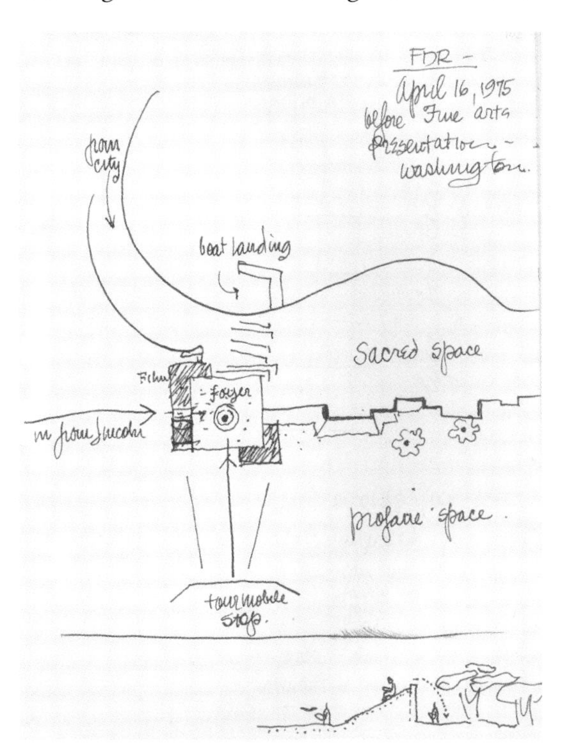

3 Lawrence Halprin, annotating a 'sacred space' in a notebook for his scheme for the FDR Memorial, Washington, DC.

in Bruce Chatwin's *The Songlines* (1986), navigated the landscape by following a trail of words and musical notes left by some totemic ancestor and used these 'Dreaming-tracks' as 'ways' of communicating with each other. The Bible also tells of visions in which ladders ascended to heaven from a particular place, or spots so holy that voices instructed shoes to be removed (Genesis 28 and Exodus 3). When such places needed signalling to a wider audience, to those to whom no voice spoke nor ladders appeared in their dreams, some physical mark or re-ordering took place. Whoever established these sites – the sacred groves of ancient Greece, for instance – were perhaps priests or devout citizens, but they were practicing an early form of landscape architecture, 'gardens' to the extent that something was contrived and concentrated in them.

Early Chinese parklands were privileged enclaves for their rulers. All had specifically religious justifications and were designed to celebrate 'immortals', to commemorate illustrious mortals and, more largely, to represent famous geographical and mythical locations.[3] Throughout these extensive and elaborate landscapes there were many acknowledgements of sacred space. Burial sites were particularly regarded as sacred zones (and not just in China). Some were vast, accessible only by certain persons, the result of huge artificial constructions – walled enclaves, tumuli and artificial hills – and could be peopled, as at the tomb (illus. 5) of the First Supreme Emperor of Qin, with thousands of terracotta figures, recently unearthed; later emperors used actual hills for the topography in which to set their tombs and tumuli.[4]

In 138 BC Emperor Wu expanded his 'Supreme Forest' to include both untouched wilderness, innumerable buildings, 'divine ponds and numinous pools', exotic animals and plants; this conspectus

4 Ian Hamilton Finlay, 'When the winds blow', inscription on pan-pipes in his garden at Little Sparta in the Pentland Hills of Scotland.

made his realm a replica of the whole universe. Though not without immediate political ambitions – Wu was not above using his magical lake as a training site for his navy, such a site represented the magical belief that 'by artificially making a replica of something one wields power over the original object or place'. Wu also constructed a garden for his Jianzhang palace, designed as a landscape in which three artificial islands represented the homes of the immortals, whom Wu thought dwelt beyond the sea and might now visit him at his palace; a soaring platform was prepared as the meeting place to which the immortals could now be lured. Other emperors followed with gardens at their palaces, some with lakes, platforms and above all mountains – the last being special possessions of rulers, with designations like 'spiritual' or 'holy' mountain, and one with a 'Hostel of the Immortals' on its summit. Two such sites survive with later additions and changes as public parks: one the lake of Beihai belonging to the imperial palace in Beijing, the other at the Summer Palace to its west.

These very early Chinese enclaves were primarily dedicated to sacred observances and rituals; though they were planted with trees and shrubs, graced with birds and other animals, it was only later that any specifically aesthetic response dominated their building or reception. Even then, when

5 Terracotta warriors in the mausoleum of the 'First Emperor' Qin Shih Huang near Xian, Shaanxi province, China (211–206 BC).

aesthetic appreciation came to play a significant part, the idea of such a garden would be less sacred than somehow special, significant or beyond the commonplace, something that 'answered' to and educated the mind's cultivation. But for that kind of significant, rather than sacred, landscape we would need to wait for much later Chinese creations of scholarly and aesthetic designs (see chapter Ten).

The sacred groves of ancient Greece, to which Blake refers, would have been much smaller and far less grandiose than early Chinese parklands, but may still be considered early examples of landscape architecture, locations that were amplified and augmented to give effect to local perceptions of significance (illus. 6). Among these 'improvements', according to the later *Guide to Greece* by Pausanias, who recorded many of these sites in the second century AD,[5] were walls, pathways, altars, statues of deities or the muses, and different groves of trees, about which Pausanias was particularly curious; some groves were of cypress, fruit trees, oak (*quercus robur*) or pine; some groves had just one species, others a mixture, while others contained just a single, unusual tree. There was even a stadium in the grove of Asklepios at Epidauros, presumably for religious games, although there is some testimony that less sacred recreations could be allowed. There were laws on persons who could be admitted and those who were banned: only women were permitted in the grove of Demeter in Arkadia, but excluded during annual festivals of Ares at Geronthrai. Springs and other physical formations like rocks or caves could also be identified and elaborated for worship or ritual, and some were included in specific sanctuaries. The diversity and profusion of the sites that Pausanias records is a clear testimony to the deliberate formulation of landscapes for special purposes.

All of these sites received some form of organization: some deliberate interventions designed to augment the particular location. One grove, high up on Mount Helikon, required some grading of different levels for its rituals. Another particular site, established in the sixth century BC and gradually developed into an elaborate landscape, survives to this day: Delphi. Its location and its formal landscaping were dedicated not just to establishing a site for Apollo himself, but also to celebrate the conquest of the old earth powers by the young Olympian gods.[6] Even today, this extraordinary site speaks of its powerful encounter

6 Two coins stamped in Greek: a grove with a temple and colonnade (AD 217–18, from Palestine); a grove with *períbolos* and entrance gate (AD 244–9, from Zeugma).

7 The 4th-century BC shrine of Apollo on the mountainside at Delphi, Greece, seen from the theatre dedicated to Dionysus above the shrine.

of the great sun god with the irrational, chthonic forces of nature (illus. 7)

Over several centuries Delphi established many types of sanctuary, temples, treasuries, gymnasia, a running track and pools, a theatre, fountains, altars and bases for sacrifice, and houses for welcoming guests from all over the Mediterranean world as well as from the city of Delphi itself. Yet despite its varied and monumental structures, the site seems to spring from the very mountain, called Parnassus, in the shelter of which it sits. Towering above the shrine of Apollo are the twin cliffs of the Phaedriades, the Bright Ones, and from their monumental rock face seem to come the very building blocks of the landscape (though other building materials here were transported from elsewhere to celebrate the shrine). Today, when only remains of these various edifices still stand, it is above all the approach to Delphi and the calculated ascent of the hillside leading to Apollo's shrine that impress.

There are two such approaches: from the Gulf of Corinth, whence Apollo's site is dimly registered across the plain; or the winding mountain pass that approaches from the east, with the shrine unseen, until finally the pilgrim, reaching the sanctuary of Athena, discovers Apollo's shrine high on the hillside to the right. If today it is under the auspices of the modern Greek Ministry of Culture that we enter and visit Delphi, as one begins the actual ascent this modern circumstance is slowly forgotten as one surrenders to the landscape's fragments and the awesome cliffs ahead (best to see the site early, though, before the coaches arrive). Originally the wild natural scenery would have been cluttered with the constructions of 'jostling treasuries and memorials'; today, by contrast, it is the landscape itself that holds sway, the very ruins themselves, of course, speaking to the destruction of man-made buildings by the forces of nature. But as one ascends, the peaks of the Phaedriades reappear, then the Apollonian shrine appears, but then disappears and finally shows itself, its columns and fragments set against the broken and forbidding crags, and we can perhaps share in the original reception of the site.

As you climb, now above the temple, you see its surviving columns small against the huge valley below and you arrive at the theatre dedicated to Dionysus. This wild and drunken deity who roamed the mountains behind shared his shrine with Apollo from the late archaic period: again, we have the merging of the wild and the irrational with the organizing and bright sun god. From the theatre, tucked into the amphitheatre of the hillside, we see the whole landscape perform itself literally, its drama palpable, the site now 'an entire landscape constructed around itself'. In this conspectus are embraced the built artefacts, the physical elements (the Corycian cave, or the Castalian spring where pilgrims would wash themselves after the long journey from the east), and the wider and even more numinous landscape that surrounds it. It is perhaps no surprise that some early Christian monuments found the site so impressive that they too established their own shrine when the Apollonian oracle was abolished.

If we jump from the mountains of central Greece to the valley and cliffs of Yosemite in North America we enter, clearly, a very different world. The Native Americans had none the less observed a similar instinct as explained by Blake, animating and naming every cliff, rock and spring of this valley. Their marks, however, were invisible upon the land when the first white adventurers discovered the valley, their toponyms ignored or ill-comprehended; many of the Americans who first 'discovered' Yosemite thought it was 'unstoried, artless, unenchanced' (to quote Robert Frost's poem 'The Gift Outright'). Faced with this apparent emptiness, the explorers struggled to find apt descriptions for this imposing and sublime landscape, and then, physically, to make it amenable to visitors.

Among the many responses to Yosemite were conventional or even familiar terms like 'garden' and 'park'. It was also a 'cathedral', a 'sculpture gallery', 'a horse trough in granite', 'in one word – a chasm'; but none of these lent themselves sufficiently to grasping, reformulating or transforming the Yosemite, as did the notion of a garden or park. Gardens may even have held intimations of the lost Garden of Eden, now perhaps recovered in distant California; John Muir saw the 'level bottom [of the valley] to be dressed like a garden'. Other Americans, too, has sensed that this and other American wildernesses could be 'great and almost boundless garden spots of earth', a 'pristine beauty and wilderness . . . a magnificent park' (George Catlin). Emerson also agreed that 'interminable forests should become graceful parks, for use and delight'. And in the form of a garden or park was what Yosemite was made to be.

Stanford E. Demars's *The Tourist in Yosemite* felt that it was reminiscent of the 'beautiful parks of Europe, especially those of England and France' and that it could soon be improved by cutting 'away more of the trees and shrub', as he insistently argued, to give 'the Valley a more park-like appearance – it could be made a beautiful park', giving the 'impression of a park as much as possible'. And so inevitably it acquired the familiar interventions of other public parklands, though in America these were themselves very recent creations: paths, roads, ladders to reach waterfalls ('practical footways for ladies'), trails to higher locations and eventually climbing routes, with their climbs named after famous pioneers. A circuit road would allow, argues Hans Huth, 'all the features of Yosemite [to be] enjoyed from a carriage'. There were hotels and hostels, campsites and spectacular events at the famous Camp Curry that featured the evening 'rite' of the 'Fire Fall', when a cascade of flames was precipitated down into the valley from the cliffs at Glacier Point, an event that begun as early at 1870. The focus of the Valley relied much on these various insertions, on the associations and aspects by which its features could be apprehended and understood, on how its special associations could be imaged and portrayed.[7]

To what extent it was a sacred place ('a cathedral', after all, where you read the 'Scripture of Nature') or simply something special or significant (like a gigantic 'horse trough') depended upon individual responses, and how the park itself was utilized and how visitors responded to it. For many, the park could be the occasion for excursions far, wide and often lonely: the climbers on the cliff in modern times manage better than most visitors to achieve a solitary communication with the mountains, as do those who penetrate further into the surrounding hills. Yet even for those who simply gazed at the mountains from the valley floor, the mountains themselves were the key to its sublimity, and sublimity was akin to the sacred in both the broadest and the more theological sense. Albert Bierstadt's 1864 painting of the valley lake (illus. 8) captured much of this appeal: the soaring cliffs, the famous waterfall, the placid and undisturbed lake, the deer, and a telltale fallen tree trunk à la Salvator Rosa. But its geology, too, with the ancient carving of its precipices, and the valley floor scoured by glaciers and levelled with their accumulated deposits, and rich now from the annual burning of the undergrowth by

8 Albert Bierstadt, *Yosemite Valley*, 1864, oil on canvas.

9 Mirror Lake, Yosemite National Park, skating scene, c. 1910s.

the Ahwahneechee people, spoke of the sublime processes of a long and awesome history. Even the uncertain early explanations of how to explain its geology worked to heighten the mystery.

On the other hand, even within the mountain valley, the commercial was both unavoidable and even by many desired: though John Muir thought 'Nothing dollarable is safe, however guarded', the entrepreneurial spirit was fully exploited. In that way, then, the Valley became not much different from Central Park, a central parkland within the western mountains. There were entertainments, skating parties on the frozen lakes in the winter months (illus. 9), guided visits to the famous showpieces like the Falls, accommodations, both grand and modest, and the opportunity to take refreshments. And if Yosemite was indeed a huge sculpture gallery, this central parkland also provided a museum for its collection of natural sculptures. Posters for the Southern Pacific Railroad championed indigenous dwellings like the cedar lodges, the majestic and almost eagle-like profile of Half Dome, downhill skiing, motor-car excursions to the Mariposa Big Tree grove, and shots of tourists taken at 'Inspiration Point' (nowadays '*Old*' Inspiration Point).

Whatever entertainments were provided within the Valley, especially for those who did no more than follow the circuits around the valley floor, there was however one truly sublime moment: the approach, the first apprehension of this place, as travellers crossed its threshold for the first time. The local native Americans called the Yosemite Valley 'Ah-wah-nee', the place of a 'gaping mouth' (illus. 10) – an opening into the fastness, but also perhaps apt for the open mouth of astonishment that greets visitors. Earlier writers thought the 'entry' of Yosemite was 'guarded' by 'mighty sentinels' to 'watch and ward over the secrets of the gorge' (another invocation of Eden, forever guarded by the angel with a fiery sword). Today's visitors can approach via a 'suburb' that is called 'El Portal', after which they enter the valley through a tunnel in the rock. From the earliest days of sacred spaces, sanctuaries were set apart, walled, even guarded, and the admission of a general public carefully monitored or even controlled; such places were also designed for the elite, for the special members of a given society who would appreciate and value the 'immortal' visitors or who could recognize the numinous. But Yosemite was increasingly, as a public park, opened to everyone who could reach it; it has been more and more overwhelmed with admissions, though moves have been sought to limit access and permits for overnight wilderness camping are now required. So, however crowded and 'trippery' the valley became, even at times very sordid with its plethora of establishments and camps, there was the awesome moment when the valley was first seen. Every visitor would see it differently, to be sure, depending on a prior familiarity with the place itself, with similar adventures elsewhere, or some other confrontation with places that made this 'unique'.

There are many such moments from the extensive literature on Yosemite. In 1864, accompanied by Frederick Law Olmsted, William Henry Brewer rode the 30 miles, camping overnight, from Mariposa to the valley entrance. His record reads:

> Trail as usual through forests of fir and p. contorta in the wet places. At last to Inspiration Point, a name not at all pedantic. View by far the finest of all about the valley. Day very clear and all spread about – the grand Cathedral, the grander Tutucanula above the emerald bottom. Beyond the rugged peaks and barren slopes towards Mono. Then descended and the rather tedious ride up the valley to Olmsted's camp just opposite the Fall . . .'

The tone, as befits a diary, is dry; but the moment of arriving at the threshold ('not at all pedantic') is palpable, and the following descent after the initial confrontation even 'tedious'. His onward travels are equally dry, but Olmsted himself records how the valley was 'sublimely beautiful and more beautiful than I had supposed', the valley 'sweet and peaceful as the meadows of [Shakespeare's] Avon'. Other initial impressions also characterized the quality and meaning of the place: in *My First Summer in the Sierra* of 1869, though published in 1911, Muir writes of making 'haste to high ground, and from the top of the ridge on the west side of Indian Canyon [I] gained the noblest view of the summit peaks', and later, at 'the brow of that massive cliff that stands between Indian Canyon and Yosemite Falls . . . the far-famed valley came suddenly into view throughout almost its whole extent'.

Yosemite, garden or park, is in a quite literal sense what the French call a high place ('un haut lieu'). As Bonnefoy explains in his *Entretiens sur la poésie*, this designation or 'practice' of place, namely identifying and knowing placefulness, is an old and essential obligation practised by many, whether on the 'threshold' of their hermitages, 'in a forest seclusion, a valley between rocky slopes, [among] the fragments of the Edenic Garden'. Place is the mirror where human 'truth' can be glimpsed. But, says Bonnefoy, it is constantly oppressed by a 'destructive tourism' that takes 'yesterday's virgin forest, peaks, the immensity of deserts, and in the name of national parks, zoological reservations, extends everywhere the unrealities of Buttes-Chaumont, no, even of Disneyland, repainting the sand, yellow, and the sky, blue'. Yet the traces of sacred, precious places, where a territory has been

10 Yosemite Valley today.

explicitly consecrated, can provoke once more the chance of 'capturing the absolute', what Bonnefoy calls (accepting the word itself) 'religious'. He does not make for easy reading or any easy identification of the problem. But he holds out the chance of something special, something even sacred, in our apprehension or dreaming of place.

Many today are dubious alike of the sacred as of any special sense of place, what has been called its *genius loci*. The two, though not identical, are intimately connected. The modern philosopher and landscape critic André Roger may well insist that 'En lui-même, le génie du lieu n'existe pas' (in itself, spirit of place doesn't exist), but he sneaks it back in as a cultural construction, jettisoning any supernaturalism and in the process making fun of a writer like Maurice Barrès, who in his *La Colline Inspirée* of 1912 wrote of 'des lieux où souffle l'esprit . . . qui tirent l'âme de sa léthargie, des lieux enveloppés, baignés de mystère, élus de toute éternité pour être le siège de l'émotion religieuse' (places where the spirit breathes . . . that pull the soul from its lethargy, places wrapped and bathed in mystery, elected for ever to be the site of religious emotion). Anyone who recalls E. M. Forster's short story 'The Road from Colonus' will surely understand what Barrès, for all his incantatory prose, was getting at: that places do reach out and seize one with an emotion that is – Blake would happily acknowledge – spiritual without being religious. And Hegel, too, invoked this same understanding of noumena: 'ancient Greek . . . demanded the meaning of springs, mountains, forests, storms; without knowing what all these objects said to him one by one, he perceived in the order of the vegetable world and of the cosmos an immense frisson of meaning, to which he gave the name of a god, Pan'. Some faint intimation of this understanding undoubtedly explains why in some eighteenth-century gardens, like Rousham, a statue of Pan could still haunt the shrubbery.

A no less theological argument than Bonnefoy's or Barrès' sustains Philip Sheldrake's more deliberately Christianizing discussion of *Spaces for the Sacred: Place, Memory, and Identity* (2001). His plea is self-confessedly theological, yet much of his often telling argument does not require one to buy into his ultimate beliefs. For him, place is 'space that has the capacity to be remembered and to evoke what is most precious'; it is the result of relationships between actions, conceptions and physical experiences; since we no longer dwell in 'pure nature', we must exist in what the anthropologist Clifford Geertz called realms of 'mediated meaning', where we need to understand what the meanings are. Narrative then becomes essential, for we account for ourselves by explaining our historical consciousness and its role vis-à-vis place. To illustrate this, Sheldrake appeals to the thirteenth-century philosopher and theologian Duns Scotus and to his articulation of *haecceitas, quiddity* or 'this-ness' of the world; it is an appealing doctrine, taken up also by Stephen Dedalus in *The Portrait of the Artist as a Young Man.* As Sheldrake's examples for the powerful particularity and focus of place, he takes two poems by Gerard Manley Hopkins, 'As Kingfishers Catch Fire' and 'Duns Scotus' Oxford'. The religious impulse of both Scotus and the Jesuit Hopkins is plain, but not (for this reader) their point of rest or destination. Things, says Hopkins, have their very own peculiar quality ('each mortal thing does one thing and the same'); Oxford, too, is charged with the innate fullness of its being, regardless of its 'base and brackish' suburban contexts. Scotus, then, for Hopkins becomes the 'rarest-veined unraveller' of things and places.

The 'sacred', then, occupies a considerable space in contemporary commentary, not least perhaps because it seems so elusive or even missed in the world. Nor does much of it, certainly, focus upon landscape architecture, but this theme is never far away, whether in discussions of the Vietnam Memorial in a volume entitled *The Sacred Theory of the Earth*, a deliberate echo of Thomas Burnett's work of 1681, or in discussions of the 'sacred and the profane' in cultural landscape photography in twentieth-century America. Yet there is also, unsurprisingly, a strong current of scepticism, which David Robertson's book on Yosemite, *Real Matter* (1997) samples, even if he does not totally accept it. Robertson often journeys to Yosemite, reads about its earlier visitors, takes his students there and has them chronicle their associations, notably to answer the question 'How do I make nature happen to me?' He wants to discover the meaning that is found in those mountains, and he takes his title from a remark by Gary Snyder repeated in Jack Kerouac:

> The closer you get to real matter, rock air fire
> wood, boy, the more spiritual the world is.

Yet that 'closeness' seems like Tennyson's 'Ulysses',

> all experience is an arch where through
> Gleams that untravelled world, whose
> margin fades
> For ever and for ever when I move.

This frustration is echoed by earlier visitors to Yosemite, like Fitz Hugh Ludlow ('Climb forever, and there is still an Inaccessible'). Ludlow meant an actual physical impasse, but as park-like interventions persisted (paths, ladders) these could be overcome. Yet what many understand as the 'specific powers in the landscape' still seemed to elude them. Robertson sets out to explore how that elusiveness might be formulated and even pinned down, what John Muir called 'the heart of the world'. Robertson is particularly fascinated by the ways that, for a decade or so in the 1980s, the National Park Service chose to signal visitors' approach to the 'inaccessible': signs alerted them to 'Listen to water sounds', posting quotations from Muir's pronouncements ('a step closer to wilderness'), as well as instructions on danger, or on 'staying alive'. But Robertson falls back, like the theological Sheldrake, by relying on the Book of Hosea and its injunctions from Yahweh to 'woo' the lover and 'escort her into the wilderness'.

I am myself less mystical about this, more attuned to the 'real' and to the dialogue between the landscape and our human interventions; this still, of course, begs the question as to how each of us does, or can, 'listen to the water' or, with Ian Hamilton Finlay, 'venerate the sound of the wind'. Walt Whitman said that architecture is what a building does to you; so, to an even larger extent, does landscape architecture; even if some of its elements have not been deliberately constructed, it still 'does' something to you. And what *you* in return do 'to' or 'for' a landscape (as also for buildings) is also part of that relationship. Yosemite is a national park as well as a natural phenomenon. What is 'does' is a result both of how it exists in itself (its 'this-ness') and of how the landscape has been transformed, adapted to afford access to views and facilities, providing it with amenable consumption. With Yosemite, as with Chinese parklands, Grecian sacred groves or Delphi, what the unmediated place 'does to you' is as crucial as what you have done to it. It is this dual response that we need to track into other garden worlds.

2 Hunting Parks to Amusement Parks

The world's largest cruise ship, Royal Caribbean International's *Oasis of the Seas*, has 'a huge park full of trees in the middle of the boat'. So, to, does the sunken courtyard of the Bibliothèque Nationale in Paris. Apparently the ship lacks opportunities for hunting, but ensures other entertainments; in Paris, one cannot get access to the forest, and the amusement lies elsewhere in its enormous collections. However, on the scale of inventions from Chinese or Assyrian hunting parks to Coney Island, *Oasis of the Seas* surely occupies a significant if unexpected place. Once upon a time, hunting parks were royal preserves (though on dry land!). In more democratic times, the place of resort is likely to be more varied, and unlikely to be linked to the activity of hunting: hence Vauxhall Gardens, Coney Island (though this sported shooting booths), Disney World and its other theme parks, grounds of travelling fairs (more shooting booths), and now cruise ships.

The history of elite and popular amusements interacts with the history of gardens in many ways, although sometimes obliquely. One of the earliest activities of nomadic hunters is discussed by both ancient and modern writers largely in terms of how the hunt was conducted – different kinds of open ground where prey was found, the often detailed discussion of hounds, and different kinds of human participation. Any landscapes specifically set aside for, yet alone designed for, that activity are rarely addressed. Nonetheless, for both royal and aristocratic pursuits, specific grounds were often identified and laid out, usually with boundary walls to keep in the prey. These developments came in part because specific sites were deemed more convenient (nobles could hunt in the vicinity and did not have to venture far afield), in part because prey could be moved or raised in the enclosures. The excuse for these precincts and pursuits was the pleasure and exercise of the hunt itself, as well as the use of hunting parks to serve as a supplier of food: at one extreme, hunting parks were simply, if exotically, open-air larders; at the other, they were an elite activity marked by the prestige of the hunt and, often, by stocking preserves with exotic and local beasts collected or gifted by powerful neighbours. Even today, landed estates in Britain stock their property with fowl (pheasants and grouse) for the pleasure and the enrichment of those who want to pay for shooting there. Poaching and fowling, in return, became the sport of those less advantaged who wished to benefit from a nearby private estate or open ground.

Hunting parks set aside for the chase, whether large or small, could be enhanced and supplied

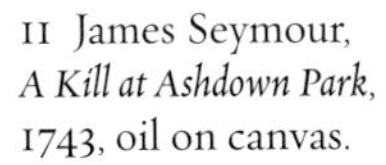

11 James Seymour, *A Kill at Ashdown Park*, 1743, oil on canvas.

with a variety of both permanent and temporary buildings: stands for spectators who were not in the chase itself, and banqueting tents. Hints of these refinements are found in both imagery and some literary accounts. But it was also the appeal of parkland per se, rather than any specific organization and decoration of parks for hunting, that would interest later proponents. Hunting parks became a key ingredient in the rise of English landscape gardening in the eighteenth century, for it was to the well-established custom of hunting during the Norman occupation of England that Horace Walpole looked back when, seeking to find some early *English* design to oppose to the over-organized and largely French design of contemporary gardens, he pointed to what he called 'contracted forests, and extended gardens' of Norman parklands. He was shrewd and patriotic enough to realize that ancient hunting parks in England were the proper ancestor of new parklands. Indeed, many landscapes that came to be developed by 'Capability' Brown were on the sites of former deer parks, like Woodstock, where Henry I had built a rustic pavilion and kept lions, leopards, camels and a porcupine. Walpole specifically singles out the early history of Woodstock/Blenheim as an example of an English hunting preserve. But eighteenth-century hunt paintings also make clear that the chase could be conducted within private parklands or estates (illus. 11, 12). Walpole wrote:

> It is more extraordinary that having so long ago stumbled on the principle of modern gardening [landscaped parkland], we should have persisted in retaining its reverse, symmetrical and unnatural gardens. That parks were rare in other countries, Hentzner, who travelled over great part of Europe, leads us to suppose by observing that they were common in England.

12 John Wootton, *A Hunting Group*, 1729, oil on canvas. Henry Hoare II of Stourhead directs the chase across his Wiltshire parkland.

Yet under the entry for 'Les Parcs' in the 1765 *L'Encyclopédie*, there was a similar account of ancient deer parks, which suggests that Walpole's patriotism was somewhat exaggerated, however much the hunting park did indeed serve as a model and even the former site of new English landscaping.

So it is worth examining this most ancient form of landscape to understand its place, not only just in Walpole's 'English' narrative, but in an expanded world of both elite and popular parks. Whereas hunting was a prime motive for aristocratic parklands, the later role of popular gardens was focused on a much wider range of activities that included musical events, drinking and eating, and the simple pleasure of seeing and being seen (if the site attracted more notable visitors, all the better). Sometimes these popular resorts might imitate effects from the parklands of nobility and gentry – pavilions, particularly, and some attention to the arrangement of scenery. But some sites, of course, had nothing that could usefully be understood as a garden: fairs, like Bartholomew Fair, or those held in other towns to delight local inhabitants, or travelling shows and circuses that toured the countryside, did not have specific or formal locations until very late and then they were largely conceived as county or state fairgrounds. But there had always been places of popular resort, beer gardens and inns with the rudiments of some outside facilities, and it was out of these gardens that some of the famous amusement parks emerged, like London's Vauxhall Gardens, Spring Gardens in Bath or Gray's Gardens on the banks of the Schuylkill in Philadelphia. In their turn, these popular hangouts spurred the creation of public parks during the nineteenth and twentieth centuries – newly invented arenas where bourgeois citizens could suitably disport themselves in the surroundings of 'nature' (see chapter Fifteen); but there was still a role for places that invented and catered for the more boisterous and uninhibited world of fun, with no more than a cursory gesture towards landscape. Central Park in New York and Coney Island in Brooklyn can usefully serve to reflect these two somewhat different extensions of nineteenth-century parkland; the more recent diversions of Disney's worlds seem to bring entertainment or amusement and landscape or scenic pleasures somewhat closer together.

Without attempting a detailed history of the enclosure and decoration of these particular prototypes, some glimpses into hunting in different cultures reveal how much this could become a significant contribution to the world of ancient and modern 'gardens'. The fame and the skills displayed by their hunters are not, however, the story here: neither the lion hunt depicted on Achilles' shield, fashioned by Hephaestus, the smith of the gods; nor Hercules and the Nemean Lion, though Hercules nevertheless stands sentinel on the roof of William Kent's Praeneste Terrace at Rousham, with the lion skin over his shoulder; nor yet the boar hunt in the forests of Mount Parnassus described in Book 19 of the *Odyssey*. But it is the landscape itself, with its manipulation and even decoration, in the ancient world of hunting parks in China, the Near East and in the classical and early modern Europe, that is the topic.

There is some considerable information about ancient imperial hunting parks in China. The Yellow Emperor (Huangdi) kept dragons in his park, though later rulers had to be content with more earthly beasts to provide material for ritual sacrifices.[1] In 138 BC the Han Emperor Wu enlarged an imperial park, the Supreme Forest, outside his capital with a circumference of four hundred *li* (a Chinese mile = 558 metres). This contained many palaces, ponds, a huge range of plants, exotic stones, sceneries that represented distant and sacred places and a population of rare animals from faraway places. The park was thus presented as a symbolic universe, with planting and animals placed in different areas depending on their place of origin. Among its various pursuits – entertainment of guests, religious sacrifices – there was a famous annual winter hunt.

Such imperial parklands persisted, but rare animals for the hunt were not the only resource. There is even the sense that the Chinese appreciated simple communion with wild things: Emperor Jianwan reported in the fifth century that 'unselfconsciously birds and animals, fowls and fish, come of their own accord to be intimate with men'. The profusion and the considerable extent of imperial Chinese domains allowed a similar mix of hunting preserves with parkland for all manner of plants and animals and with a variety of ceremonial locations: platforms were used for ritual observances, for greeting of the weather or for the consultation of heavenly constellations, and upon these were eventually built palaces. Some included representations of other famous scenic spots like Mount Xiao (Henan province) or Mount Hu (Anhui province).

We have, then, in these enclosed realms, insertions that were both simple and complex, renditions of other localities and other topographies (rockeries, islands), and used for a variety of recreations including, apparently, farming, forestry and mining as well as hunting. The imperial gardens reached a peak in AD 581 when Emperor Yang constructed his Xi Yuan (western garden), with some of the same imagery as before (a lake, three artificial hills) and with over a dozen palaces and a meandering canal. It was the succeeding Tang dynasty, enjoying more peace and prosperity, and the Song, though much less stable, that increased the number of parklands, some of which could occasionally be opened to the public and others that were extraordinarily vast. In Genyue, an imperial park in Beijing, a huge labour force took six years to construct Longevity Hill (Wan-Shou-Shan), clothed on different sides with thousands of plum trees and medicinal herbs; Genyue also contained representations of five famous mountains and gorges from the Yangtze river. The last of the most spectacular productions came in the Qianlong emperor's reign (1735–96); they included the Yuanming Yuan (Round Bright Garden), to the west of which lay the large park of Jing Yi Yuan, where a specific part in the south-east was noted as a hunting area; there was also the imperial summer resort of Chengde to the north-east of the capital; an early base for hunting had been established in 1677, with ten thousand square kilometres enclosed four years later, and this Jehol district soon became a primary hunting parkland provided with villas or hunting lodges.[2]

In the Middle and Near East, hunting was also a consuming activity. The Assyrian king Sennacherib (reigned 704–681 BC), moved his capital to Nineveh and surrounded it with hunting parks, as well as well-planted gardens and orchards, and reliefs in the British Museum show pavilions, palm trees, streams and other plants. A later ruler, Ashurbanipal, was particularly proud of his hunting skills, and bas-reliefs from his palace of *c.* 650 BC show the hunters (with horses and chariots) and the prey, but with little sense of any topography associated with the event. He certainly hunted in a park in Nineveh, and had animals trapped and brought to him for the kill, including lions. Such events were designed to show off the prowess of the ruler, and his exploits could be watched from a building decorated with images of the hunt. A relief in the palace of Ashurbanipal does show tame lions and lionesses in a garden of date palms and grapevines, suggesting that a hunting ground was also part of some decorative parkland (illus. 13). Another shows Ashurbanipal crouching in a pit prepared to shoot his prey, and in another deer are being driven into nets in a landscape dotted with trees; but whether these were nets set up within a designed place or simply out in the open landscape is not clear. Nets for catching animals and birds were a common and traditional device, and were described at great length by Xenophon, along with other appurtenances and the breeding and use of hounds. Netting was a tradition that persisted well into the Renaissance, and we shall see Medici gardens in Tuscany in which netting of prey and other sport and entertainment were featured.

In Greek and Roman times hunts were also frequently depicted,[3] but again with little indication of where the hunt would have taken place (illus. 14) When animals abounded in the wild – lions, boars, wolves and lesser beasts such as hares and deer – the hunter presumably sought out the prey wherever it was available, and there was little sense that hunting need have taken place in any designated

13 A gypsum relief in the Assyrian North Palace of Ashurbanipal at Nineveh in modern-day Iraq, showing lions in a parkland of palms and pine trees; 645–640 BC.

location. The royal hunts of Alexander the Great were chronicled by Plutarch and appear to have taken place in Iran, yet a Latin chronicle writes of Alexander hunting also near Samarkand in 'great woods and parks', which were walled and had 'towers for shelters'. These enclosures seem to be been further embellished with 'many springs of water that flow all year long'. It became a mark of especial prestige to dedicate an enclosure, often huge, to hunting and to the collection of special prey; the collection of beasts to be hunted also recommended itself as a convenience – the ruler didn't have to go out and find game and have to return empty-handed. The Persian parks were described by Xenophon in *Anabasis* as 'a great paradise full of wild beasts' that he had witnessed on his eastern travels and which were chronicled in his account of Persian wars in the very early fourth century BC, *Cyropaedia*. This account of Cyrus the Great (mid-sixth century BC) and his legacy tell of

14 A mosaic from Villelaure, Vaucluse, France, showing a Roman hunt with nets and hounds; the suggestion of landscape gives no sense of where such a chase was taking place.

arranged hunts, where the prey was driven into a suitable location where it was hunted on horseback, but also, since he so relished the chase, he hunted animals reared in parks. It was the younger Cyrus, with whom Xenophon served, who is credited with planning trees and laying out a beautiful orchard with his own hands, and Xenophon is famous for having translated the Persian word *paradeiso* into the Greek, from which we derive our term 'paradise'. The term seems to have been applied generally to both hunting parks and enclosures that contained trees, flowers and birds, and they made a considerable impression on Xenophon: he noted the younger Cyrus's appetite for parklands as well as his power, destroying both a Syrian palace with a 'park, great and fair, that contained all the fruits of the season' or another with 'fair palaces and parks full of trees and game'. Xenophon himself, a retired general eventually exiled from Athens, enjoyed a large estate near Sparta, where he eventually bought a small property beside a river, where he erected a small temple and where there was 'all manner of beasts of the chase' (as reported in his *Anabasis*).

Reports and detailed analyses of hunting continued throughout Greece and Rome, with Roman nobles learning from Middle Eastern and African practices how to hunt in the grand manner.[4] Tyrants like Domitian enjoyed parklands and the slaughter there of a 'hundred wild beasts of different kinds', but this practice was opposed by those like Trajan who still chose to chase prey in the wild. Nero maintained both tame and wild animals in the park of the Domus Aurea, and Varro wrote of an orator with a walled parkland of 32 acres, in which – Orpheus-like – he summoned the beasts by playing on a horn; in another he reports a 9,000 acre park in Gaul where couches and banquets were served to those who watched the games. Literary accounts refer to game reserves of various sorts: avaries, fishponds and ponds for snails, tame collections of boar, bear and deer; the elder Pliny reports an imperial elephant park. But where the hunt and hunting parks intersect with the theatrical appeal of the amusement park was in the public staging of hunts and fights within circuses and ultimately the Colosseum. Part ritualistic, part victory celebrations, these soon became a theatrical display: lions, bears, a hippopotamus, crocodiles pitted against fighters chosen for their skills in the sport and presented for the satisfaction of the populace; elephants had been used in fighting, and their appearance, goaded by fighters with javelins, were a conspicuous display in the arenas. This marks perhaps that moment when hunting, as an elite pursuit, converged with the world of popular 'amusements' (if that is the apt word) and gladiatorial 'hunts' were performed before thousands of noble and plebeian fans.

For designed and elaborated parkland for hunting, however, more relevant and more exciting gardenist material can be seen in a cluster of sites in the late Middle Ages and early Renaissance parks, now reasonably well documented: the late thirteenth-century Burgundian park of Hesdin in the Pas-de-Calais in northern France; the ideal hunting park proposed by Filarete in the twentieth chapter of his *Treatise*, written in the 1460s; the very varied uses of Medici gardens around Florence a century later for a variety of different hunts and amusements; early hunting parks around Rome in the sixteenth century.

Hesdin was well known through a celebratory, but non-specific, poem by Guillaume de Machaut ('Remède de fortune'); but detailed archival records show that construction and improvements proceeded throughout the 1290s, and that it was

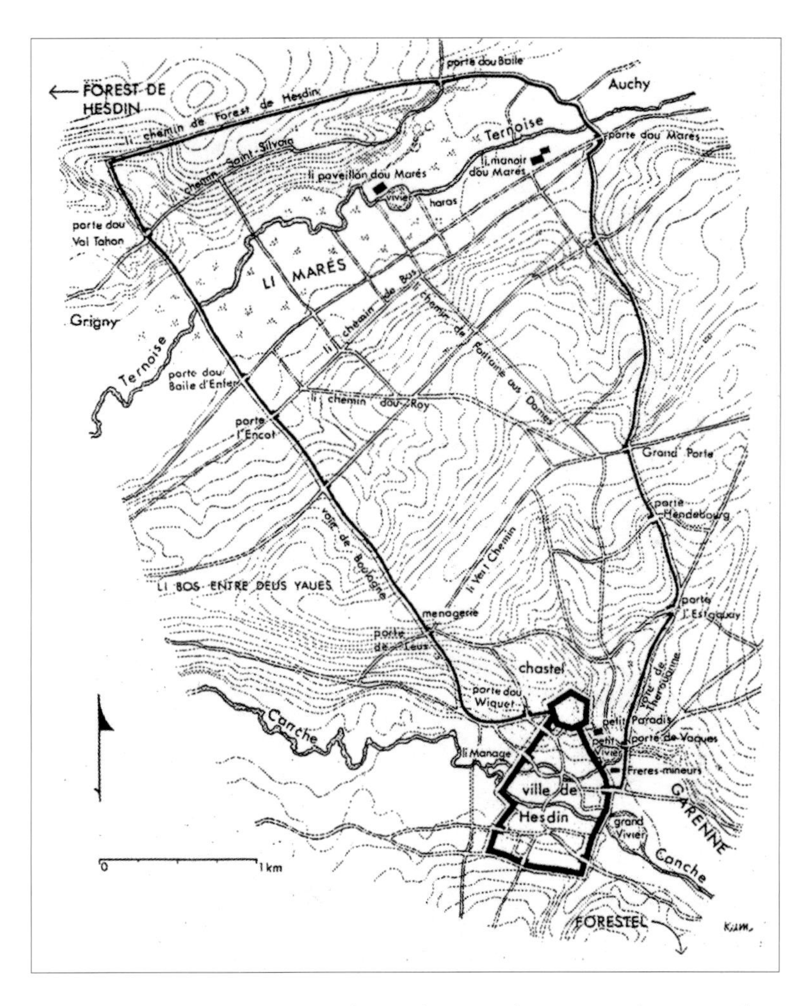

15 Anna Hagopian van Buren, map of the Park of Hesdin, Pas-de-Calais, France.

finished and stocked with peacocks and swans by 1306.[5] A modern map of the site (illus. 15) shows the parkland stretching for four kilometres to the north of the town and its castle; Hesdin Forest, still extant, exists to the west. Between the town and marshlands around the river Ternoise, the park is criss-crossed with roads, with gates and posterns opening into a landscape of 'hamlets, lodges, fish ponds, orchards, gardens, barns, stables, paddocks, mews, aviaries and a menagerie', the last situated by the 'gate of the wolves'. The castle enjoyed its own garden, one of which was called 'li petit Paradis' (a curiosity is that the pronunciation of 'Hesdin' drops the 's' and, ignoring the 'H', makes the French sound like 'Eden'). At the far end of the park alongside the river were other pools, springs, more orchards and vineyards, a stud farm, an earlier manor and a pavilion (both dubbed 'dou Marés', of the marshes). The pavilion itself was an ample establishment for hospitality, with seats, a crenellated fountain, shrubberies and bowers or gazebos, and a rose garden with a stone tower, and the whole place was decorated with *automata*, most probably wooden statues or marionettes, some of which contained water-spouts. The castle itself also contained similar clockwork devices. There

16 A 16th-century painted copy of an earlier tapestry shows a wedding party at Hesdin in 1431.

was even, later, a fifteenth-century moveable 'folly', a 'dining house on wheels that could be rolled into the park and turned to face the sun'. In all, Hesdin was a place of entertainment, and the view of a fourteenth-century wedding party, while not accurate, suggests some of the layout; its pavilion by the river is perhaps a feature based on merely verbal accounts that conflated various elements of the marshlands (illus. 16).[6] Animals were certainly a feature of the park, and the wedding party shows hounds in the far distance; there are records of hunting along the river to the north.

The treatise by Filarete (Antonio di Pietro Averlino) is utopian, literary and couched in narrative or fictional form; but its thrust is instructional.[7] His lord and patron asked him to invent a hunting park or 'game preserve, with woods here and there and beautiful spots'. It would be large (ten by five miles) and walled around, with planted groves (various species are suggested), a 'small mountain' with laurels on its summit and a hermitage, ponds for ducks and herons, and with compartments to keep the carnivorous animals segregated from those destined for the chase (these species are also listed). A castle with an interior courtyard was then proposed at the centre of the park, with a walled garden around it. This detailed description of the site then proceeds, unexceptionally, to take up details of an elaborate hunt that would follow; but Filarete's specific attention to its spatial organization and design are the key elements of his architectural project. For a plausible visual equivalent, from a manuscript some 20 years later, we might turn to a drawing by Francesco di Giorgio Martini (illus. 17): here within a walled enclosure are a little mountain, groves, a semi-circular courtyard, a fish pool, pavilions, orchards, a canal and a circular pavilion on a mound. While it presumably exhibits his particular concern with giving regular forms to gardens, the similarity to Filarete's proposal and the mention of animals and their disposition is nonetheless an intriguing parallel.

The countryside round Rome was increasingly used for hunting in the late fifteenth century, even by churchmen, who found ways around canon laws that forbade clergy to indulge in sports, and who could ignore the early denunciations of hunting by Saints Ambrose and Augustine.[8] Information is mainly devoted to hunting lodges, their accommodation (examples of which survive) and to the basic spaces designed for the chase, with some attention to their walls or boundaries, modelled on the large hunting parks or *barchi* from northern Italy. There were platforms erected to watch the hunting, and open-air banquets, music and dramatic performances were held at the lodge at La Magliana, west of the city, which often acted as a seat of government (a Chequers or Camp David for the papacy). Bishops of Viterbo used a hunting park at Bagnaia, where the Villa Lante was later to be built, but meanwhile they enjoyed a hunting lodge that still survives in the *bosco* adjacent to the villa gardens. The Villa Farnese at nearby Caprarola also maintained a hunting park, walled, with woods, meadows and woods, a lake for fishing and a collection of domestic and exotic beats. At Tivoli, a *barcetto* had fountains with its fishponds, and another in the valley below was enclosed with two miles of walls, largely as a preserve for wild beasts and roebucks. These sites were thought of as having decorative as well as functional uses.

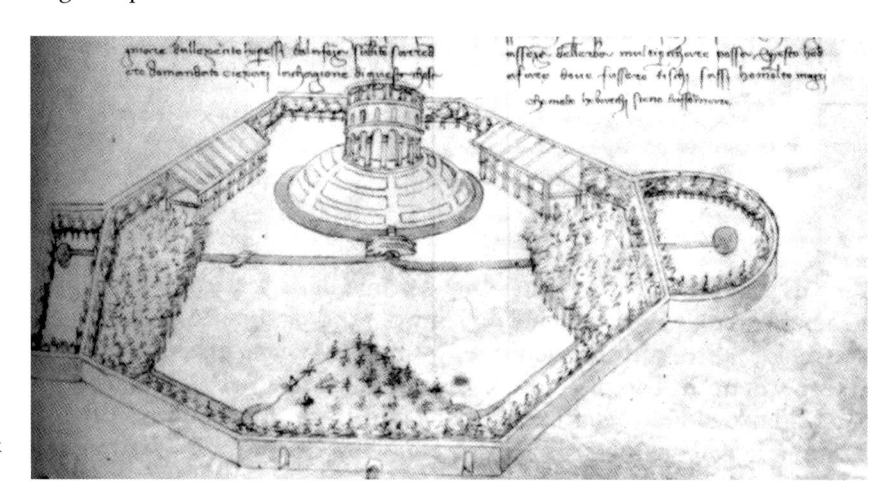

17 Francesco di Giorgio Martini (1439–1502), design for a 'barco' or park at Urbino, Italy.

With Medici properties in Tuscany we have much more elaborate information on garden facilities for the hunt.[9] The Medici owned several country farms and villas, and these were used variously for both private hunting and for the entertainment of visitors: in particular, the Villa Castello, to the north and near enough to Florence for frequent ducal visitations, and the Villa Pratolino, northwards on the road towards Bologna. At Castello the central garden was surrounded by several enclosures for hunting or fowling, notably in the woodland or *selvatico* on the upper terrace (illus. 18); here small birds and animals were chased by professional hunters as entertainments for guests who were feasted nearby. To the east (right) of the central garden was a larger area consisting of a fowling grove (*frasconaia*) and a rabbit warren (*ragnaia*), in both of which hunting was practised with fine nets hung from poles and formed into tunnels into which the prey could be chased. Again, a pavilion in the eastern area was placed for observation of these events. A further area, immediately in front of the villa and overlooked by its balconies, was an open space, the kind identified by Alberti as suitable for exercises in chariots or on horseback; but at Castello it was also used for peasant dances, where citizens enjoyed the hospitality of the location.

The small, local hunting pursuits at Castello need to be contrasted with the larger park of Pratolino, which was contained within an extensive reserve called 'La Bandita di Pratolino', stocked with different beasts, including gazelles and ostriches, reserved for the Medici, and enclosed with a stone wall; it is the lower half of the park below the villa that is represented in Utens's late sixteenth-century lunette, with its irregular pools cascading down each side of the park, a central *viale* and narrower pathways (*viottole*) through the woods (to serve as cover for the animals). Hunts were staged and presented with banquets, watched by spectators and, with all the routines of hunt protocol, managed within the different spaces of the parkland. The whole area – especially the central area depicted by Utens – is a landscaped space dedicated equally to hunting and to the delights and amusements of fountains, with a Mount Parnassus, fountains, grottoes and herms (see chapter Nine). This was a hunting ground that was also fashioned as an elaborate Mannerist garden.

In these elite domains hunting had come to coexist with an abundant garden world, where delight and pleasure extended the focused activities of the hunt. But popular amusements in garden spaces, however, relied neither upon the aristocratic hunt nor upon any elaborate provisions of landscape. Activities like bowling, maybe archery, swimming, fishing, cockfights, dog fights and bear-baiting, among other athletic events, as well as the less energetic pleasures of eating and drinking, all required places where these events could be held and enjoyed. Thus the amusement park, as we know it today, had its origins in small, local and often very modest sites. Many places of popular resort like inns, dance halls and tea and beer gardens bothered little with providing or enhancing their gardens, but they did provide places of resort for the less well-to-do where outdoor spaces were clearly a valued 'extra'. Such small gardens sometimes survived even the most ambitious urban redevelopments: after Baron Haussmann in the mid-nineteenth century had cleared away much of the medieval clutter throughout Paris, there were still opportunities to enjoy small gardens that his modernization had spared: Gauguin painted these garden booths, called *guinguettes*, in Montmartre.

London was not unique in its provision of popular garden amusements, but as Warwick

18 Giusto Utens, *The Medici Villa at Castello, Tuscany*, 1599, oil on canvas.

Wroth's survey of *The London Pleasure Gardens of the Eighteenth Century* (1896) makes clear, it was a city with many 'sweet gardens and arbours of pleasure' (at Islington Spa, for example, in 1684); some of these coexisted with gaming rooms, tea-houses, coffee shops and dancing rooms, duck hunting (at Duckling Pond House), ponds set up for anglers and swimmers, and the consumption of salubrious spring waters that were billed as 'a powerful antidote against the rising of the vapours, also against the stone and gravel'. Such pleasure gardens were scattered throughout London and its environs (Wroth has a map of these[10]); many often survived only for short periods, or were taken over, their names changed, as new entrepreneurs sought to draw in fresh custom with different amusements, festivals and events. But the range of locations and their mixed entertainments over the seventeenth, eighteenth and nineteenth centuries was truly impressive, and offered a garden world somewhat at odds with the development of a landscape aesthetic in the 'English' garden.

Essentially, the pleasure gardens of London and its surroundings invoked garden settings, whether contrived or actual, limited or sometimes agreeably large, and even drew visitors in by their glimpses of some surrounding countryside; some, like the Apollo Music Room, sufficed with a painted country backcloth. The Spaniards Inn, still standing to the north of Hampstead Heath, was originally laid out with grass lawns and gravel walks and a mount that offered extensive views into neighbouring counties. These displayed occasional garden effects that were derived from earlier and more sophisticated elite domains – an enchanted grotto, with a water-mill that, activated by its proprietor, represented fireworks and a beautiful rainbow: there were several places called 'Merlin's Cave', presumably named after Queen Charlotte's similar structure at Richmond, designed by William Kent in 1735, its exotic waxwork figures and astrological symbols also copied in other locations; images from about 1760 show an ornamental pond and Grand Tea Room on Duck Island in St James's Park, with distinct reminiscences of both Burlington's Ionic Temple by the canal at Chiswick and William Kent's landscape sketches; large pools, gravelled walks, avenues of trees, groves or even quincuncx could be found at the Mulberry Garden at Clerkenwell, Pancras Wells, or Marylebone Gardens; The Spaniards' garden was decorated with coloured pebbles that depicted famous places (Tower of London, the Colossus of Rhodes, Salisbury cathedral's spire) along with 'an odd association of things earthly and celestial'. Cuper's Gardens, south of the Thames opposite Somerset House, was famous for its fireworks, but its layout, as recorded in the mid-1740s, mingled orchards, regular groves, a pool and carefully modulated meandering paths.

The reputations of all these gardens ranged from the 'scandalous' (Pancras Wells in 1722) to 'genteel company' (Shacklewell Green coffeehouse in the 1790s). They were generally visited by what might be called a 'mixed company', such judgements depending as much upon the particular commentator or advertiser as upon the fluctuating fortunes of its clientele. Several leisure gardens drew much attention and prestige from notable patrons: Islington Spa, or the New Tunbridge Wells, received royal patronage in 1733, when the daughter of George II came regularly to drink the waters and was greeted with gun salutes and a thronged crowd. But it was Vauxhall Gardens (illus. 19), the most famous and visited of all the major amusement gardens, that earned the especial accolade of royal favours, since it was both rented from and frequented by the Prince of Wales; Ranelagh Gardens also benefited from a considerable patronage by eminent personages, such as Lady Mary Wortley Montagu, Sir Joshua Reynolds and Oliver Goldsmith.[11]

The significance of Vauxhall Gardens as a famous and early amusement park depends largely on its longevity (from the Restoration until its closing in 1859) and on its proprietors' and owners' skills in responding to a changing clientele; careful manipulation of landscape effects, entertainment and social display over the years appealed to precisely this range of a truly 'mixed' company. Thus it managed to combine three elements of public gardenhood. First, there was its parkland or garden effect, however forced and factitious it seemed to become as the years went on. Pepys saws citizens eating its cherries off the trees, Joseph Addison pronounced on its 'prospect of fields and meadows', a French visitor liked its 'pretty contriv'd plantation' in the 1710s and Jonathan Swift went there to listen to the nightingales. Though many found its 'rurality' unimpressive, much language was expended on its 'groves' and 'green-wood', and the garden itself comprised walks, trees, lawns, sculptures and on the northernmost edge of the site a representation of 'Rural Downs', with a statue of John Milton by Roubiliac on a hillock.

Then there was its fashionable aspect – its appeal to a sophisticated audience who enjoyed its parade and self-performance, its masquerades and opportunity for promenading (Walpole estimated that there were 'above' 2,500 visitors on a visit in 1736). Thomas Rowlandson offers a typical scene in 1785 with his marvellous watercolour (found in a junkshop in 1945 – illus. 20): here is a mix of the fashionable and the coarse among the auditors; the singer is Mrs Weichsel and the conductor, James Hook; also shown are the Prince of Wales (wearing the Garter Star), his lover, the actress Perdita Robinson, and Georgiana, Duchess of Devonshire. James Boswell himself captured the essence of these events in his remark that Vauxhall was 'peculiarly suited to the taste of the English nation' by being a mixture of 'curious show', 'gay exhibition' and 'music . . . not too refined for the general ear'. Boswell also added 'good eating and drinking'.

Finally, there was the sheer entertainment value – the much applauded musical performances, for both vocal and orchestral music gave Vauxhall a privileged place in the repertoire of 'English song', and many performers made their debuts at the gardens before appearing at Covent Garden and other London theatres. But there were also the nightly illuminations that Walpole specifically admired, and the somewhat tacky events that offered the 'motley crowds' a variety of painted prospects – 'a view in a Chinese Garden', other painted and illuminated backdrops (as in any theatre) and a famous landscape with a Cascade where strips of tin mimicked the flow of water. As landscape and garden taste advanced, many found these items merely 'puppet-shows', and by

19 *A general prospect of Vaux Hall Gardens, showing at one view the disposition of the whole Gardens*, 1754, copper engraving by J. S. Muller after Samuel Wale.

20 Thomas Rowlandson, *Vauxhall Gardens*, 1785, watercolour.

1770 even Walpole found trees more beautiful without being hung with lamps; yet its repertoire of fake ruins, meandering paths, Chinese effects and other 'oriental' pleasures offered some intimation of other picturesque sceneries elsewhere.

This careful combination of effects would eventually find their own separate forms, as public parks and gardens developed during the nineteenth century. Parklands became one of the most characteristic garden forms of the century (see chapter Fifteen), and were often pastoral enclaves where the civic virtues of an urban bourgeoisie could be learnt and displayed. On the other hand, there continued to be places where the emphasis was emphatically on amusement, on the sheer fun of entertainments – peep shows and popular songs – without much attention or expense expended on anything like landscaping. Another aspect of this particular amusement-directed parkland can be traced in the world of international and national exhibitions (see chapter Sixteen), which combined the promotion of science and industry with the excitements and challenges of education, and usually some component of gardenhood and landscape.[12]

Vauxhall, in both name and its reputation as a garden theatre, enjoyed international fame. It was imitated in France and in the United States, and its name was given to a Moscow underground station. Vauxhalls appeared in or were proposed for Philadelphia, Boston, Charleston in South Carolina and on four sites in New York City;

often the name was simply applied to any open-air amusement sites, such as farms and woodlands that recalled the early rurality of the original Vauxhall or New Spring Gardens. Bowling, fireworks, waxworks and sculpture, concerts, dances, balloon ascents and theatrical performances were generally accompanied by some garden effects, promenades or nursery gardens, where refreshments were always served and mineral waters sometimes available.

Many of these American ventures were short-lived, faced local obstacles or generally could not deliver enough support. But they also encountered rival and more successful developments that catered to new and more nuanced cultural activities. Central Park would be one key instance of a park that, while it catered for amusement, also provided a various landscape of promenade, meanderings pathways, lakes, hillocks and even 'wild' areas like the Ramble. It was designed to commit its citizens to an education in the appreciation of the countryside, now epitomized within the park, to enabling less wealthy folks to appreciate both the importance of healthy living (fresh air, exercise, free milk) and the chance to encounter suitable religious and artistic objects (the Bethesda Fountain, the statues of eminent musicians and writers, and eventually a zoo and the Metropolitan Museum of Art, both established within its boundaries). It was, in short, a pleasure garden designed for all Americans (at least those who could afford to get there), where they could learn and parade their own special identity. It was genteel, a model of social order, an education for an emergent middle-class nation.[13]

Yet such parklands, and their promotion by one of the most prominent landscapers of his age, Frederick Law Olmsted, did not eliminate more popular venues; they were simply displaced onto a new development of amusement parks, where the opportunities for the mass marketing of attractions and pastimes were deployed. Of these, Coney Island would be a prime American example; but its shows could be found nationwide, where the same events, even if less extensive, could be enjoyed on seaside piers, parks and beaches in a variety of cities, from Willow Grove outside Philadelphia, to New Orleans, Denver and San Francisco.

Coney Island was a collection of three parks,[14] one of these, Steeplechase Park, being enclosed. All were provided with stupendous scenographic events, but little planting or landscaping. Panoramic views did seem to suggest extensive parklands (illus. 21), and some of its attractions – boat rides through hills and waterfalls, or scenic railways like 'coasting through Switzerland' – invoked various mimetic landscapes, and the electric lights of Luna Park transformed it into 'an enchanted garden'. The early years of Coney Island saw a cluster of hotels, like the Brighton Beach Hotel, with flower beds and a certain air of gentility. But what most came to characterize Coney Island's separate parklands was a proliferation of follies or *fabriques*, exotic and exaggerated items, which filled the spaces and the visitors' imaginations. Luna Park was a 'land of trellises, columns, domes, minarets, lagoons, and lofty aerial lights', for unlike Vauxhall it could now boast electric illuminations. While some of these, like the Ferris wheel imported from Chicago, the Iron Tower from Philadelphia or the 'Trip to the Moon' cyclorama from Buffalo, had their immediate origins in recent exhibitions (see chapter Sixteen), they all traced their descent from the proliferation of Picturesque pattern books in the early and mid-nineteenth century, designed for the decoration and stimulation of both private and public gardens

21 Elevated promenade at Luna Park on Coney Island, New York City, *c.* 1900.

and parks (see chapter Fourteen). There was an Iron Pier (1903), an Elephant Hotel (1883) constructed out of wood and tin with spiral stairways in its hind legs, an elaborate grotto at 'Dreamland' supported by a high naked sculpture of the 'Creation', beacons, bridges, trick benches and chairs (like Renaissance water jokes) and – this being the age of excited mechanical devices – a whole host of other contraptions. And by 1900 the spread of tramways, steamers and trolleys permitted a visitation that far exceeded the numbers that could reach the pastoral enclaves of Central Park after its creation.

The vogue for Coney Island began in the early 1890s and flourished in the years before the First World War. Its earliest beginnings, interestingly, had popular hunting opportunities for duck and snipe, and clam fishing along the beaches, though these rapidly succumbed to a plethora of amusements and pastimes, some of which by the mid-1890s were so infamous that it acquired the nickname of 'Sodom by the Sea'. The subsequent transformations of this mixed and decidedly louche site by the turn of the century, while still celebrating all the excitements and amusements of popular entertainment, established a carnival atmosphere of good humour, laughter and much communal participation. If visitors to Central Park needed to be well-behaved, visitors to Coney Island would find restraints upon behaviour and social comportment less oppressive; as a writer in 1901 reported, 'Coney Island has a code of conduct which is all her own'. If the subsequent history of entertainment parks, like California's and Paris's Disneyland, and Florida's Walt Disney World, somewhat sanitized the vibrant and somewhat tawdry popularity of Coney Island, their new landscapes reaffirmed the exoticism of the scenery and seductive *fabriques*.

22 Pergola and water channel in the House of Loreius Tiburtinus, Pompeii (a reconstruction of the original garden from AD 79).

3 Ancient Roman Gardens and their Types

With the Roman garden, we begin to have sufficient and exciting material on a whole range of gardens – villas, suburban and rural sites, seaside palaces and mountain retreats, sculpture gardens and even sacred groves. Part of this excitement simply relates to our ability in modern times to have uncovered – literally – so much evidence of Roman garden-making, as well as to our knowledge of Roman literature, so much more extensive than what we know of Greek and Hellenistic gardens. Yet part of our response also relates to our deep-seated recognition that the world of Roman gardens, even sometimes its visual and stylistic forms, still speaks to our own sense of how gardens serve our continued needs and desires.

Recent publications, including accounts of the practical world of Roman horticulture, signal something of this continuing inspiration: indeed, a comparison of Roman horticultural implements, illustrated in Linda Farrar's *Ancient Roman Gardens*, makes us realize how similar are our own gardening tools, so little has the practical work of horticulture changed over the centuries.[1] Romans came early to write about gardens, mainly in a practical vein; around 37 BC Varro wrote his *World of the Countryside* (*Rerum Rusticarum*). Virgil touched upon garden matters only obliquely, leaving it to other poets to do this, which was indeed taken up by Columella a hundred or so years later in his twelve books of *De Re Rustica*; the tenth and eleventh books, in both prose and hexameters, deal with the garden (other texts have been lost, and a later writer, Palladius in the late fourth century, wrote one on gardening that was still subordinated to agriculture).

Columella, who was originally from Spain but came to farm near Rome, makes a useful introduction to what he himself called a truly 'modern' reoccupation. He attends to the annual cycle of planting, to the usual activities of choosing the soil, laying out beds, digging, spreading manure, sowing and weeding, and he also signals the activity of a female farm manager in organizing the transfer of food from the garden to the table. In his verses Columella, properly invoking Priapus as the suitable garden deity to whom the hexameters are dedicated, waxes suitably heroic, with some nice moments of bathos, exhorting heavy work with plough and hoe, celebrating the maturing of ground with what the 'privy vomits from its filthy sewers', relishing the different origins of his plants and enjoying at the end the juice of the wine-press. (John Henderson has a wonderful reading of this poetic and often mock-heroic bravura of gardening skills and horticultural reverie.) Yet Columella, for all his pyrotechnics, is a down-to-earth gardener and farmer – insisting on living hedges, on beds that are no wider than somebody weeding can deal with from each side, on the varieties and sowing times of everything from cabbages to asparagus. From other sources we may learn how plants were laid out, embellishing the open spaces between colonnades (Vitruvius), the handling of different trees and shrubs, their fragrances and colours (the elder Pliny), and the cultivation of gardens for the making of ceremonial garlands (Varro) – but for more sense of the larger purviews of garden form and usage, we need to look elsewhere, beyond the garden shed.

Roman garden layout continued to shape later design as well as planting practice: the atrium became the Christian cloister, while architectural features, like peristyles, terracing (*opus pensile*, or hanging gardens), grottoes, pools or *vestibula*, and furniture, like trellis-work, pergolas and seats and tables, continued to sustain the deployment of domestic space today. The Roman villa would be the model for many subsequent generations of those who wished to establish a place in the country, complete with both the necessary amenities of elegant living and maintenance. For the villa, we have archaeological discoveries, along with detailed and instructive writings – most notably those by the younger Pliny, whose literary texts were endlessly reviewed to yield specific site plans for subsequent architects.[2] For the more modest gardens, we are lucky to have the remains of Pompeii and Herculaneum – for when the lava and ash from Vesuvius buried those cities in AD 79, it embalmed a whole host of gardens, modish as well as lavish, which modern archaeology has further exposed to our view.[3] And this is where it is useful to begin.

The sheer range of gardens at Pompeii – in luxury houses, modest households, temples, baths, restaurants, inns and hotels, schools, shops, theatres, funeral parlours, market gardens, vineyards and even brothels – testified to the Roman delight and reverence for and usefulness of the natural world (illus. 22). This was translated alike into painted scenery, garden layouts and the creation of sacred landscapes both depicted and designed, all arguably asserting a particularly rich understanding and perception of nature and its inhabitation by numerous spirits. Yet there seems to have been little inclination to confuse the illusion or artificial worlds with the reality. However lifelike were the wonderful images of dolphins, dogs, rabbits, an octopus, orioles, thrushes, partridges, snakes, strawberry plants, *viburnum*, even images of the hunt, and many water effects (illus. 23), there was no confusion of these paintings or sculptural objects with either the actual elements of the garden world, its planting, its water and its birdlife, or with the larger world beyond the garden and the city. Gardens were much loved, their plants cultivated and animals valued, but these were enjoyed alongside gods, goddesses and every imaginable chorus of satyrs, dancing fauns and cupids. Every house and its garden had their favourite gods, altars and shrines (*lararia*), be they devoted to Dionysius, Priapus (good for the garden's fertility), Venus, Diana, Hercules, Pan, Pomona, Mars or Diana. There was therefore, as Michel Conan argues, a wonderful sense of unreality alongside the lively apprehension of real nature: space was dilated – even tiny gardens suggested images of much larger territory with their frescoes; time too was suspended in gardens, which were filled mainly with evergreen plants (ivy, myrtle, bay, rosemary). Space and time were effectively blurred, too, by the presence among

23 Mosaic fountain with a garden scene, originally from Baiae, 1st century AD.

human inhabitants of deities and other spirits. Once aqueducts were established during the time of Augustus, water features became a central part of gardens in Pompeii, represented in atrium pools, jetting craters and water channels, as well as being replicated in painted imagery. Indeed, everywhere in Roman gardens, water – all its forms – was prized more highly than anything (Cato put only vineyards higher). Statuary and pots were everywhere, often locally produced for garden use; but – again – also imaged on walls and illusionistic niches, so that the garden was intensified by this mirroring of real and imagined garden scenery. Everything we have learnt about the buried city of Pompeii speaks of the interpenetration of its gardens with the busy world of daily life; opportunities for outdoor living and especially dining made the garden setting a fundamental part of existence, telltale fragments of which have emerged from modern excavations to make palpable the life there. Outside the small spaces of Pompeii, as we shall see, larger villas would be able to display a much more elaborate mix of countryside (*rus*) and town life (*urbs*).

The eruption of Vesuvius claimed the life of the naturalist Pliny the Elder, whose nephew became famous for his praise and description of his own country villas. If Pompeii suggests how small gardens could be, even given the illusion of space that offered no extensive explorations on foot, and allowed the interpenetration of gardens with images of a larger nature, then the villas of the younger Pliny were certainly gardens that invited exploration and the exchange of buildings and landscape within a very mixed ensemble. Pliny wrote about his villas in the early years of the second century AD, and two in particular were extensively described: at Laurentinum, near Ostia at the sea, and in the Tuscan foothills of the Apennines. There is some lively debate on whether these were in fact accurate representations, topographically grounded, or whether they were more rhetorical, deploying an array of classical *topoi* and invoking clusters of familiar garden forms and functions without any precisely defined sense of their configuration. Pliny's own recognition at the end of the Tuscan letter that he was emulating other famous ekphrastic descriptions, like Homer's shield of Achilles, may suggest that he was less focused on a specific topography than concerned with an ideal design. Since neither site has been wholly unearthed nor even properly identified, Pliny's two letters must be read rather for their insights into how a Roman would have experienced and enjoyed the formal designs of his villas, rather than for their prescriptive authority or factual accuracy; this did not, however, prevent many subsequent architects from extrapolating the written texts to construct ground plans of the later buildings and gardens (see illus. 33).

Several themes are rehearsed as Pliny describes for two different friends the pleasures and entertainments of his villas. The coastal villa was easily reached on horseback from Rome, and was therefore something of a nearby retreat, while the distant Tuscan villa afforded much more ample accommodation for guests and considerable space for gardens and landscaping, of which Pliny makes much. It is clear that Pliny delighted in his movement through the buildings and grounds – following different paths, different routes through *diaeta* (inner sanctums), interior courtyards, sheltered walkways, walks that skirted a garden, woodlands, a hippodrome or race track and even a garden area that astonished by presenting itself as 'a spot of ground, wild and uncultivated'. (The exact sense of what that 'spot' would be is, of course unclear – except that by contrasting it with other elements of the property it is clear that this particular area must have been kept deliberately untended and yet admired. As I also am relying here upon a mid-eighteenth-century translation by John Boyle, fifth earl of Cork and of Orrery, there is also the chance that his version is coloured somewhat by later landscaping ideas).

The exchange between inside and outside was enhanced by the views from the villa buildings themselves – various sights of the sea near Ostia ('so many different views', 'three different oceans'), and of the whole surrounding landscape in Tuscany. All of these considerably augmented what the smaller gardens of Pompeii would have provided, with its paintings and sequences of imagined landscapes. Even so, a bedroom in Pliny's Tuscan portico, shaded by 'the verdure and gloom of the neighbouring plane tree', contained a painting of more trees where birds sat amid the painted branches. The variety of different landscapes ('prospects . . . finely diversified') created rich impressions of distant vistas. But it was the Tuscan villa complex that gave Pliny his best experiences by setting the garden work in and against the 'situation of the country'. In a famous phrase that was later echoed by other writers, not least for its vocabulary, Pliny wrote of 'an amphitheatre of immense circumference', which – as the Latin put it – was 'formed only by the hand of nature' ('amphitheatrum aliquod immensum, et quale sola rerum natura possit effingere'). That this *natural* scenery was, as it were, *formed* by nature gives a delightful perplexity to the experience of the scenery; a few paragraphs later Pliny also looks 'upon a real country' that seemed 'a landscape painting drawn with all the beauties imaginable'. The effect, though, is of a diverse but still mediated landscape rather than insertions of a simply 'natural' terrain. Within these mingled impressions of a large agrarian and worked landscape of woods ('stocked with game for all kinds of hunting'), meadows and ploughed fields were set the much more regular and ornate garden work: crafted hedges, pallisades and topiary of 'various figures' of animals, and his own name and that of his gardener in topiary boxwork (the Roman term for gardener is *topiarius*).

Pliny, like many villa owners, was eager to spread the fame of his properties, for them to be visited – or even just 'seen' in the ekphrastic descriptions of his letters – as well as to be places for his own and his guests' enjoyment. Whether his letters were or were not accurate descriptions, they were obviously visions, in several senses, of an ideal property. Other Romans, too, notably those in their extensive outposts, were proud to fashion their own gardens, some patrician, others more workmanlike. Excavations of these have uncovered a variety of scattered sites throughout the Roman Empire, from Morocco to Turkey, from Tunisia to England and Germany, and their remains speak to both the care and precision with which properties were laid out and a more precise configuration of actual examples than Pliny's letters permit. Inevitably, it is the outlines and remains of buildings that receive more attention than garden layouts; yet archaeological work, along with some inspired explorations of pollen samples, tree root holes and other palaeobotanical evidence, have disclosed more than we might imagine of these lost gardens. Models, too, and reconstructions (illus. 24), have allowed some sense of sites that are readable now only at ground level; these can also remind us how important must have been the contrast of building elevations with the gardens.

Here we can look further to a selection of elite examples, the evidence of widely scattered sites throughout the Roman Empire, and some sense of public gardens; for all of which we need to draw upon literary texts, archaeological sites and both Renaissance and modern reconstructions.[4]

Three famous imperial gardens and villa layouts, in even their imperfect states, reveal a huge effort of grandiose display. The Emperor Tiberius, after his withdrawal from Rome in AD 27, particularly enjoyed his seaside villas, of which he possessed twelve on the island of Capri alone, such as the Villa Jovis with its terraces down steep cliffs and its (now very ruined) grottoes. A similar exploitation of a natural cave on the coast near Naples became a banqueting hall at Sperlonga (illus. 25): the cave was artificially enlarged, equipped with pools and gardens that extended out towards the water and with a sculptural display that, aptly, represented scenes from the *Odyssey*, including

Ulysses and his companions blinding the giant god Polyphemus (a sculpture that survives).

Another emperor, Nero, established his Domus Aurea (Golden House) in a valley between the Palatine and Esquiline Hills, a site over part of which the Colosseum would later be built (illus. 26). The Domus itself had its own garden courtyards with a rustic grotto, but the main effect was the park itself, approximately 125 acres in extent, and designed to imitate a large rural property within the city walls. An earlier aqueduct was tapped to create a large water body, fed by a series of descending fountains; this pool, as Suetonius was to write, 'looked like the sea, surrounded by buildings which gave the impression of cities; beside these there were rural areas with ploughed fields, vineyards, pastures, and woodlands, and filled with all types of domestic animals and wild beast'. This mixture of wild, solitary countryside with more cultivated and used scenery was devised, according to Tacitus, 'to create a semblance of what Nature had refused to do'. By all accounts Nero was unimpressed; yet after his flight the parkland was opened for public visitation and may then have inspired more awe. The park itself no longer exists, though there are some remains of the buildings; huge caverns of masonry are what we see today.

The largest villa complex ever established in the Roman world must have been that for the Emperor Hadrian at Tivoli, when he ruled between AD 117 and 138. It is also well discussed,[5] and its enormous site more readily comprehended by modern visitors, helped by somewhat indulgent restorations, most notably of its statuary and water works. It is however vast, more a town

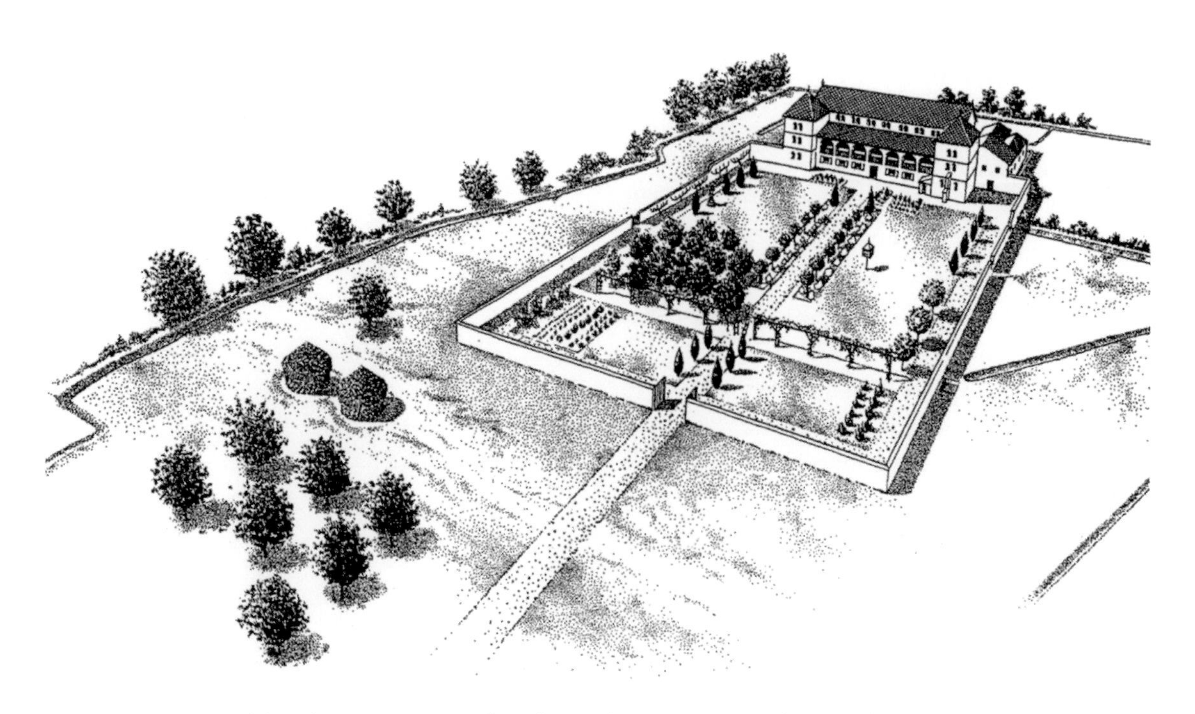

24 A reconstruction of the 4th-century Roman villa and its gardens at Frocester, Gloucestershire.

25 The cave and pool at Sperlonga, Italy, redeveloped by the Emperor Tiberius c. AD 30.

than a villa, a piecemeal layout with no overall configuration; yet it may well have inspired later villa-builders to select various elements from it for their own layouts. Three aspects are of particular significance: it followed the lie of the land; it exploited both statuary and flowing water; and it authorized its own *genius loci* by commemorating a series of other sites elsewhere and far away. These elements, even in the Renaissance, were less visible and constructed than they are now, still served to make the nearby Villa d'Este one of the glories of a revived antiquity (see chapter Six). Here the Cardinal d'Este employed among others the antiquarian Pirro Ligorio to exploit the steep site, to spread its available water supply through a host of different effects and to create, arguably, a scenario that proclaimed the prestige of both Tivoli and the Cardinal while also acknowledging the influence of the neighbouring site of Hadrian's villa.

98 RITRATTO

De la Casa di Seruio Tullio. De la Casa di Nerone, & del Tempio de la Fortuna.

QVella parte di questo Monte Esquilino, la quale è vicino à Suburra, di sopra à la Chiesa di s. Lorenzo in Fontana, era da gli antichi chiamata Cliuo Vrbico, nel quale già era vn Boschetto di

26 A crude engravings of what purported to be the gardens and park of Nero's Domus Aurea of AD 64–68, from Totti's *Ritratto de Roma antica* (1627), a typical guidebook for the early curious tourist.

Hadrian's particular legacy in terms of his villa layout and its association seems to have been to commemorate the holdings of the Roman Empire, which had been expanded hugely under his predecessor, Trajan. Hadrian had himself travelled widely, especially in Greece, which he much admired, and he was seemingly determined to recreate or visualize his travels, particularly those in Rome's eastern empire. A Latin text claims that Hadrian

> fashioned the Tiburtine Villa marvellously, in such a way that he might inscribe there the names of provinces and places most famous and could call them, for instance, the Lyceum, the Academy, the Prytaneum

(council house), Canopus, the Poecile, the Vale of Tempe. And in order to omit nothing, he made an underworld.

Later, sixteenth-century antiquarians continued the fashion by labelling other parts of the Villa, and this particular instinct to associate individual parts with other, distant places is a practice that continues to be much imitated in later gardens.

The eastern valley of Hadrian's Villa is convincing enough as a Vale of Tempe – though miniaturized, as would be the case with many later imitations of original Roman and Greek sites, like the Praeneste Terrace at Rousham in mid-eighteenth century England. The usual identification of the statue-lined canal as the Egyptian Canopus (illus. 27) – the now-vanished canal that linked Alexandria to the town of Canopus – is less plausible, as is the partly destroyed *trinclinium* at its head: the latter, a dining area, might have reflected the licentious goings-on in Egypt as well as some recollection of the various deities of the Nile that were worshipped at Canopus, but it is doubtful that Hadrian himself viewed this canal in this way. The canal was lined with caryatids based on those of the Athenian Acropolis, as well as a sculpture of a crocodile that has also been unearthed there. Hadrian may well have reflected on the places that he knew from his travels without needing to replicate or specifically name them; it is we, or his successors, who have

27 The Canopus at Hadrian's Villa, Tivoli.

28 Modern model of Hadrian's Villa.

felt the need to label the different parts of this huge property, perhaps to anchor its enormous variety and to give some focus to how we might imagine the scattered holdings of the Roman Empire.

That Nero's Domus Aurea after he fled was opened to the public recalls another Roman, as recorded by Plutarch in his life of Julius Caesar. Mark Antony in Shakespeare's *Julius Caesar*, seeking to arouse the mob to revenge Caesar's assignation, invokes the terms of his will:

> Moreover, he hath you left you all his walks,
> His private arbours, and new-planted
> orchards
> On this side Tiber; he hath left them you,
> And to your heirs for ever: common
> pleasures,
> To walk abroad and recreate yourselves.

That there were public gardens in Roman cities is certain, though our records of them are less clear than for villa gardens or the remains of Pompeii and Herculaneum. Innumerable gardens or little orchards were attached to inns and taverns, to baths, to funerary gardens and to tombs, most notably the mausoleum of Augustus in Rome, where the tumulus was ringed with evergreen trees, and its surrounding parkland, as recorded by Suetonius, was opened as a public park (illus. 29). Sacred woodlands dedicated to spirits were also open, but the most frequented gardens were those in various Roman *portici*. One of the first such constructions, by Pompey the Great in 55 BC, was built as an addition to a theatre: derived from Hellenistic prototypes, an area enclosed by covered walkways offered shelter from rain or hot sun and could be used as an assembly point by folk using neighbouring buildings. The enclosed gardens of Pompey's portico had water, statues and 'an avenue thick-planted with plane trees in trim rows', according to the poet Propertius. An inscribed plan of this site on an engraved marble map of Rome, the *Forma Urbis*, was once displayed for public view in AD 200; it marks items that could be lines of pollarded trees, water basins, a double woodland (*numen duplex*) and exedra seating along the outside edge; written sources by the poet Martial and the elder Pliny mention a host of different plants, some dedicated to the relevant deities. Another famous portico was that of Livia, the wife of the Emperor Augustus, which may not have been attached to a building, but existed as a quasi-public garden in its own right.

Such porticoes could be added to a variety of different buildings besides theatres, and were used for a variety of social events, including (according to Ovid) amatory assignations. A large example, partly restored, exists at Mérida in Spain, with a runnel or water basin along three sides of garden and in front of the portico itself. The town of Conímbriga in Portugal had several small gardens, including a portico inside baths, while others are known in Tunisia, Turkey, Athens (a library complex established by Hadrian, where readers could consult their scrolls in the attached garden), at Vaison-la-Romaine in France (where colonnades overlooked a partly excavated garden with a fishpond and central island), and perhaps in a temple at Colchester in England. When the poet Ovid was exiled to the Black Sea, he remembered fondly some of the gardens back in Rome: he recalled the gardens attached to Agrippa's baths or *thermes*, their grass, lakes, water channels and beautiful gardens ('pulchros spectantis in hortos'). But he could also find versions of Roman features that were built rather than simply recollected in banishment; they were a familiar sight around the Mediterranean, for wherever the Romans went, their best and favourite buildings and gardens were repeated over several centuries. In ancient Carthage and Ephesus, as in Libya, Morocco, Tunis, Croatia and Portugal, excavations have revealed columned courtyards, patterned floors of glass or mosaic, fountains,

29 Mausoleum of Augustus, the model in the Museo della Civiltà Romana, Rome.

some lined with jets of water, hunting parks, seaside villas, sculpture, frescoes (at Leptis Magna, Libya, a painting shows people boating on lake in front of a colonaded villa front with a garden behind, illus 30), pools with marine images and several mosaics with mansions set in parklands full of trees. Northwards, villas with spaces for garden courtyards may be seen near Orange and Saint-Rémy-de-Provence, some near the Pyrenees having extremely large sites; country places near Trier and Cologne in Germany often were more simple, with layouts that consisted mainly of rural necessities (orchards and vegetable plots). England is most famous for the excavated palace of Fishbourne, near Chichester, a place of nearly ten acres, with regular gardens (sometimes termed 'formal'), peristyle courtyards, kitchen gardens and what may have been a 'natural' area. Other excavations have located villas with pools beside the Thames (below Cannon Street station), and a late country villa at Frocester in Gloucestershire. Given the more northern climate, planting especially and topography needed to be modified, while retaining much of the usual formal vocabulary and infrastructure of earlier Roman villas.

Throughout Roman culture there was a continuing debate, even criticism, of opulent and luxuriate landscaping. Indeed, it became a familiar *topos* of Roman garden commentary: from Varro objecting to display and unnecessary opulence, to Martial mocking an owner who lays out various gardens but forgets that he also needed rooms for eating and sleeping, to the elder Seneca's attack on the inversion of natural order – 'living at odds with nature' – and the younger Seneca's contempt for every kind of frivolity, including constructing 'natural' landscapes in the very midst of the countryside. Increasingly villas undertook vast

30 Fresco from Leptis Magna, Libya, 2nd century AD.

31 Remains of Horace's modest villa at Licenza.

32 Jacob More, *View near Horace's Villa*, 1777.

infrastructural reworkings of the land – Plutarch told how one consul suspended 'hills over huge tunnels' and reorganized the sea so it surrounded his dwelling, while others reworked land forms, created terraces, removed rocks and established planting where nothing had grown before; 'total landscapes' could be invented and inserted onto available land. Particularly popular were elaborate fishponds, aviaries (even Varro had one) and hunting preserves – so crucial to elegant eating and living were these amenities. Such condemnation and satire were clearly fuelled by such grandiose landscaping as Nero's at the Domus Aurea, but it was clearly not applicable to all Roman villas and gardens, especially those belonging to more modest households or those with no pretension to ostentation (illus. 31). Early Latin writings on horticulture and farming continued to inform these activities, and many outposts of the empire were established with much more practical amenities; what has been called the Romanization of barbarian culture must have diminished or modified some of the more extravagant practices of imperial and elite place-making, even while it adopted some of them.

The spread of gardens throughout the Roman Empire, around and beyond the Mediterranean, ensured that the culture of gardens took hold wherever the Romans billeted their troops or subdued other nations, even if these far-flung gardens did not mirror those in Rome. But many did, especially in climates for which the use of gardens for outdoor living was essential. This surely explains something of the enormous vogue for gardens in later cultures, as hints and memories of Roman garden pleasures and their literary celebrations and archaeological remains permeated them. When Alexander Pope in the early eighteenth century reworked Virgil's *Eclogue* VII ('muscosi fontes, et somno mollior herba / et quae vos rara viridis tegit arbutus umbra') and Handel set his lines to music, they spoke of

> . . . blissful seats,
> The mossy fountains and the green retreats!
> Where-e'er you walk, cool gales shall fan
> the glade,
> Trees, where you sit, shall crown into a shade,
> Where-e'er you tread, the blushing flowers
> shall rise,
> And all things flourish where you turn
> your eyes.

In England in Pope's time, there was little systematic exploration of Roman remains, as had happened earlier in Rome during the Renaissance; there was nonetheless a determined effort to replicate or reimagine gardens of ancient Rome (and increasingly, Greece): at Castle Howard, with its temples and mausoleum, Pope with his grotto and obelisk, or Rousham with its antique recollections of Roman sculpture, pyramid and *cryptoporticus*. Pliny's letters on his two villas and other writings were translated in 1728 by Robert Castell in *The Villas of the Ancients Illustrated*; he extrapolated plausible plans for the ground plans and estate and garden layouts, and wrote extensive commentaries on the Latin descriptions, arguing that ancient Rome's attention 'as well for the Pleasures of a retir'd Life as the Conveniencies and Profits of Agriculture' would be apt for English culture. Its dedication to the Earl of Burlington and its substantial list of subscribers supports this relevance to English usage, especially since Castell's two proposed maps for each of Pliny's estates permit extensive areas of natural and irregular layouts (illus. 33); some of his readers would find such layouts exemplified, or imitated, in their own country estates.

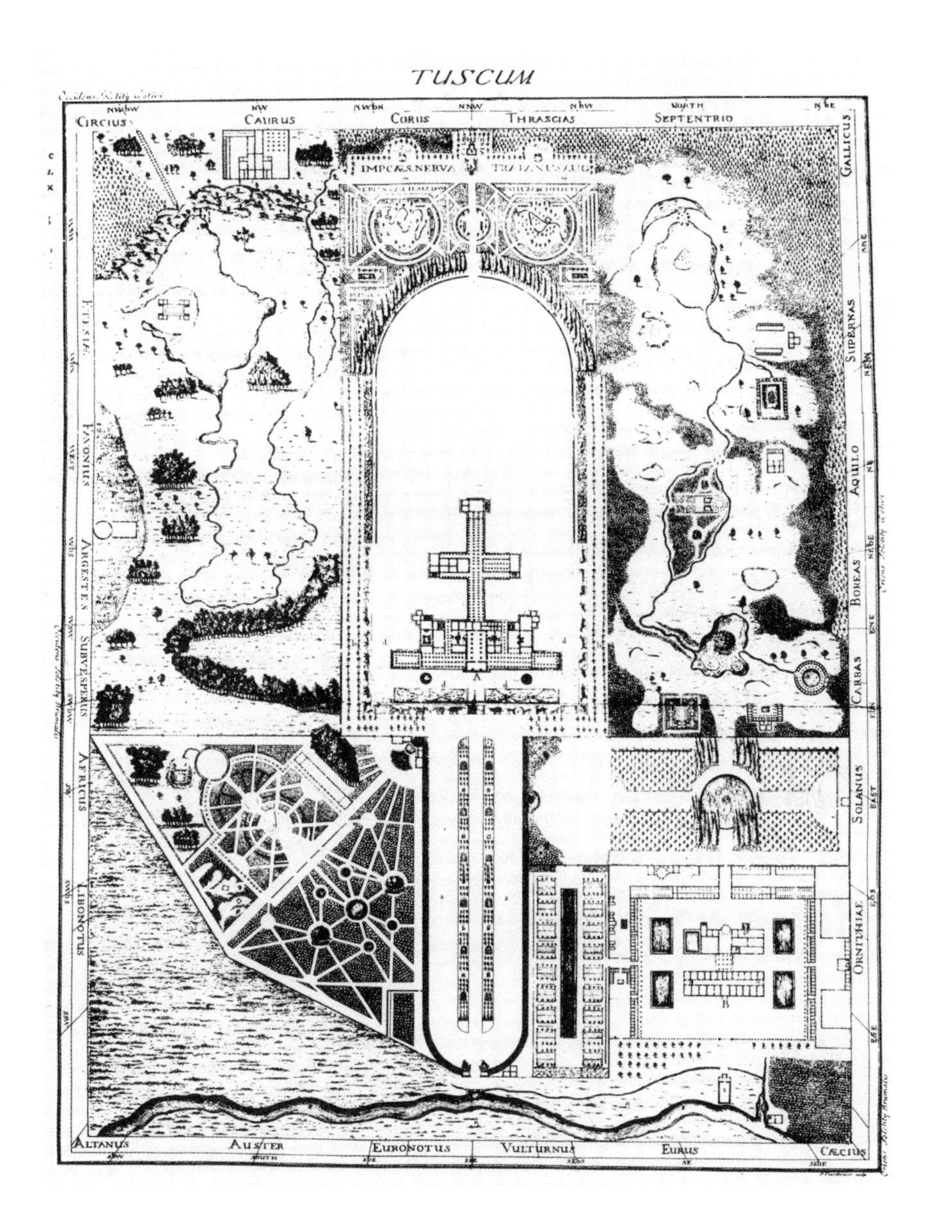

33 Imagined reconstruction of the younger Pliny's villa estate in the Tuscan foothills, engraving, from Robert Castell, *The Villas of the Ancients Illustrated* (1728).

A study by Victoria Emma Pagán, *Rome and the Literature of Gardens* (2006), was published in a series on 'how Classical ideas and material have helped to shape the modern world'. It enjoys a provocative cover, where the gilded and guillotined head of *Apollon Terroriste* by Ian Hamilton Finlay at Little Sparta peers at us through an opening in the garden wall of the House of the Gilded Cupids at Pompeii. It provokes, not least because Finlay used his vision of the classical world to disturb and augment our contemporary habits of landscape. But in fact Pagán's book works better at showing how it is our modern ideas and readings – Finlay's, J. M. Coetzee's *Life and Times of Michael K.* or Carolyn

Forché's poem 'The Garden Shukkei-en' – that shape our approach to the Roman world of gardens, rather than allowing, as the series itself claims, the use of classical materials to shape our own ideas of gardens.

Indeed, garden habits over two thousand years continually colour how we view Roman gardens. For all the superb and documented recovery of ancient sites, like many at Pompeii, we see columns that have been re-erected, roofs replaced, pergolas rebuilt, water channels filled, plants reintroduced and fine pieces of sculpture and other artefacts that have been removed to the safety of museums, and copied for display in spots that they did not originally occupy. There is nothing wrong with such restorations, especially when they are premised upon painstaking efforts by scholars and curators. Yet this 'afterlife' of Roman gardens is refracted through two millennia of other garden art and experience. The hedges that line the broad central path at Fishbourne were replanted with box along lines of bedding trenches discovered there, which alternate semi-circular with rectangular recesses, a familiar Roman pattern, probably occupied by statues (though not here). The hedges look wonderful, and the archaeology for them is doubtless compelling; but they also speak – perhaps unwittingly – of Baroque parterres, and above all of maintenance that makes them look wholly modern. What we bring to 'authentic' old gardens, faithfully restored, are our own techniques of maintaining them and our own habits of mind.

Freud compared the human mind to the city of Rome, with its palimpsestial layers, which can be uncovered by a chance remark or probing enquiry. We still delve into these memories. Revisiting Roman gardens, below the city's modern surfaces, sometimes exposed to view by archaeologists, sometimes revealed by a literary fragment, sometimes confronting a statue or other artefact that once decorated those spaces, we ourselves recover a world of gardens that still resonates today. Certainly, we are unlikely to want to recreate those gardens – though the J. Paul Getty Museum at Malibu, California, has done so, as did Lord Astor, a former US Ambassador to Italy, when he bought a site at Sorrento that contained fragments of a first-century garden by Agrippa Posthumous and where he augmented a lush, even overgrown modern garden with a selection of Roman sculpture and sculptural fragments. Nonetheless, whether it is around our own swimming pools, besides fountains, under pergolas, with items of garden figures (even if gnomes came later), places for recreation and dining (the barbecue pit), or even in miniature intimations of larger landscapes where we fantasize the possibility of the country within the town, we are still in touch with the gardens that the ancient Romans were among the first moderns to create and publicize throughout the Western world.

4 Islamic and Mughal Gardens

The world of Roman gardens encountered Islam, or rather Islam encountered Roman gardens, around the Mediterranean basin, when the Roman empire declined. Yet when Islam was born in the seventh century it flourished alongside the remains of Roman culture, and among Jews and Christians from both Rome and Constantinople. This multicultural aspect is worth emphasizing. If we visit, for example, the surroundings of Palermo in Sicily, we find the remains of a seat of an Arabic governor transformed into a residence and estate for a Christian ruler, the Norman king Roger II (*r.* 1130–1154). Little survives of the Favara Palace (from the Arabic for 'fountain jet') grounds, but the Christian kingdom flourished amid emphatically Islamic buildings and gardens, and life there was ruled by an administration that drew directly upon a mixture of Byzantine and Arabic laws.

Two garden remains at Palermo (illus. 34) demonstrate this cultural mix. The Ziza Palace was commissioned by William I in 1166 (much remodelled in the seventeenth century) and stood facing a pool in an enormous park. Its name is from *al-'Aziz* or 'magnificent' edifice. In its central ground floor hallway, water poured down a sloping marble chute or *shadirwan* into a channel on the floor which emptied into an exterior pool, where a small pavilion stood on a platform in the water; rooms on the second floor looked down onto the garden and its entertainments. An Arabic inscription proclaimed it as the 'loveliest possession of this kingdom' and saluted the 'great king of his century in his beautiful dwelling-place, a house of joy and splendour that suits him well. This is the earthly paradise . . .'. The other garden building that survives (largely as a ruin) is the somewhat later and smaller Cuba, used by Boccaccio as the setting of the sixth story in the fifth day of the *Decameron*, which concerns a lady captured by pirates and given to the king. Its name possibly comes from the Arabic for dome, and it contains traces of what once were magnificent *muqarnas* (ornamental vaulting). The central room contained an interior square pool, decorated with a star on its floor, and with a fountain. The commentary on these buildings, their surroundings and their provenance is complex and need not delay us; but the synthetic mix of Arabic and Christian forms, usages and visitors, must make us wary of seeing a wholly Islamic tradition at work here. And the various uses of other Islamic buildings and gardens at different historical moments, at Córdoba or Istanbul, teach us to see these primarily as places and not as icons of faith or ideology.[1]

34 A modern photograph of the *iwan* in the interior hall of La Zisa, Palermo.

A journey through Islamic and Mughal gardens begins, however, in Spain and then will move eastwards to Iran and the Indian subcontinent.[2] The world of Spanish-Arab gardens was itself influenced, not only by the legacy of ancient Rome (for private courtyards as well as public gardens), but complicated by the subsequent exchange of plant materials from the Americas to the west and by the Italianization of palaces during the Renaissance to the east. Unlike eastern Islamic and Mughal gardens, Spain provides few early visual or literary descriptions of sites. The gardens there today can also be misleading: their planting has changed and been modernized to please later tastes, its garden levels altered or diminished, as in the Court of the Lions at the Alhambra; new garden enclosures have been constructed, and the original sense of place is no longer readily appreciable – Generalife, for instance, was one of three fortified villas, in part to protect the Alhambra, and its considerable orchard and agricultural surroundings have disappeared. Its estate was a dozen times larger than the Alhambra, and its holdings served essentially as a farm to supply the other city-like complex, where official receptions were held (in the Hall of the Ambassadors in the Comares Palace). Yet even considered with a sceptical historical eye, these two gardens at Granada may still convey a strong sense of what such an Islamic city-palace and villa would have been, and of course they are still much visited today, in part because they are surviving examples, however compromised, of early Islamic garden art.

35 An engraving of La Cuba, Palermo, from G. F. Ingrassia, *Informazione del pestiforo . . .* (1576).

36 The 'Court of the Myrtles' in the Alhambra, Granada, from Alexander de Laborde, *Voyage Pittoresque et historique de l'Espagne* (1812).

Islamic gardens were first established in the rich agricultural land around Córdoba between the sixth and tenth centuries, and were indebted to both Roman agricultural practice and that of the Umayyad of Syria, who ruled there. Gardens were created for the caliph and his entourage. A characteristic if still only partially excavated and understood site, in the lands outside Córdoba, was Madinat al-Zahra, which took 40 years to build, using craftsmen from Baghdad and Constantinople and forced Christian labour, and importing vast quantities of building materials from the east. It was characterized by elaborate gardens, extensive hydraulic engineering and views across an extensive panorama, which Ruggles notes as a representation of the authority of the ruler. This generous and empowering regard over surrounding landscapes would also be found in the various *mirador* (what we might call belvederes) that are scattered through the later Alhambra and Generalife.

The huge precinct at Madinat al-Zahra was both palace and city, with various halls, baths, mosque, a zoo, barracks, many other functions (like a mint and workshops) and an autonomous administration. It was disposed down three terraces, each separately walled, on the side of the Sierra Morena mountains: the upper garden was for the palace inhabitants, the second devoted to gardens and orchards, the third to houses and mosque and probably the zoo, itself moated, with an aviary and compounds for exotic animals. The gardens were irrigated and enhanced by water brought from the hills by bridges and underground tunnels, stored above the gardens for use there, and then distributed to farm lands below (some aqueducts at this level were of Roman origin). We need much imagination to visualize this site as it must have been, with pools and many water channels, and arcades opening onto gardens, pavilions and courtyards. For a better grasp of these gardens and courtyards we should go to Granada.

The Alhambra (from the thirteenth century onwards) is perhaps overwhelmed by its fame, the uncertainties of its building programme or development (including the garden enthusiasms of early viziers from a Jewish family), and the

37 Generalife, Granada, Acequia Court. The planting is modern.

romantic lore that later accrued around it (see illus. 36). But the fortress site on its hilltop is impressive, and Islamic commentators praised the fertile region, the landscape of which, though altered, is commanded on all sides. If the Alhambra was a series of enclosed courtyards with small gardens, the gardens of the Generalife descended a hillside and opened it to the countryside. Its main feature is the Acquia Courtyard, a long channel of water, with a fountain at its centre and four quadrant gardens along its length; at one end is a pavilion opening onto a *mirador*. A wonderful water stairway descends the hillside to this main complex; while it might remind one of the *catene d'aqua* at the Villa Lante or Caprarola in Italy, its delicate paving, the handrails on either side, scalloped for the running water, and the intimacy of the pathway are quite different.

The Alhambra is most famous for its two gardens, the Courts of the Myrtles and of the Lions, set at right-angles to each other within adjacent palaces, the first linked to a council room and the sultan's state room with a fountain at its centre, the second leading to the Hall of the Two Sisters. The Court of the Myrtles has an oblong central pool, once planted with oranges as well as myrtles, as noted by the Venetian ambassador Andrea Navagero in the mid-1520s, and with narrow strips of water sunk into the pavement; at each end of the courtyard a circular basin sends a short jet into the air (illus. 38). At one end the courtyard opens into the Hall of the Ambassadors, richly decorated, and on three sides its alcoves have windows that give onto the larger landscape. Inscriptions around this chamber appeal to the cosmos, the heavens and the God who is 'observer of all things', religious parallels to the sultan's power over what he surveys and to the praise of perspective by the fourteenth-century poet from Granada, Ibn Luyun, on 'the gaze [that] may freely roam in its contemplation'.

38 Alhambra: detail of the water spout in the Court of the Myrtles, with the Comares Tower reflected in the pool.

The later, fourteenth-century Court of the Lions is divided into quadrants, once laid out with deep, sunken beds of vegetation, arcaded around the sides with slender columns like a forest, and surrounded with richly carved stucco and perforated screens that seem to blend the interior garden with its surrounding buildings. Jets spring from recessed pools and flow into channels on the garden pavement, where the eponymous and famous lions hold up the basin inscribed with praises of the 'abundant' blessings of the fountain. References

39 Alhambra: Court of the Lions.

have been proposed to link this basin with King Solomon's gardens and with the throne supported by lions that he made for the Queen of Sheba, but the centrepiece maintains its own unique presence in this garden courtyard. The many inscriptions laid into the fabric here and in the Generalife, many of which were written especially to celebrate the Alhambra, are more relevant to the meaning of the building for its Islamic occupants.

Though even a cursory glance at ground plans for Islamic and Mughal gardens (in books by Ruggles and *Il giardino islamico*) will show how varied was their layout (for reasons at once topographical and social), it is possible to identify some basic forms (illus. 40). Most were set out in a variety of grids, in squares and rectangles, within and around which were cultivated enclosures, laid out less rigorously; all were enclosed, what is

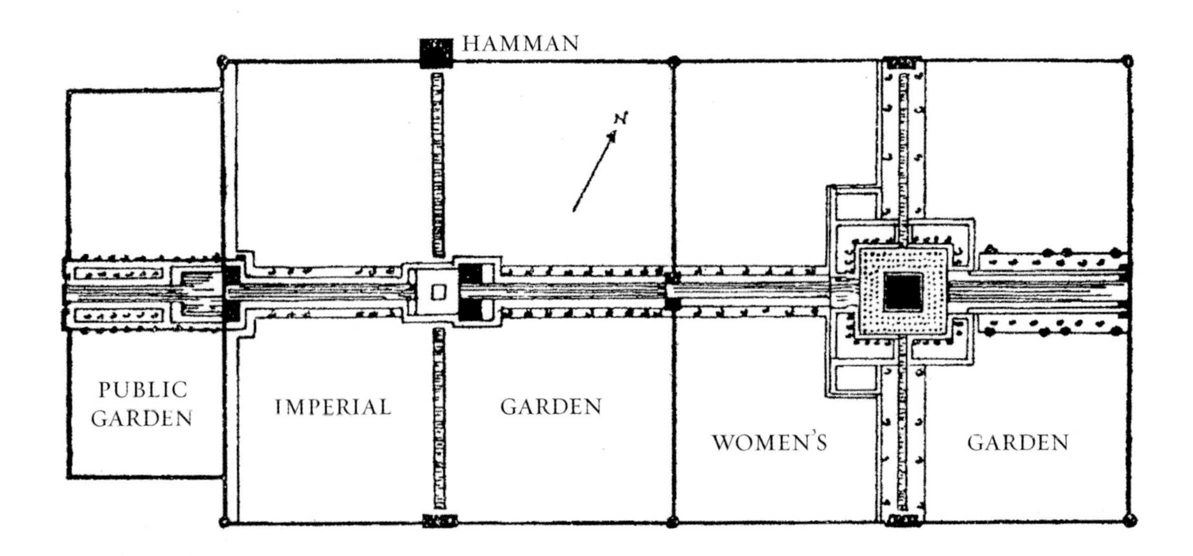

40 Shalamar garden near Srinagar, Kashmir.

termed a *hayr*, derived synecdochically from the word for a pavilion and used also for enclosures, including for menageries; sophisticated irrigation and hydraulic machinery fed central pools and distribution systems to cool visitors and water the grounds; walkways followed the irrigation channels, and these paths defined the garden; many of the planting beds were sunken below the walkways, so that strollers along the paths could see across them, or, seated below, be immersed in the foliage and perfumes. In one instance in Spain, now lost, crossed aqueducts supported on arches allowed visitors to walk beneath them.

We often assume that the quadripartite layout (the *chahar bagh*) was automatic in these gardens, though in practice there was some variety, and the term itself can also be applied to gardens with different formats. A Persian agricultural treatise of 1515 did describe the planting of a *chahar bagh*: it was walled, and consisted of two water bodies along which were paths, lined with flowers and rows of poplars; one of the channels led to a pool or water tank in front of a pavilion, surrounded by a rich carpet of fruit trees, regularly planted, and beds of clover.

Another quasi-myth that frequently appears in discussions is that the quadripartite water channels corresponded to the four rivers of Paradise; but this notion emerged very late in literary commentary and did not authorize its original form or meaning. References to paradise as a wonderful, pleasant, even celestial place are, however, different, some referring to real gardens irrespective of their layout, some to paradises at the beginning and end of time, like the paradise reserved for the faithful. Ibn Luyun, in the midst of a treatise where he focuses on the location of villas, soils, seasonal climate and tools for agriculture and gardening, talks warmly but in the end very vaguely about an enclosed garden layout:

> With regard to houses set amidst gardens
> an elevated site is to be recommended, both
> for reasons of vigilance and of layout: and

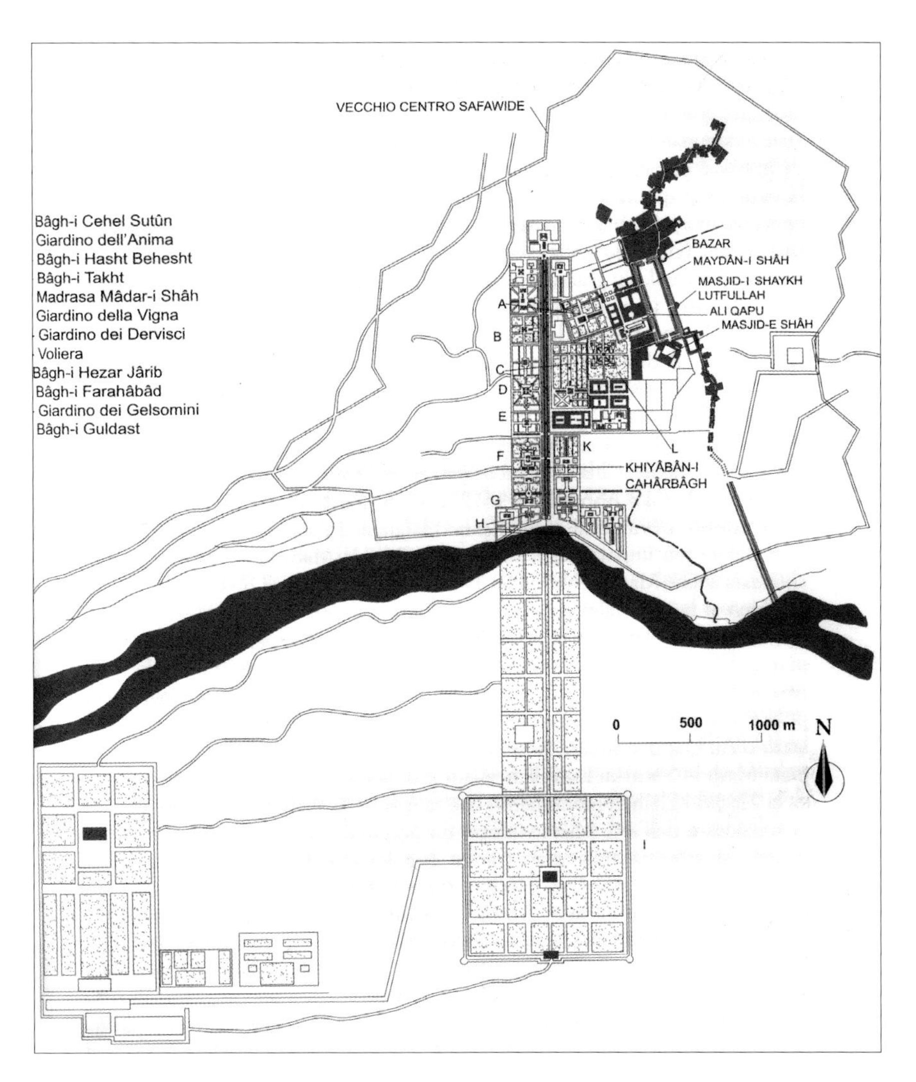

41 The Chahar Bagh, Isfahan, Iran.

> let them have a southern aspect, with the entrance at one side, and on an upper level the cistern and well, or instead of a well have a watercourse where the water runs underneath the shade. And if the house have two doors, greater will be the security it enjoys and easier the rest of its occupant . . .

He was particularly sensitive to questions of security, and the central raised pavilion he describes later in the poem would also have had the advantage of being visible to all who approached it through the garden. He continues:

> Then next to the reservoir plant shrubs whose leaves do not fall and which [therefore] rejoice the sight; and, somewhat further off, arrange flowers of different kinds, and, further off still, evergreen trees, and around the perimeter climbing vines, and in the centre of the whole enclosure

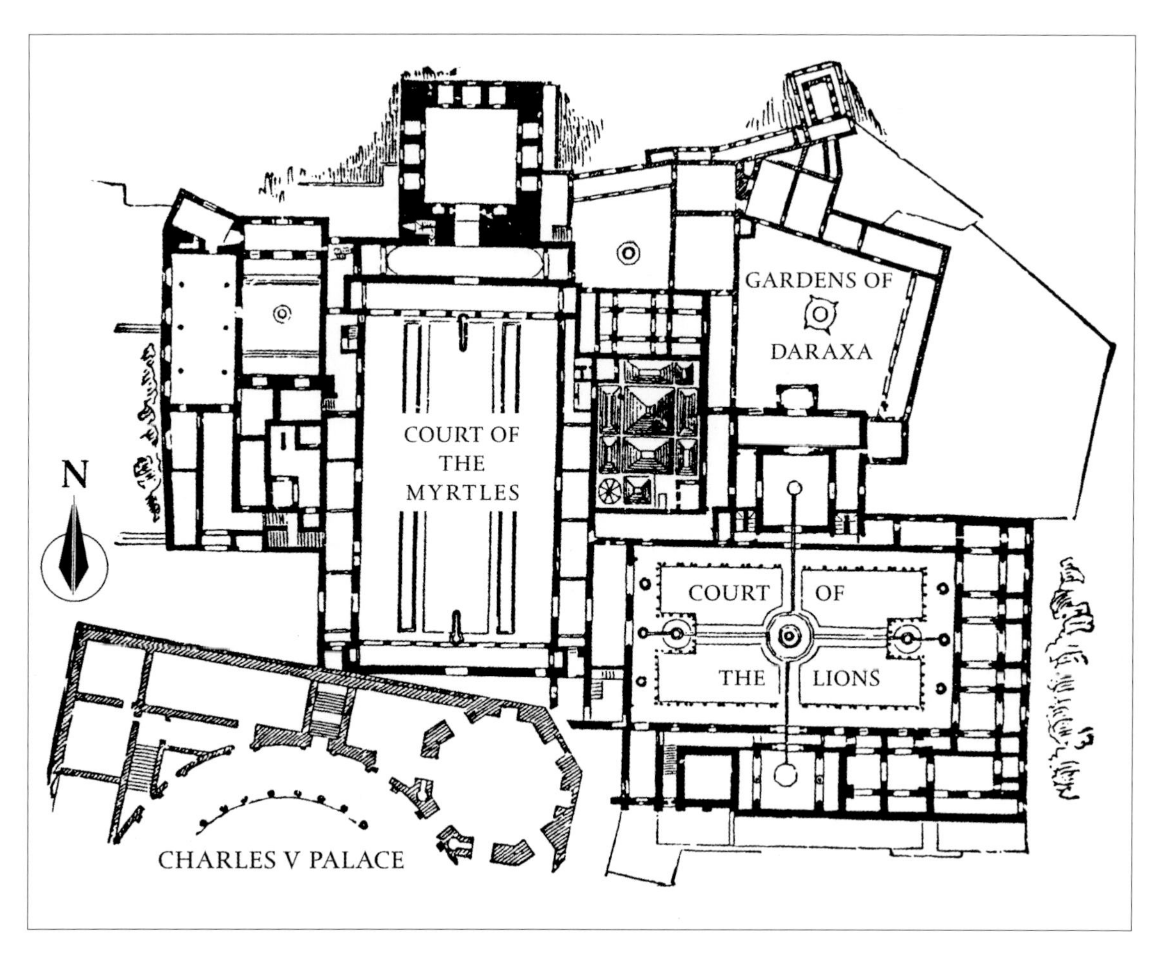

42 The central areas of the Alhambra in Granada, Spain.

> a sufficiency of vines; and under climbing vines let there be paths which surround the garden to serve as margin . . . In the background let there be trees like the fig or any other which does no harm; and any fruit tree which grows big, plant it in a confining basin so that its mature growth may serve as a protection against the north wind without preventing the sun from reaching [the plants]. In the centre of the garden let there be a pavilion in which to sit, and with vistas on all sides, but of such a form that no one approaching could overhear the conversation within and whereunto none could approach undetected. Clinging to it let there be [rambler] roses and myrtle, likewise all manner of plants with which a garden is adorned. And this last should be longer than it is wide in order that the beholder's gaze might expand in its contemplation.[3]

The Arabs were particularly skilled with irrigation and hydraulics, in particular machines

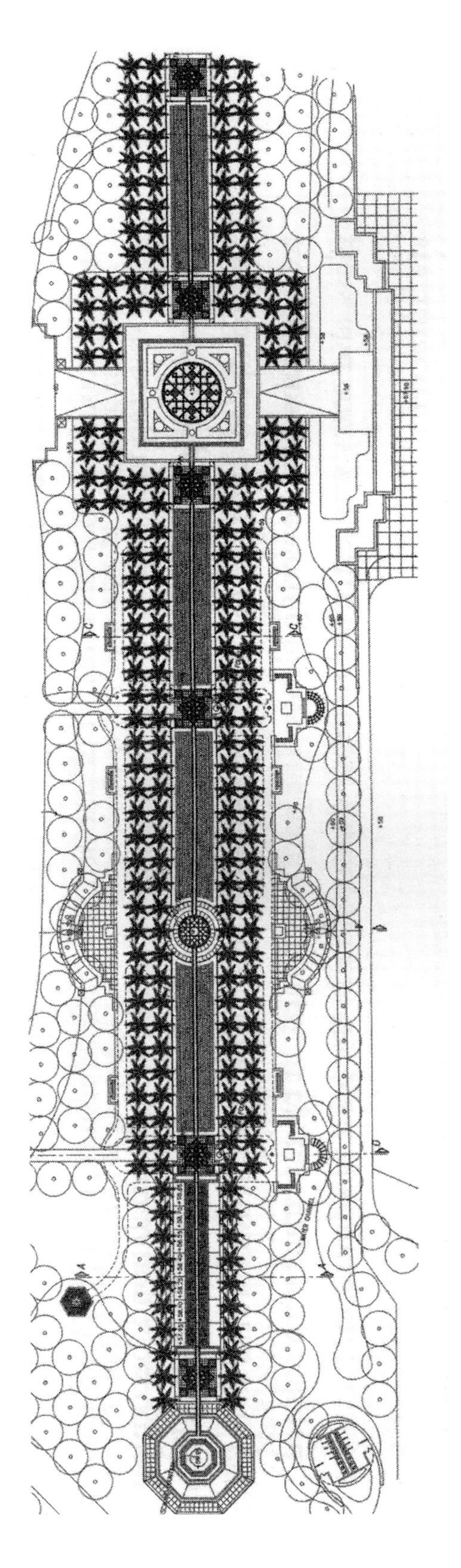

for delivering water like the noria, a mechanism for drawing water in vessels from a well for distribution throughout a garden. This item was well illustrated in several manuscripts, and large functioning ones can be seen today at Hamah in Syria. A much smaller version also waters the eponymous garden in Provence, where its modern designers have suggested a variety of Islamic references, a small oblong pool and pavilion, and vases and benches that give a particularly modern colour to the garden (see illus. 247). Pavilions and water were the standard features of almost all Islamic gardens, and both could be varied and often richly adorned. But as well as regular pools and channels, there were serpentine watercourses – early examples of the meandering rills that featured in English landscapes for their 'naturalness'– such as those that were much prized in a Córdoba garden in the eleventh century and later at Málaga.

The gardens that came to Spain and Sicily had their political roots and technical skills in the Near and Middle East, and eventually found new territories in what are now India and Pakistan. Before Islam, the Arabs had found water and shade at oases; subsequently, both for agriculture and, more slowly, in the making of gardens it was the discovery and manipulation of water resources that determined these sites, abruptly contrasting their green and flourishing worlds with the surrounding landscape of desert and mountain. Garden worlds may have had some symbolic significance, but more crucially they were obvious and visible expressions of a cluster of ideas (beauty, fruitfulness, order, security and elite patrons) contrasted with a bleaker world (chaos, lack of cultivation, enemies and those excluded from the garden). The agricultural

43 Central section of the Aga Khan Trust for Culture's Azhar Park, Cairo.

landscape of Islamic culture, though these days largely built over and unrecognized, surrounded and was the basis for their gardens, which then came to flourish in their midst. The love of flowers, apparent alike in Turkish/Iznik tiles, Indian silks, the decoration of miniatures and carpets, marks this world of careful cultivation; even when wandering tribes settled for only one or two nights, they spread their carpets to make an instant garden, and garden carpets were much prized (illus. 44). For the creation and cultivation of gardens was the mark of an established regime and the expression of a ruler's power was displayed in the building of gardens, or even imaged by the spreading of carpets during his travels.

Iranian excavations have shown that there were garden spaces as early as the seventh century AD, and gardens were a prominent feature of Persian miniatures from the fourteenth century, as was the use of gardens as places of entertainment; the Persian language, Farsi, is replete with imagery of flowers. A miniature illustrating a book of poems, contemporary with Timur (Tamburlaine), shows a visitor on a horse at a gatehouse, with many flowers inside and outside the courtyard, and a stream seems to be used to irrigate the outer garden area (illus. 45). The gardens at Samarkand, Timur's capital, were described extensively by a Spanish envoy, Ruy Gonzáles de Clavijo.[4] He was received by Timur in the Garden of Heart's Delight,

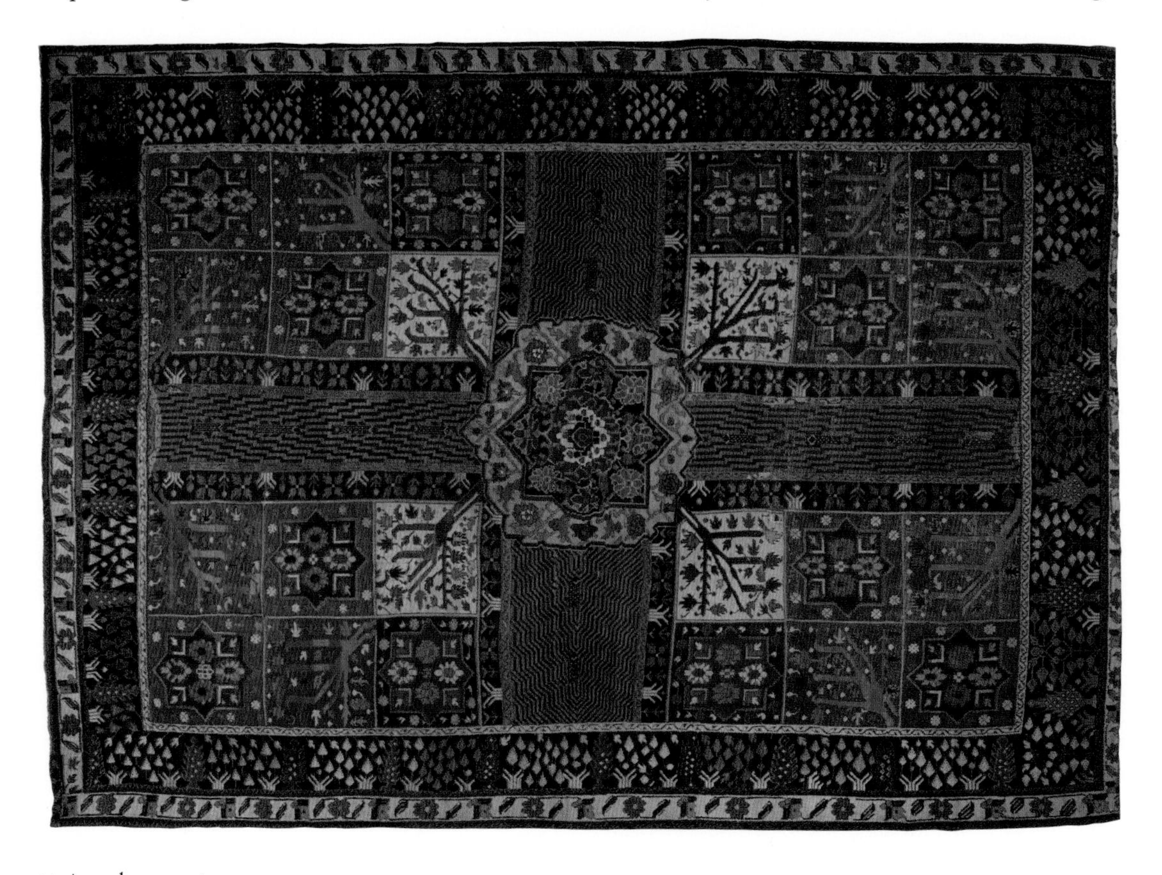

44 A garden carpet.

45 Miniature from the reign of Timur, 1396, watercolour.

established in meadows outside Samarkand, but refers to many other gardens and writes of walled gardens of fruit trees and vineyards, irrigated by streams and decorated with pavilions in the form of tents or awnings, which permitted breezes in the hot climate yet were shaded from the sun. Other travellers who saw these gardens were Marco Polo in 1300 and Sir John Chardin and Sir Thomas Herbert in the seventeenth century: their descriptions, eloquent in Herbert's case about the plants and water features, are enthusiastic yet none the less general and repetitive ('grandeur and fragrance'; 'flowers rare to the eye, sweet to the smell and useful in physic').

Over these visits, and the awe with which these gardens were viewed, hung memories of the famed Hanging Gardens of Babylon. As early as the late third century BC it was described as a palace with lofty stone terraces, that closely represented 'mountain scenery', and two hundred years later it was still imagined to have been invented by a king 'to imitate, through the artifice of a planted garden, the distinctive landscape of Persia'. The hold of these hanging gardens on both architecture and landscape imagination has been enormous, and graphic recreations of them sought to capture their exotic and inspiring engineering, and to envisage exactly how it was reared; even terracing, adorned with flowers and trees, held out the hope that the mythical place had been re-animated, like the Isola Bella on Lake Maggiore (see illus. 109)[5].

But it was real Persian gardens visited by the young Babur that may well have inspired his gardens in Afghanistan and India. The culture and cultivation of flowers and gardens were widespread in the cities of Iran, like Isfahan, Tabriz, Shiraz, Tehran or Herat (now in Afghanistan), where Timur's son made his capital and which Babur also visited. Herbert saw Isfahan during the reign of Shah Abbas I, who created the Chahar Bagh, which exists today as a busy avenue through the city (illus. 41). It was over 2,000 metres long and flanked on either side by enclosed gardens, all with romantic names like the Garden of Flowers or of the Nightingale; it was lined by plane trees and a central watercourse ran down the middle over descending *chadars* (water chutes, scalloped to give the water texture as it flowed down) until it reached the Zayanden River; the whole promenade was linked to other estates and by tree-lined avenues to further gardens on the other side of the water.

Though there are earlier records of Indian garden lore, in Hindu epics and their later illustrations, the eastward movement from Iran of gardens and garden makers was most famously exemplified by the Emperor Babur and, after his death in 1530, by his Mughal successors. He built at least ten gardens around Kabul, as well as his own burial ground, but his work is best known from the famous pair of miniatures that portrayed Babur's Garden of Fidelity (Bagh-i-Wafa) (illus. 46): it was halfway between Kabul and Peshawar, but the site cannot now be identified with any precision. It was built in 1508, but was only illustrated over 80 years later, and the image was undoubtedly influenced by gardens that Babur would go on to build in Agra, Delhi, Lahore and Kashmir; the artist would not have been able to record the planting in the early sixteenth century.[6] It was drawn to illustrate a memoir of Babur, where he wrote of the Garden of Fidelity with its rich plantings (oranges, citrons, pomegranates, with little distinction between ornamental and utilitarian plants) and water that runs down the sloping hillside to a reservoir; some of his words are inscribed on the miniatures. What we see in the miniature is a lush enclosure set within a bleak and rocky landscape, where a deer lurks, ready perhaps to be hunted; the garden itself is laid out according to the instructions of its architect, who is holding a large red-coloured diagram as he directs the workers in their planting, watched by Babur on his right; the gardeners dig and plant in the flower beds, fruits hangs on the trees and birds hover, perhaps to eat the fruit: water courses through the channels and empties into a pool. Horses and visitors arrive at the doorway to the garden, and other guests have already entered and await its inspection. Pleasant as this garden retreat may appear, it was also a centre where politics and administration were conducted.

The sense of precision and immediacy of this garden should not deflect us from thinking of it as the *idea* of a garden conceived of by late sixteenth-century artists. Babur himself reflected much on the places where he established gardens and about the order that a garden would bring to them; yet this image may well be a mode of celebrating the idea of the place rather than what we know (or the artist knew) of its original layout. Babur's *Memoir* talks about a hillside at Bagh-i-Wafa, so it may well be that the garden was laid out in terraces (which the artist avoids, the better to celebrate the garden's order); around the central *chahar bagh* (Babur writes) was apparently a large meadow filled with fruit trees; the artist only hints at this, making the scale of the quadripartite segment disproportionally large. (Another image in the British Library, Or.3714, folio 173b, however, shows a more expansive meadow below the pool and its water courses down the pathways.) The artist's dedication and devotion to the symmetry of the Garden of Fidelity may also be shaped by Babur's own distaste for the disorder and muddle of the region's landscape. 'Formal geometry', says James Wescoat, 'became an emblem of political and cultural control' by Babur and a successor, Akbar.

The narrative of Islamic and early Mughal gardens is handicapped in a variety of ways. We are too much impressed with images that may be more dedicated to the idea of a garden than a specific reality (no harm in that, if we can recognize the image for what it is); those that show reception or visitation are perhaps more useful and accurate; Babur did write of drinking parties, though he would later abandon alcohol. Furthermore, archaeology of gardens has not yet advanced, and anyway gardens always present different problems: pools and water channels, terraces and the footprints of pavilions are more easily identifiable, but planting eludes us, or has been overlaid with subsequent taste. And above all, we tend to lean upon later sites – like the Taj Mahal, most famously – to direct our responses. There is nothing wrong with that either, except (as in all garden history – think of England in the early eighteenth century: see chapter Ten) that it is misleading to read the past because of what followed it and when it seems to be a clearer example of what preceded it; that proleptic narrative risks losing moments of real interest, when societies were reformulating their ideas without yet formatting them. Sometimes earlier examples have a way of getting lost in our certainties of what emerged. And India had a long tradition of gardens before the Mughals, which included pools, arbours and shaded walks, long before there was contact with Islamic culture; then, around 1200, Sultan Firoz developed many gardens around Delhi, probably largely in the form of fruit orchards, all irrigated, and there were accomplished gardens before Babur in Central India (Malava and Baihmani kingdoms).[7]

The Mughal empire in India dates from AD 1526 to 1857, when the British deposed the last emperor. During that time new gardens appeared on the plains of India – at Agra – and on the hillsides of Kashmir. Babur had seen water wells as the solution to providing sufficient water in Hindustan, and he needed them to maintain a supply of flowers, many of which specimens he brought with him from Kabul, while he noted many and gathered many more in travels through India (oleanders, banana plants and much citrus fruit). His successor, far less involved with garden activities, died before (as tradition suggested) he had designed and built his own burial ground: so his widow created one in Delhi on the pattern of an Iranian *chahar bagh*, with a raised mausoleum on a central platform with arcades below, its

46 *Babur Laying Out the Garden of Fidelity*, from a copy of his memoirs from *c.* 1590, watercolour, ink and gold.

درختهای انار همین است گرداگرد حوض تمام سبزه زار
بالای عین باغ همین است در وقت زرد شدن برگهای بسیار
عمل بشنداس چهره نامی نانها

47 View from the dome of Humayun's tomb, looking east to the Yamuna floodplain, Delhi.

quarters further divided into nine units; water channels run down the paths between them and feed small square pools between them; the whole was protected with high walls. It still survives and was restored in 2003 (illus. 47).

But the emperor who, after Babur, most extended garden culture was Akbar, who ruled from 1556 to 1605. He promoted Babur's work through illuminated manuscripts and established his own gardens at Agra and in Kashmir (which he conquered in 1586). His own tomb was vast; a nineteenth-century drawing depicts its multiple gardens, courtyards, pools, fountains and buildings, and tiny gardeners work along the base of the drawing (illus. 50). It was completed by his son, Jahangir, who seems to have inherited his great-grandfather Babur's garden enthusiasm; he was especially enamoured of Kashmir and its rich floral culture ('the flowers of Kashmir are beyond counting and calculation'). But its springs and mountain waters were equally impressive and were exploited, though still in regular forms, in long channels, with cascades and waterfalls descending the hillsides; this tended to break the pattern of a central building with subsidiary pools around it (ideal for flat terrains), and often

48 Humayun's tomb, Delhi, 1560s.

49 An entrance gateway into the gardens surrounding Humayun's tomb.

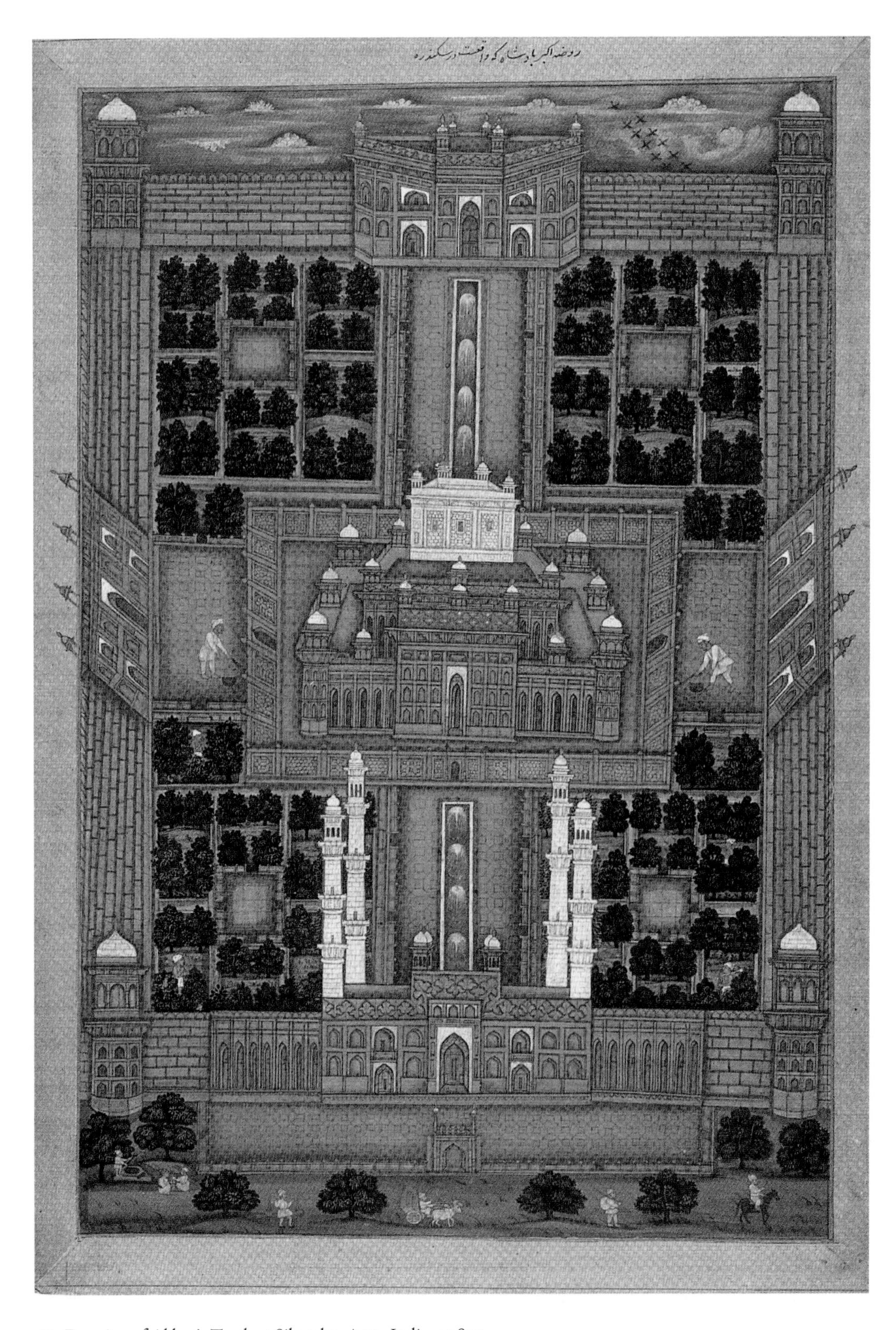

50 Drawing of Akbar's Tomb at Sikandra, Agra, India, *c.* 1870.

placed a pavilion at the head of the water channels, as at Nishat Bagh or Achabal Bagh in Kashmir. This organization was most famously followed at the Taj Mahal, the tomb built for Mumtaz Mahal, wife of the next Mughal ruler, Shah Jahan. Yet its placement here, on the banks of the Yamuna river, allowed it to extend its view across the river to another pool and garden at the Mahtab Bagh. The irrigation demands on these and other Agra waterfront gardens have substantially diminished the hydraulic river's resources at the Taj; the grounds too have been much reworked while the country was under British rule.

The life and forms of the Islamic and Mughal garden extend even to our own time, and still encounter some of the same cultural heterogeneity that marked the early gardens of Spain and Sicily. We may end, then, with three examples: the mixed vision of British and Indian culture in Edwin Lutyens's gardens for the Viceroy's Palace (now the Rashtrapati Bhavan) in New Delhi; and two truly modern examples, in Cairo and in Bradford, England. Egypt is not a surprising site for a new Islamic garden; nor, upon reflection, is Bradford, with its large immigrant population. Yet they both use modern parklands in which to position recollections of Islamic garden art.

British rule in India produced a fascinating mixture of garden cultures, where local traditions of forms – and, necessarily, planting – vied with importations of British social, horticultural and design practice.[8] But one of its most distinguished architects, whose work with gardeners like Gertrude Jekyll contributed a fresh inspiration to gardening back in Europe around 1900 (see chapter Eighteen), was called to work at New Delhi when Great Britain inaugurated this new capital in 1912. Lutyens built the Viceroy's House and laid out its 15-acre gardens inside even larger grounds of 250 acres. Its debts to Mughal garden patterns are clear, and the architect has done his researches well: a *chahar bagh*, but with a grass island for social receptions where a tomb might have been; elements like orthogonal channels, bridged with flat slabs,

51 Maunbari Garden, Amber, Jaipur, India, 17th century. A large, stepped garden, with the topmost section laid out as a *chahar bagh*.

52 Amber palace garden.

textured surfaces in the manner of the chadars, fountains with overlapping disks to 'suggest lotus pads' and steps leading down to the water, 'like ghats [stepped embankments] along a sacred tank'. Yet the site seems over-busy, overwhelmed with its detailing. In contrast, both the grandeur of the site (an imperial version of those Islamic sultans who surveyed the landscape) and the ingredients culled from British traditions – its lawns, its eight tennis courts either side of the axis, the pergolas and the triple, concentric terraces of the circular garden to which the axis leads at its far end – must have (somewhat) solaced the British (illus. 53).

Al-Azhar Park in Cairo, opened in 2005, is an astonishing collection of Islamic motifs, but set within a modern parkland of meandering paths, grading hillsides, playing fields, children's playgrounds and eating places, intended largely for a low-income neighbourhood. A 'Processional Walk' stretches down the centre of the park and displays sunken beds, axial water channels, fountains, polychromic stonework and a geometry that melds a certain modernist purity with the Islamic ingredients. From a hilltop, with views over the city, an axis descends through a series of *chahar baghs* and finally, after shifting to the right, this

53 Looking west over the gardens of Viceroy's House, New Delhi, India, 1912–29. This photograph of Lutyens's gardens is by Lady Beatrix Stanley, wife of Sir George Stanley, the British Governor of Madras from 1928 to 1934.

54 Lister Park, Bradford, England, 2001.

second leg moves past a series of sunken gardens towards a cafe that juts out into a lake. A promenade on the northern edge overlooks the site and uses the Ayyubid wall, recovered from being buried in the ground. Both the archaeology and the design of the park were funded by the Aga Khan Trust for Culture, with a master plan by Sasaki and Sites International as the landscape architects.

Just outside Cartwright Hall at Lister Park in Bradford are the Mughal Gardens (illus. 54). As in Cairo, these are set, somewhat oddly, within modern parkland. But their dialogue with local worthies (Cartwright and Lister) and with the habitual layout and decoration of late nineteenth-century public parks adjacent to them makes this insertion of Pakistan recollections more acute. Perhaps, like the British replaying their memories of British gardens in Asian landscapes, the immigrant newcomers to Yorkshire will find these gestures to Pakistan reassuring: the water channels and slab bridges, the fountains (even though modern security requires pools to be protected with railings, and the planting smacks, necessarily, of local horticulture). Yet in just a few generations, Bradford will respond not just to that nostalgia, overlaid or even forgotten, for Pakistan, but, perhaps more importantly, to an understanding of how inclusive and varied can be a world of gardens and how soothing, too, can be an appeal to formal design.

55 Harold Peto's water channel at Buscot Park, Oxfordshire, 1904.

56 The miller, his mill house and his property, from René d'Anjou, *Mortifiement de vaine plaisance*, *c.* 1470.

5 Western Medieval Gardens: From Cloister to Suburban Backyard

The Western medieval garden was, by and large, an enclosed space. There were other landscapes, notably parks (often for the hunt, see chapter Two, but even those might have been walled to keep in the prey) and large agricultural fields, which were also enclosed within walls, as the *Très Riches Heures du Duc de Berry* make clear. Otherwise, most medieval gardens were relatively small in scale and were securely bounded. Enclosures helped to focus the mind, shielded unwanted vistas of wild and terrifying wilderness, protected the elite from intruders (human and animal alike) and even in much more modest sites still maintained appropriate security against theft and animal invasion. Yet if they were physically restricted, some of the ideas and associations with medieval enclaves were immense and unbounded. It is frequently said that the Middle Ages were thoroughly 'mystical' and 'religious', and much consideration went into the making of gardens; but that did not exclude some very pragmatic and utilitarian endeavours.

Medieval gardens were, of course, as culturally specific as for any time and place; yet paradoxically the essential and longstanding dedication to personal space in gardens is still an ambition that matters today. Protective zones continue to enhance today's suburban backyards (no prying neighbours, keeping children under parental surveillance) and the privileges of private or closed communities (corporate headquarters or academic institutions), but they do tend to provide far less extensive associations than earlier sites. Imagery and ideas in our own gardens may be worlds away from the high walls, turf seats and flowery lawns of medieval gardens. But the instinct for keeping and elaborating one's own plot of ground, a space of personal satisfaction, is the same as ever, and the pleasures of peace and quiet, as well as activity and gaiety, are essential to all gardens.[1]

Medieval gardens were more varied than one tends to imagine – not just the familiar imagery of cloister, lordly retreats and ladies' bowers, or the symbolic or allegorical gardens of Genesis, the *Roman de la Rose* and the *Song of Songs*, but also, though far less visible in both writing and in paintings, the small local holdings of the farmers, millers and those who worked on and for the land. However, those vernacular places have mostly not survived, although other sites may do so, at least by recalling lost gardens – notably in monastic spaces. A profusion of modern recreations of medieval gardens, in France particularly, seems to have revived our thinking about these places. But historically we are dependent on verbal descriptions and visual imagery, with most of our more lively information coming from the later Middle Ages. We should not underestimate how much the basic, familiar enclosure, or versatile courtyard, survived through the centuries. Indeed, the survival of the Roman atrium must have been a powerful formal precedent in the creation of Islamic compounds and monastic cloisters.

This chapter may best be approached as a commentary upon a dozen images and relevant texts.[2] This will display the range of sites that should be emphasized during the period as well as the assumptions that sustain them. This *range* of medieval garden types needs to be emphasized. The important ones, the ones most likely to be acknowledged in image or text, were elite: royal, noble and ecclesiastical sites. They were to be found in towns, attached to manors, castles and palaces, and in a variety of spaces designed for monastic use – herbal and medicinal gardens, individual gardens for Carthusian monks, retreats for important priors and, above all, the communal, principal cloister. But imagery also suggests the frequent incidence of smaller holdings, plots for the essential cultivation of useful foodstuffs. The miller in this miniature (illus. 56), though not as stout and brawny as Chaucer describes him, may also have made a good living ('he hadde a thombe of gold, pardee'), which allows him to keep a small herb garden, a pollarded orchard (both of which are enclosed) and a solid mill house.

Another image presents a more varied scene: labourers are ploughing in the fields, corn is

57 Gardens and fields, with cultivation, from the 14th-century 'St Omer Psalter'.

being cut and (somewhat implausibly, given the harvesting season) a sheep is being sheared, along with other agricultural activities (illus. 57). Around what is presumably a manor house to one side, a residual garden has been laid out with a wattle border around a fine tree; grass is being scythed, a man is digging in the oblong beds, a woman walks in what might be a 'flowery mead', and there is a fruit orchard of two trees, also enclosed by wattles. The long vista allows an extensive view into the countryside, mainly (one imagines) to show off more of the property and its activities and to show how the different segments of the land make up the larger property. It also implies how closely the garden and agricultural work are part of a shared labour force, some of whose workers seem to be eating inside the adjacent farm buildings. Yet there is an appreciable difference between the larger, workaday activity and the more relaxed leisure of the grass for walking, the espalier against the house and the tree for shade.

Other images make clear how much a garden was both valued and yet expendable when a property was attacked or under siege. A wonderful scene of St George battling the dragon, with the princess awaiting her rescue, is watched by a crowd of folk who have retreated into the safety of the castle to watch the confrontation; below the castle walls are enclosed gardens that have also, for the time being at least, been evacuated (illus. 58). Setting manor houses in moats (illus. 59) would serve to protect a property as well as castle walls, but equally necessitated that the garden be located outside the safety zone. Other images show that bears have climbed over the fences to steal the honey from the beehives. In peaceful times, such gardens would be maintained and even, as in this illustration, enlarged under the direction of the lady of the house, who would most likely be charged with supervising the garden as opposed to the agricultural spaces. Other images make clear how much the gardens were both set apart from other living areas and protected with high walls; their doorways signalled a special point of entry, liminal thresholds across which the privileged or the invited could safely pass. Town gardens, too, kept the curious neighbours at bay, through in one particular scene the doorway into the street is wide open – whether to admit the four gardeners, or to allow the curious to observe its elegant series of square beds, which were the conventional format for the layout of flower gardens (illus. 60).

58 Detail from Bernardino Martinori, *St George and the Dragon*, late 1420s, tempera on panel.

Social diversity was of particular concern to one of the best-known and widely circulated authorities on garden-making, Pier de' Crescenzi. His *Liber ruralium commodorum* was written in the early years of the fourteenth century, first in Latin, and then translated into Tuscan Italian in 1350; published editions appeared from 1471 onwards and in all came to number about 60.[3] Since his treatise survives in more than 100 manuscripts in four languages, some later versions of which are illustrated, it was clearly the premier instructional guide to gardening. In his eighth

59 'March', a leaf from a Book of Hours, 16th century.

book ('On making gardens and delightful things . . .'), Crescenzi distinguished between gardens for kings and 'other illustrious and wealthy lords' and those for people of 'moderate means'; yet the materials and techniques he discusses range across a broader agenda of garden-making, and could have been applied to those whose main task was propagation of foodstuffs for themselves. But rank and status nonetheless determined the extent and display in gardens and its

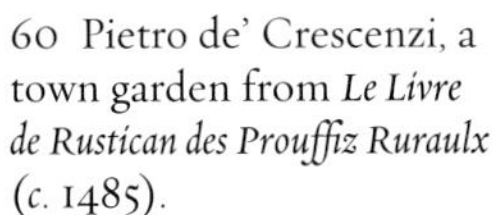

60 Pietro de' Crescenzi, a town garden from *Le Livre de Rustican des Prouffiz Ruraulx* (*c.* 1485).

amusements and pastimes – and things have not changed much today!

Crescenzi's treatise is a veritable encyclopaedia of garden practice and theory, dealing with farms as well as orchards and gardens, with plants and their cultivation and pruning, the preparation and use of fields and their soils, cultivation of vines and other trees, herbs (he lists these and other vegetables in Book 6), farm animals and animals for hunting, meadows and groves. Book 8 is dedicated to making 'delightful gardens'. Like all gardeners, he is concerned with three major elements: a site needs to be selected and prepared, its spaces organized and furnished with proper amenities, and decorated with whatever ornamentation the owner's rank and means allowed.[4] But the modern idea of a garden designer, producing ground plans, sketches or even models, would not have applied;[5] gardens were the work of local workers and household stewards, and it was precisely for such garden-makers that Crescenzi spelt out his various instructions.

He is precise and leaves few details to be assumed: thus gardens are 'made only of plants, others of trees and yet others of both'; clearly for him the plants and plantings were the all-important ingredient.[6] Soils are rigorously explained and their different qualities described; the precise arrangement of flowers and herbs are listed ('rue' is recommended on account of its beautiful colour and because 'its bitterness drives away poisonous animals'); he urges a level ground, surrounded with ditches and hedges of thorn and roses (a hedge of pomegranates could be used in warmer climates); there should be no trees in the middle of turf, not least because people will brush against the spiders' webs between them; turf seats and shade trees are recommended, but also 'shaped' or topiary trees, and pergolas formed 'in the manner of a house or tent'; above all there should be a supply of water, a pure spring ('if possible') conveyed to the centre of the gardens, the centrality of which allows for the irrigation of the whole garden. For wealthy owners, he urges walks and bowers, to be measured out and demarcated 'for delight', as well as a variety of cabinets and other shelters. There will also be 'walls of living trees' that resemble fortifications or palisades, and trees on the top of embankments, all devices that will continue to be used much later, in Baroque gardens or along the city ramparts, well into the eighteenth century.

Crescenzi, then, is a wholly practical garden-maker, and his gardeners are encouraged to use line and measure in their layouts on the ground. Nowhere does he do anything more than hint at the pleasure or delight which his garden instructions and proposals should ensure. We may note, for example, his insistence on the need for a good water supply and irrigation. Yet the remark that water 'may be diverted into the middle of the garden, because its purity produces much pleasantness' suggests, if it does not even conceal, other themes. Anyone who has laid out a fairly extensive garden will appreciate the availability of convenient water,[7] but to many in the Middle Ages that centrality of a 'pure' and 'pleasant' water supply implies other associations, recalling both the biblical description of the Garden of Eden (Genesis 2:8–10) and the praise of a central fountain in Solomon's *Song of Songs* (4:12–16).

Genesis tells the familiar story of how

> the Lord God planted a garden eastward in Eden; and there he put the man whom he had formed. And out of the ground made the Lord God to grow every tree that is pleasant to the sight, and good for food; the tree of life also in the midst of the garden, and the tree of knowledge of good and evil. And a river went out of Eden to water the garden; and from thence it was parted, and became into four heads.

It did not escape the medieval mind that the lost Eden had contained, not only every beast, but 'every tree' that was both easy on the eye and suitable for the table; so that fallen mankind would strive to recover what Crescenzi terms 'an abundance and variety of good trees' in subsequent gardens. But the Bible also notes that the land of Eden provided water for the garden – that is, it drew its water from the larger surrounding territory – and that after irrigation the waters were dispersed into four rivers outside ('thence it was parted . . .'). There is a wonderful manuscript illustration that takes the biblical description literally: it shows a handsome fountain that dominates a walled garden and presumably waters the ground (after a fashion, since there is not much planting inside the seemingly tiny space) and thence is diverted or 'parted' into four rivers, on which boats are sailing (illus. 61).

But part of the surprise in this image is the delightful but wholly implausible image of the dominant fountain itself, drawing its water magically from the earth, and then jetting it from four spouts into the rivers (placed so that we see them,

61 Manuscript illustration by Maitre de Boucicaut for the *Voyages* of Jean de Mandeville, 15th century.

irrespective of their regular spacing around the fountain rim!). But the artist has nevertheless collapsed his understanding of Genesis into his reading of the *Song of Songs*, where a prominent fountain is celebrated:

> A garden enclosed is my sister, my spouse; a spring shut up, a fountain sealed. Thy plants are an orchard of pomegranates, with pleasant fruits; camphire, with spikenard and saffron; calamus and cinnamon, with all trees of frankincense; myrrh and aloes, with all the chief spices: A fountain of gardens, a well of living waters, and streams from Lebanon. Awake, O north wind; and come, thou south; blow upon my garden, that the spices thereof may flow out. Let my beloved come into his garden, and eat his pleasant fruits.

Again, we notice the plenitude of the garden – 'all the spices' of its trees and shrubs.

But the love song that celebrates the beloved, fragrant, pure ('a spring shut up') and ready for the lover came to be read as a paean to the Virgin Mary, the fount of all virtue and the mother of Christ. Hence the two texts came to be entwined, and the original garden from which man and woman were banished – and which is often pictured without any fountain or irrigation system (illus. 62) – is now blessed through the devotion and symbolic fountain of the mother of Christ (illus. 63). Two walled gardens, one about to be abandoned, the other inhabited and watered, now

62 Banishment from Eden, from the Cologne Bible (1479).

coalesce, so that Solomon's fountain comes with hindsight to adorn the garden of the fallen world. Mary is endlessly depicted – as in this woodcut for the *Song of Songs*, but also in innumerable representations of the Virgin – in the presence of a fountain, symbol both of her own fruitfulness and of the nurture of the church. But the presence of fountains, often elaborate and very conspicuous, that occur in many other garden scenes that have nothing to do with the Virgin Mary, nevertheless impart to other amorous activities a suggestion of her original role as the fountain of youth.

The illustration of Emily in her garden does not contain any fountain (illus. 64), but its fruitfulness and ripeness are insistent, and that she is mistaken for Venus by one of her lovers in Chaucer's tale and is initially described as a flower (fair as 'the lily upon his stalke greene' and with the 'rose colour strove her hew') also lends this representation a suitably chivalric air. The artist's choice of this incident from Chaucer's 'The Knight's Tale', which has *its* source in Boccaccio's *Il Teseida*, concerns the rivalry of Palamon and Arcite for the love of Emily, and it is presumably these imprisoned rivals who spy on her as she sits in the garden with her back to them. Their rivalry and ultimately tragic contest of wills – Palamon survives and marries her, Arcite dies – suggests that the garden is not without its dangers. Walled to shut out everything but the blue sky, protected, too, by Emily's sitting on a turf bench in the midst of the garden and sheltered by a trellis of roses, she is still spied upon; and the artist's need to depict her as seated rather than walking up

63 Illustrating the *Song of Songs*, in the *Speculum Humanae Salvationis* (1435).

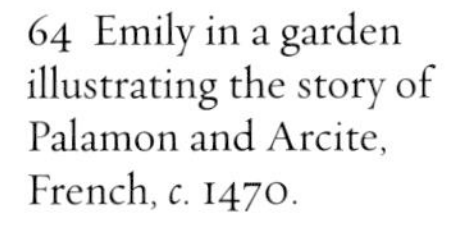

64 Emily in a garden illustrating the story of Palamon and Arcite, French, *c.* 1470.

and down (as she does in Chaucer) makes her presence more vulnerable.

Such small enclosures were, as has been noted, often walled or, in this case, provided with an archway that makes Emily's presence in the inner sanctum particularly notable, and her eavesdroppers the more sinister. But the importance of privacy and accessibility is well presented by the narrative of the *Roman de la Rose*, the denouement of which is pictured as the arrival of the lover in the enclosed and secretive rose-garden. Translated by Geoffrey Chaucer, this long and elaborate French narrative of how a lover seeks, learns to understand and eventually wins his beloved makes much of both its garden settings, with their symbolism (the rose above all), and the very physical designs and constraints of actual medieval gardens. Here, in a modern version of Chaucer's English by F. S. Ellis (1900), are fragments of this poem:

> A garden spied I, great and fair
> The which a castled wall hemmed round.
> The wall was high, and built of hard
> Rough stone, close shut, and strongly barred,
> Enclosing round a garden vast.
> Mine eye with eager joy I cast
> Upon a wicket, straight and small,
> Worked in the stern, forbidding wall . . .

Once entered, the garden is elaborately described and enriched with exotic and various trees, and its flowers give the narrator no little pause in attempting to identify what plants might be missing from the rich array:

> no light task some flower to name
> That was not found thereon, each came
> To lend its beauty, blue periwinkle
> 'Twixt rose and yellow broom did twinkle,
> With violets, pansies, birds-eye blue,
> And flowers untold of varied hue
> Sweet-scented roses red and pale.

There is inevitably a 'fountain clear' beneath a glorious pine tree that beggared belief, for no

> mortal eye e'er seen,
> In any garden as I ween,
> A pine so tall, straight-grown, and fair.
> And in a stone of marble there
> Had nature's hand most deftly made
> A fountain 'neath that pine tree's shade.

The fountain, like the illustration of the fountain in the Garden of Eden that springs miraculously from the earth, is described, despite its obvious artifice, as being 'made' by 'nature's hand'. Thus are the twin roles of skill and nature ultimately blurred in this garden art.

The final climax of the poem, where the lover succeeds in winning the rose, is depicted in one famous manuscript as taking place in a small enclosed rose garden where, like any pilgrim with his staff, he finally arrives at the large and luscious rose. But to arrive at this denouement, the lover must negotiate many challenges, most of which are presented, allegorically, by his arrival at some castle gateway or firmly closed doorway, where inscriptions alert him to the next barrier he must overcome or where gatekeepers accost him with varied advice and directions. A typical scene is depicted in an illustration in a British Museum manuscript (Harley 4425, illus. 65): the lover is met at a threshold by a woman who holds the key that will open the subsequent doorway; as with many images that narrate a sequence of events in a single frame, the lover is next seen inside the entry, having gained access, and is making his way down a garden path towards a garden that he can now see, that could not have been glimpsed over the outer wall; yet even here, he must once again pass through an elaborate archway, its door now open, that finally admits him to this garden, an early promise of his final success.

The scene is eloquent of many gardens, its allegorical furnishings drawing on both extant examples of elite gardens and a whole repertoire of visual depictions. It is enclosed, and the view beyond its walls blocked, save for the blue sky and the crowd of birds. The inner garden with its flowery mead is separated by trellis work from another area with raised beds; the trees bear

65 Courtly company in a garden, illustrating the *Roman de la Rose*; Flemish, late 15th century.

66 The cloister of Sant'Apollonia, Venice.

fruit; music is played, and ladies read poems (this recalls another illustration when the Latin poet Ovid, dressed now in medieval clothes, recites his verses from the *Ars amatoria* to three pairs of lovers gathered beside his desk in a flowery grass plot). The conspicuous and elaborate fountain of the *Roman* illustration occupies the centre of the turf, decorated like a 'flowery mead', and the water, having presumably watered the garden, runs in a channel under the garden walls and out into the neighbouring territory.

The most recognizable medieval garden today is still, as it was then throughout Christendom, the cloister, some still in use and cherished, others, in England especially after Henry VIII's dissolutions, preserved in ruins. The cloister was a mode of self-protection and a focusing of the religious mind: turning away from the outside world, with the divine territory of the sky as the only exterior landscape, its enclosure ensuring that the 'life within' was where the focus would remain. Apart from the actual rituals within the church itself, it was the symbolic centre of the monastery or abbey. Cloisters come in all sizes – my particular favourite is the very small cloister, now paved, in Venice of Sant'Apollonia (previously Santa Scholastica, illus. 66). The surrounding walk allows its users to air themselves in rain or heat, with shade provided by the arcade, and in larger ones this gives ample scope for exercise; the cross-paths would be derived from garden walks, and may be embellished with flowers, and the central well – the most feasible and pragmatic source of water for the community – honours both the Solomonic song as the image of the Church and images of Eden's water supply, recovered now in our post-lapsarian world. Some fountains, however, can be as elaborate as medieval images showed them

67 René d'Anjou, in his house with a chequer-board garden beyond, a manuscript of *Le Mortifiement de vaine plaisance*, c. 1458.

to be: that at Monreale, outside Palermo, sits in its own enclosed garden at one corner of the cloister and is handsomely decorated, while the central Venetian well at Sant'Apollonia is plainer (actually the wellhead gives access to a cistern below where rainwater was collected, a necessary activity in the salty lagoon). Big monastic institutes would, like manor houses, have areas for the supply of food and pharmaceutical herbs for use in the infirmary and to cater to the sick who came to them for medicine. Some large institutions – for example Carthusian monasteries, of which there are good remains at Mount Grace Priory in North Yorkshire – provided special gardens for the abbot or prior, while individual monks enjoyed a garden attached to each of their cells.[8]

There is, in many ways, no reason to connect medieval gardens with anything we might think of today. Our associations for gardens do not allude to Genesis, the *Song of Songs* or any symbolism of the church or Virgin Mary. Gardens are no longer only a site of elite pastimes, and while they can still today be a *locus* of lovemaking and even courtship, there are other places that welcome such activities today – even the public park, where the illusion of being alone with one's lover and being ignored by others around one is still richly appreciated.

68 Medieval garden at Daoulas Abbey, Finistère, Brittany.

69 Withy fencing at the reconstructed medieval garden of the Templar Commandery in Coulommiers, Ile-de-France.

70 A Franco-Burgundian garden, from an illustration from the second half of the 15th century to Pietro de' Crescenzi's *Livre des prouffitz champestres et ruraulx*.

But in many ways the medieval garden continues to provide a diverse set of uses and associations. Above all, many gardens, private certainly but also institutional, are both enclosed and their entry sometimes limited. Crescenzi's agenda of items by which to decorate a garden are still valid: seats, though now wooden or iron rather than turf, are necessary; shade is useful, and the various devices – pergolas, trellis work, different formats for flower beds – are still used. We use pavers to cover parts of the garden that may receive hard wear or eliminate maintenance, and chequerboard patterns, like those in the manuscript of *Le Mortifiement de vaine plaisance* (illus. 67), are still an appreciated format. Indeed, the modern and ubiquitous garden centre usually provides all of these materials and ideas, especially the gateway or – much used as an entry into private gardens throughout today's suburbs – the archway, decorated with a climbing vine, that signals an entry into somebody's personal space.

It is not only the English for whom the home and its gardens are their castle. Privacy seems even more obligatory in a world where different devices – mobile phone and Blackberry – can intrude. Neighbours still like to think their own family patch is by invitation only, that children and indeed adults can be safe and can enjoy their own personal space and behave without prying eyes. As Crescenzi argued, gardens are still where we take delight in growing flowers, vegetables and fruit trees, and where, even for him, there was little sense of any obligation to acknowledge larger issues and beliefs. True, community gardens provide opportunities for those without the chance of owning or renting their own private enclosure, but even here they have a kind of owners' pride and relish in doing one's own thing in one's own space. And gardens may still, especially those with space and variety, provide places where the pleasures of intimacy or the obligations of security can be secured. The *New York Times* (12 May 2010) noted that Hamid Karzai was taken into private gardens in Kabul for one-on-one talks with a visiting American Vice-President and again by a Secretary of State in a Washington garden. This recalls the ancient use of Mughal gardens with their central and raised pavilion (see chapter Four), where eavesdropping could be prevented and intruders or visitors spotted before they get too close; but then my Philadelphia neighbour also takes his phone into the garden to receive his calls.

6 The Renaissance Recovery of Antique Garden Forms and Usages

Garden-making is an atavistic affair, and horticulture is deeply committed to both its tools and its procedures. Indeed, it has been somewhat hard over the centuries to 'invent' new gardens, whether as overall layouts, or by stressing individual elements. It is not only the modernist landscape architect who, determined to 'make it new', struggled to devise new forms, new materials (much easier) or new experiences (also easier, since new behaviour allowed fresh responses to old forms). It is both astonishing and even amusing to see some modern landscape architects seemingly or deliberately forget past endeavours in their wish to push the envelope of their profession; they are not the first to have done so.

Turning a blind eye to earlier work can give any architect the advantage of fresh creativity, even if, with hindsight, historians may trace the latent antecedents of current work. Garden-makers have frequently decided, very successfully, to 're-invent' old garden forms and usages without any conscious acknowledgement of previous work; sometimes the re-imagining can be done with an awareness of what the past has supposedly offered them. In the history of landscape architecture some recourse to the past, or a determined blindness to its practices, has sustained much 'new' work. Earlier ideas and forms have been re-used: the atrium has been a viable garden structure – an inside/outside formula – since at least the Roman empire, through medieval cloisters, academic layouts and modern office headquarters; the *chahar bagh* (fourfold garden) perpetuates itself – not least because of the elegance of the idea – in modern reworkings of this formula; the villa culture of ancient Rome continued to inspire those with country seats and mansions; theatrical spaces continue to describe both physical layouts and receptions of place. What was claimed to be the 'new' England landscapes of the eighteenth century did not by any means spring fully armed from the heads of Kent or Brown; yet it appeared fresh and innovative and ignored or silently cannibalized earlier garden forms. In its turn, though, it provoked something of a push back during the nineteenth century, with an array of appeals to earlier forms and styles, only to be met then with the elegant inventions of the Arts and Crafts movement, which very consciously used older forms to generate new layouts and new social uses.

The Renaissance has acquired its name in part because conspicuously and determinedly it chose, from at least the late fourteenth century, to revive classical forms and uses, usually Roman at that point. That surge of academic interest and publications, archaeological investigations and building designs and practice left its mark firmly upon garden designs. And what began by being a largely Roman influence was extended and supplemented in the eighteenth century with a wave of new and now knowledgeable Greek and Hellenistic experience, which also left its mark upon the garden. But by that time a variety of other cultures had also inspired or seduced garden-makers. However, it is the Italian Renaissance garden that is primarily the topic of this chapter.

In garden matters, the Italian Renaissance did pose some difficulties, since most Roman gardens had completely vanished. Nonetheless, from the late thirteenth through at least the seventeenth century an ancient garden culture was explored through several alternative and indirect routes, and what transpired in Italy during that time elicited an astonishing profusion of some of the most beautiful and exciting new gardens in the modern world. We can still visit many of them today, though we are probably less attentive to just how these places emerged out of a revival of interest in Roman garden-making. Some have obviously changed to greater or lesser extents over the succeeding centuries – yet Villa d'Este or Villa Lante still remain as potent examples of that resurgent garden culture of the sixteenth century. Some gardens remain mere fossils of what was created, while others have been transformed by the introduction of yet newer modern styles; for into the very heart of Italy by the end of the eighteenth century came new 'English' models of gardens. This English mode came partly as a result of reacting against the spread of Italian Renaissance gardens into northern countries, though again without fully comprehending the extent to which classical Rome and early modern Italy had themselves contributed to an evolution of this new 'naturalism'.

Renaissance Italian garden expertise was grounded in a careful reading of ancient texts. These in their turn were matched to whatever modern sites had survived or were considered ancient survivals (some accurately so, others a result of wishful thinking). And then, given that gardening and agriculture were themselves such a traditional activity, ancient literary descriptions of Roman gardens and villas could be reimagined in terms of what activities still survived in general use: terracing was an ancient agricultural activity, a necessity that worked well on steep hillsides; irrigation and water supplies were a further obligation for all gardeners, and survivals of Roman aqueducts were everywhere, if often defunct; and planting itself seemed to perpetuate centuries-old habits of cultivation and maintenance. When Wilhemina Jashemski was engaged in her pioneering study and exploration of Pompeii and Herculaneum (see chapter Three), she was astonished to see how local craftsmen and gardeners continued to work in the traditions that they presumably had inherited for generations.

Visitors to Rome could consult ancient writers, and they could use guidebooks. An English translation of an Italian writer called his version *Italy in Its Original Glory, Ruine and Revival* (1660), and that was the triple perspective that most tourists adapted: they could read about ancient gardens, hope to see something of their remains, and marvel at what they took to be re-creations. Joseph Addison read relevant texts before his departure from England: 'I took care to refresh my Memory among the *Classic* Authors'. Once arrived in Italy, tourists could buy guides or even maps, or take local advice (accurate or not) that led them to visit places that they were happy to have associated with classical figures. Ellis Veryard knew that the Esquiline Hill in Rome was 'famous for the gardens of

Maecenas' because he had read about it, and at Frascati, Richard Lassels knew that 'here Cato was born, here Lucullus delighted himself, and Cicero studied'; others, like Veryard, walked out to view what he described as 'a great heap of rubbish . . . and a house almost entire, said to have been Cicero's'.[1] An early traveller, Sir Thomas Hoby, went to Cicero's villa near Mola, and John Raymond also visited 'Ciceros Grote, in which he wrote many of his familiar *Epistles*'. Whether Cicero did indeed compose his letters there was less important than the claim to have made connections with the place. If the site at Mola was 'full of bewtifull gardines', as Hoby asserts, it is doubtful that they were 'original': but they may have looked genuine. Authenticity was so much in the mind of the visitor. Addison viewed the Italian landscape in the light of 'the landskips that the Poets have given of it', thereby acknowledging a triple influence – of ancient poets upon modern painters and then upon his own perception.

Guidebooks were eventually made available to tourists, and they included images of both ancient and modern gardens as well as reconstructions of ancient sites, as in Giacomo Lauro's *Antiquae Urbis Splendor* (1612 and later editions). The ancient sites were on the whole more wishful than accurate (illus. 71); writers and readers tended to envisage ancient gardens in terms of what they knew of their own gardens. Publishers often reproduced identical images to illustrate two different sites, as in Giuseppe Mormile's *Descrittione della Città di Napoli* (1670); there was nothing unusual in this; his book derived its woodcuts from an earlier book on the Pozzuoli area near Naples of 1594. There were maps of ancient Rome that listed its gardens or *horti* in Bartolomeo Marliani's *Urbis Romae Topographia* (1534), and Lucio Fauno's much re-issued *Delle antichità della città di Roma* identified ruins via classical literary descriptions. Among the most famous ancient Roman gardens were those belonging to the poets Ovid and Sallust (illus. 71), and their locations were identified and discussed in Basil Kennett's *Romae Antiquae Notitia*, published first in London in 1696, with five more editions until 1713. The area most identified with the ancient city was the Pincian Hill, the Mons Hortorum (Hill of Gardens), which is where Sallust in the first century BC laid out his gardens in part upon the Pincio and Quirinal hills in the north-east corner of Rome. It was on the Pincio that the modern Villa Medici was established in the mid-fifteenth century, decorated with the spoils of excavations and built over the vaulted remains of the Basilica of Maxentius (the Temple of Peace); climbing the hill today in the *bosco* of the villa we can glimpse these subterranean caverns.[2]

Many early guidebooks for Rome narrated and sometimes illustrated, first, what they could of the classical antecedents of the city and then commenced a description of the modern city by visiting the Belvedere Courtyard at the Vatican. This complex served as a bridge between ancient and modern Rome for the traveller, as it does for any narrative of the Italian garden revival. Both in its forms and in its uses the Belvedere Courtyard deliberately reflected and reworked classical precedent. It borrowed formal elements from Roman buildings, gathered Roman sculpture as in a museum and paraded events in its theatrical spaces that recalled Roman entertainments.[3]

The courtyard was established in the early sixteenth century by Julius II in order to join the main Vatican complex at the south to the casino or little palace on the summit of Monte Sant' Egidio to the north. This building was probably constructed by Innocent VIII (died 1492) and must be numbered among the very early Renaissance

71 Reconstruction of the gardens of Sallust, from *Roma antica e moderna* (1750).

villas to resume the antique idea; however in architectural style it resembled medieval rural buildings with crenellations. Sketches of the Belvedere casino viewed from the north by Maarten van Heemskerck in the 1530s (illus. 72) show the building high on its hilltop, with an open loggia at ground level and an arcade above; we know that it was established in proximity to the *vigna* that Pope Innocent had purchased before his death and it was surrounded by a garden with cypress and citrus trees. It enjoyed an extensive panorama of the famous approach to the city from the west and north, with the Ponte Milvio and mountains behind. It was that view that would have greeted arriving pilgrims and ambassadors to the holy city. The interior of the little palace or villa was decorated, in the antique manner, with hunting, pastoral and seaside landscapes.

72 Maarten van Heemskerck, *View of the Belvedere, Vatican*, 1534–5, pen and ink.

The decision to rethink the uneven and steep slope from the casino back towards St Peter's and the Vatican apartments was undertaken by a new pope, Julius II, under the direction of Bramante. Against the rear facade of the casino was laid out a huge garden, with a central fountain; then matching ramps descended to a lower garden platform, with a nymphaeum or grotto placed in the massive wall between them; from that lower platform a set of straight stairs reached the lowest level (illus. 73). Down both sides of the courtyard were planned corridors, as Vasari called them, from which loggias and arcades would give onto the space below; these were reminiscent of ancient villas with covered passageways or *cryptoportici*. At the top right-hand corner of the courtyard a gentle ramp was created to reach the Belvedere villa.

The courtyard was intended for multiple uses. It suggested a huge amphitheatre, though on three levels, broken by the ramps and then by the stairs. At the very bottom was the area where different tournaments and festivals could be presented.

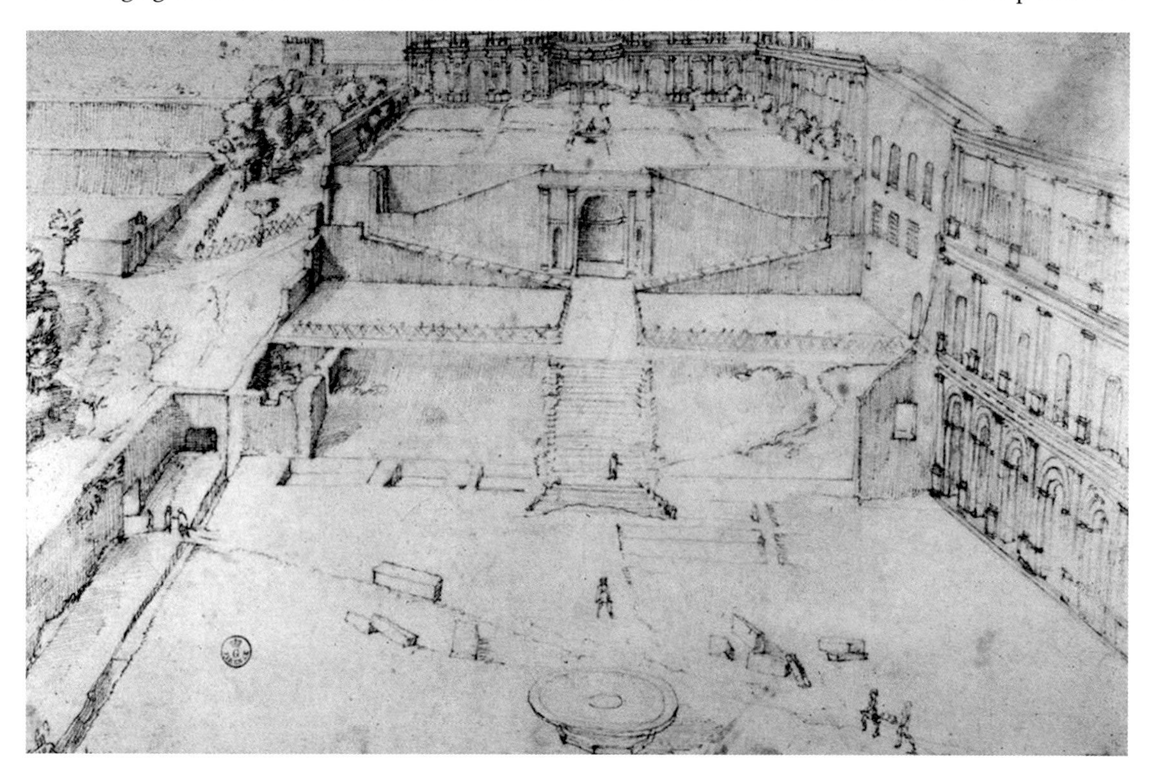

73 Giovanni Antonio Dosio, *The Cortile del Belvedere*, c. 1560, pen and ink.

74 Prospero Fontana, The Cortile del Belvedere with a scene of a naumachia, *c.* 1545, detached fresco from the Castel Sant'Angelo, Rome.

A bullfight was staged in 1509, and animals – including an elephant gifted to the Pope by the king of Portugal – were left to roam there. A pair of engravings by Antonio Lafrery from 1565 show a marriage tournament in that lower space, with views both north and south, and spectators thronging the balconies; another seventeenth-century painting probably copied that same event. But the most suggestive image of this theatrical space was a fresco painted by Prospero Fontana for the nearby Castel Sant'Angelo (illus. 74): it shows boats engaged in a naval battle in the flooded courtyard, a recreation of an antique *naumachia*. Such events had been held originally in the Colosseum, and the vogue for this Roman entertainment seemed to have triggered several modern re-creations: Fontana's fresco implies that the lower area could be flooded for such events, but in other situations it was rather the miniature imitation of a *naumachia* that was envisaged: the later gardens at the Villa Lante incorporated a similar event, consisting of stone boats with jets bombarding against the central fountain (illus. 75).

Today, the building of the eighteenth-century Museo Pio-Clementino has divided the grandiose complex and, breaking its visual unity, prevents us grasping its imposing length (over 1,000 metres) and descent (a 25-metre drop); a further piece of the Vatican library, further dividing the space, was also built in the early nineteenth century. It is only paintings, drawings and engravings that can help us to imagine this striking creation. So many other changes have occurred – the nymphaeum survives in a ruined state and is no longer part of the visual effect, and an important reference to an antique Roman temple has also been lost. This was Bramante's borrowing of a scenographic staircase from the Roman Temple of Fortune at Palestrina (the ancient Praeneste): this so-called 'double staircase' consisted of a central platform with concave stairs above, and convex stairs below.[4] The

75 The water parterre of the Villa Lante at Bagnaia near Viterbo, Italy.

76 Fresco of the Villa Lante within the Villa itself (1574–8).

77 The exedra (double staircase) in the Vatican Belvedere, from Sebastiano Serlio, *Tutte l'opere d'architettura* . . . Book III [1540] (1619).

effect – which Sebastiano Serlio engraved for his *Tutte l'opere d'architettura* in 1584 (illus. 77) – offers a stage-like experience (see chapter Nine for a discussion of the theatrical implications of this device). Another niche-like platform, though much less elegant, was later devised by Pirro Ligorio for that end of the highest garden.

Hidden behind that highest facade, aligned down the courtyard from its exedral apse, lay the sculpture garden of the Belvedere Villa, the Cortile delle Statue (illus. 78). This was set out as a garden with citrus trees and fountains, and surrounded by niches to display Julius II's collection of antique discoveries – a Venus, an Ariadne (sometimes mistaken for Cleopatra), the River Nile, Apollo and the famous Laocoön; a further garden room was adjacent to it. Its entrance was marked by a Virgilian inscription ('Procul este profane' – away you un-initiates) that announced the special and private space of this collection. The same inscription would be invoked by Henry Hoare on a temple at Stourhead in the eighteenth century, along with a copy of the sleeping nymph or Ariadne that he placed in his grotto there.

Despite its private nature, the sculpture court could actually be reached by a public staircase, the gentle ramp or 'snail' staircase that Bramante built at the side of Innocent's palazzino; it was used by visitors and its entry duly noted in local guidebooks. The space was modified in the eighteenth century, and it survives in that later configuration today. The importance, prestige and influence of this sculpture museum in a garden cannot be underestimated.[5] It is not that its dimensions or layout were of much consequence in subsequent imitations, but the display of antique sculpture in a garden setting has been influential. The proliferation of modern sculpture gardens is a direct descendent of that Renaissance initiative, and its opportunity to display massive works of art is of a piece with the papal need to accommodate large sculptural pieces (see chapter Twenty).

The discovery of Roman antiquities in and around Rome in the fifteenth and sixteenth centuries provided an endless provision of ancient pieces for collections besides that of the Vatican.[5] Often the discovery of ancient pieces was an indication of where traces of original Roman gardens might be found and studied. Leonard Barkan has charted this process in his *Unearthing the Past: Archaeology and Aesthetics in the Making of Renaissance Culture* (1999): he recounts discoveries of both fragments and whole pieces, their provenance in early literary texts, new locations for them in Italy and even abroad, the manufacture of contemporary copies and of the restorations of damaged items, and the lessons taught to artists who could now draw from these rediscovered classical images. The Laocoön was among the most famous of these discoveries in January 1506, and was immediately bought by Julius II for his sculpture collection. But extraordinary as the Laocoön was, becoming over the years both a theme for scholarly debate about the relative virtues of visual arts and literature and the object of much learned reconstruction of its entwined and anguished figures, there were many other finds, less imposing, but no less authentic in recalling the lost traditions of ancient Rome. And where originals were unavailable there soon developed a host of copyists in Italy and especially abroad, who could provide for houses and gardens a veritable pantheon of classical figures. (The Laocoön can be glimpsed on the left side of the courtyard in van Cleef's painting opposite.)

These visible signs of authentic *romanitas* were featured in gardens that also imitated ancient garden forms, like exedras, porticoes, terracing, elegant stairways, grottoes or nymphaea. Whether or not a specific feature 'spoke' clearly of *romanitas* in some garden, there was a certain display of good faith in both visitors and patrons for identifying and saluting Roman precedence: an intriguing example is the seventeenth-century view of the 'Villa of Servilius Vatia', so entitled on the engraving: yet the landscape shows no sign of a villa (unless it is the tiny castle in the background) and otherwise the scene is peopled with fishermen. It seems easy then for someone with the right habit of mind to look at any topography and imagine it as a Roman villa. Joseph Spence took his idea of a *ferme ornée* that he laid out at Byfleet in Surrey simply from the sight of 'Fields, going from Rome to Venice'.[6]

A garden nymphaeum did obviously speak of Rome – by its name (a useful way of giving a cave or grotto a special status), by the likelihood that it was occupied by the statue of a suitable classical figure and by the presence there of water (illus. 79). Water in its many forms was both in ineluctable part of Renaissance gardens, and, more importantly, a clear signal that the Renaissance was relying upon long traditions of water supply and the mechanics of hydrology. Renaissance gardens needed water; all gardens do, but none more so than those which featured elaborate and triumphant displays of water. In and around both Florence and Rome, new gardens first had to secure adequate provisions of water, and then deploy them in ways that made the presence of water truly conspicuous.[7] Today, when we assume that water is available anywhere, even if it still emerges from nearby springs, or when recycled

78 Hendrick van Cleef III, *The Vatican Belvedere Sculpture Garden*, detail of left side, 1550, oil on canvas.

79 Francisco de Holanda, *Ariadne*, 1539–40, Escorial sketchbook.

water is *de rigueur* in modern fountains, the significance and the celebration of a water supply are far less evident, and we forget to what extent Renaissance water works were sustained by traditions of Roman engineering.

Gardens like those of the Villa Castello and the Boboli Gardens behind the Pitti Palace in Florence obtained their water in aqueducts, which were the first systems to be constructed; at Castello, the supply eventually needed further conduits to bring more water from another nearby Villa Petraia. Cosimo de'Medici also proposed taking the water from the springs that serviced both these villas into Florence itself to serve a fountain in what is now the Piazza Signoria. Instead, water that was created for the Boboli Gardens, through two new aqueducts, was brought across the Arno to serve the Piazza. In both cases, water destined in the first instance for private purposes would also benefit the city – Castello's springs served local people; Boboli's water went not only to ducal uses in the Palazzo Vecchio, but also to the city's public fountains. The supply of water may not have come specifically from ducal properties, but its spectacular deployment in those villas in the first instance gave the impression that the urban water supply was a gift of the rulers or princes of the church. A similar 'gift' (as we see) came from the fountains of the Villa Lante before it descended to serve the people of Bagnaia, named after the ancient baths that gave it its name. The theatrical aspect of some of these villas seemed to offer the riches of the villa's water supply as a performance staged by the local patron for his citizens: if the Boboli hillside, scooped into an amphitheatre, was not designed or originally termed a 'theatre', none the less the design by Tribolo surely dramatized the life-giving arrival of its waters by setting Giambologna's fountain of Oceanus in the midst of the garden framed by the evergreen hillside. Oceanus, and later a further Neptune statue set about with sculptured river gods, celebrated in miniature the whole hydraulic process from springs to rivers and the ocean, and the irrigation of fields and gardens that flowed from these opportunities.

The remains of Roman aqueducts were everywhere around Rome; some were ruined, others capable of functioning, many could be resuscitated. Both the Villa d'Este and the Villa Lante found their water locally and conveyed it into their gardens. Mineral springs at Tivoli and the river Aniene fed several surviving Roman aqueducts, and Cardinal d'Este himself constructed and enlarged other underground conduits that fed both the town's inhabitants and the astonishing play of water in the gardens. Given that the villa at Tivoli was established in a place known to have been the site of Roman villas and with the remains of many – most particularly Hadrian's – still visible in the vicinity, this combination of a site rich in historical associations with both a display of sculpture, featuring classical figures and legends, and the astounding manipulation of water works within the villa garden, made the Villa d'Este a resonant example of the Renaissance recovery of classical culture. Yet these wonderful effects may well have made its modern success more potent for many visitors than its revivalist skills.

Another villa where the water supply was admired was the Villa Lante, created in a former hunting park and in a landscape rich in mineral springs. The waters were brought and collected in a reservoir outside and immediately above the top of the garden, and then released into a series of grottoes, fountains, cascades, water theatres and a water parterre; a visitor in 1578 wondered how a 'thousand' fountains, a 'whole hillside' of them, could originate in a single fountain (illus. 80). Furthermore, the town was now supplied, as noted, with its own water. The work was attributed to a hydraulics expert from Siena, Tommaso Ghinucci, whom (Montaigne noted in his journal) was 'always adding new inventions to the old'. The modern and the antique, the biblical and the classical, coincided at the top of the garden: the water first appeared in a Grotto of the Deluge, a reference to the inundation of the world at the time of Noah; but on either side of this were there two Loggias of the Muses – an equivalent to the Muses perched atop a Mount Parnassus that elsewhere (at Pratolino, for example, see illus. 88) featured in gardens and, as Muses, were the inspirations of human creativity and inventions. And the gardens at the Ville

80 Giovanni Guerra, *Cascade and Fountain of the River Gods at Villa Lante*, 1604.

Lante below were marked by this double appeal to antique effects (river gods, the *naumachia* on the parterre) and to modern engineering skills.

It was, indeed, in the late sixteenth century that Italian humanists linked the success of modern garden-making to the ideas of a classical writer like Cicero. He had lauded the primitive or 'first' world of the gods – their mountains and waters – and then had praised the infrastructures that humans made by drawing upon those resources (a 'second' nature). In their turn humanists recognized that the new successes of Italian gardening had contrived what they called a 'third nature', drawing upon the resources that Cicero had identified as the first and second natures (see chapter Seven).[8] The Villa Lante, with its exploitation of local water, its refashioning of this to serve the town itself (a new civic infrastructure), and finally the delights and affirmations of the garden, was a prime affirmation of these three natures.

Furthermore, one legacy of Hellenistic learning was found in the writings of Hero of Alexandria; his *Pneumatica*, circulated in manuscript form before being published in Latin in 1575 and in Italian fourteen years later, described various hydraulic devices that found their way into many Roman gardens (illus. 81). Pirro Ligorio read the text in manuscript and drew closely upon one of its most famous inventions for the Villa d'Este: an Owl Fountain, where an owl turning towards a chorus of birds, whose song was produced by water pressure, made them suddenly fall silent. But other devices from Hero were used at Pratolino, at Hellbrunn in Austria and at Enstone and Wilton in England, and his ancient theories sustained both Salomon Caus's *Les Raisons des forces mouvantes* (1615) and Stephen Switzer's two-volume *An Introduction to a General System of Hydrostaticks and Hydraulickes philosophical and practical*. Switzer's own career as practitioner was directed, he said, 'to the watering Noblemens and Gentlemens Seats, Buildings, Gardens'.[9]

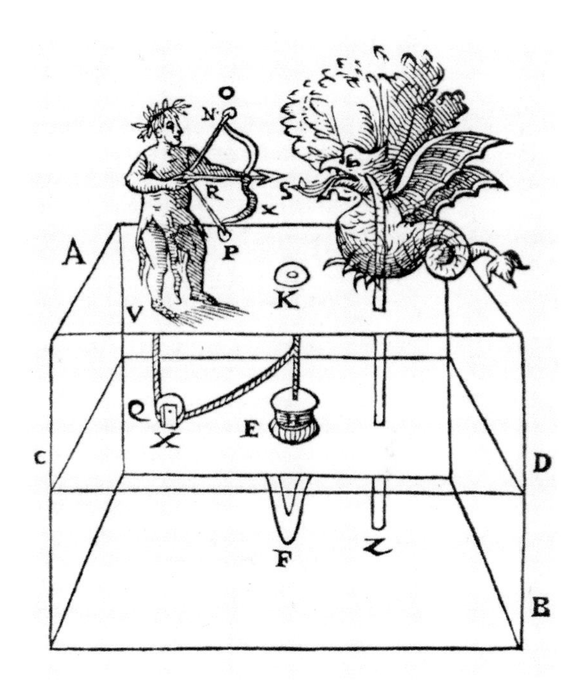

81 A theorem from Aleotti's translation of the *Pneumatics* of Hero of Alexandria (1589). If the golden apple (marked K) is lifted, Hercules draws his bow and the dragon makes a noise. Hercules was much invoked in gardens, not least for his stealing of the golden apples from the Garden of the Hesperides, a key theme at the Villa d'Este in Tivoli.

This long tradition of hydraulic expertise from Hellenistic Alexandria to Italy, to northern gardens and eventually to England, along with the parallel in the spread of antique classical sculpture and its imitations, was one of the major facets of the Renaissance garden. If there was another such revival it came only in the eighteenth century with the awareness of Greek architecture. Once James Stuart and Nicholas Revett published *The Antiquities of Athens* (1762, with a second volume 27 years later), which numbered 'Capability' Brown among its many subscribers, and which earned Stuart the sobriquet 'Athenian', and *Ionian Antiquities* (1769 and 1797), the Grecian revival in Great Britain became a 'veritable second Renaissance'.[10] If the *color romanus* had manifested itself in a whole range of sculptures and hydraulic effects, the Greek Revival found its role in buildings like the Lanthorn of Demosthenes at Shugborough, for example, or the Doric Temple of Theseus at Hagley (both by Stuart). Gardens continued to take up the Roman inheritance from Italy – the Praeneste Terrace at Rousham, the Temple of Ancient Virtue at Stowe, the ruined column at the Désert de Retz, or the Temple of Modern Philosophy at Ermenonville; moreover, Roman sculptural figures were better known, and already much copied in northern gardens. But by the later eighteenth century, pattern books for parks and gardens were chock-a-block with a congeries of other cultures, and now Greek items had to competed with Chinese, Roman, Turkish, Indian or even American *fabriques* for the patron or owner who craved 'true Taste and Eloquence'. And in England, especially, there was increasingly a distinct reluctance to colour its gardens with anything that was not somehow deemed to be indigenous; sculpture suffered particularly hard – for patriotic reasons and on account of its expense: George Mason in *An Essay on Design in Gardening* of 1768 also thought that the 'use of statues is another dangerous attempt at gardening'.

So after these northward progresses of the Renaissance Roman revival reached and flourished in England, there were many obstacles to its continuance. Financial constraints for both elite and more modest properties, changes in taste dictated by fresh perceptions of how the natural world functioned, and politics – all contributed to a fresh search for gardens rather than those that rely upon a useful and fruitful foreign past. New landscapes sought their inspiration in local topographies, and in native British architectural styles. The nineteenth century certainly witnessed an eclectic host of allusions and references, but it was the Arts and Crafts garden at the end of the century that found its own best model in local planting and the old traditions of vernacular building.

7 The *Paragone* of Art and Nature in the Renaissance and Later

Competition between the arts was a prominent topic in both Renaissance theory and practice: debates as to the specific character and efficacy of any one art form in comparison to others, in the first instance between painting and sculpture, were extended to other arts, like music, poetry and eventually garden-making. At issue in most of these contests or (as they came to be termed) the *paragone* was the ability of any art to capture and represent nature, considered the be-all and end-all of artistic productions. The English critic Sir Philip Sidney, explained this fundamental concept in *An Apologie for Poetrie* (1595): 'There is no Arte delivered to mankind that hath not the workes of Nature for his principal object'. Nature comprised the whole material world – animal (human included), vegetable and mineral; Sidney further enunciated the canonical aesthetic that nature's 'world is brazen, the Poets only deliver a golden [one]'. In short, the ambition of any artist was to exploit both his skills and the opportunities of his medium so that his subject-matter was presented (represented) compellingly. There were those who sought the ideal, the idea behind all material appearances; other artists presented the natural world in all its fullness, yet clear and sharp, purged of accidentals and contingencies.

This fundamental concern is obviously one to which garden-making needed to aspire. A garden's basic materials were the traditional four elements: the earth and the plants that grow in it, the water that irrigates them, the air (or winds), and the sun (fire) that nourishes. It was the business of the gardener to use this 'art' to exploit these elements as efficiently and productively as possible, laying out his site in ways that gave him and his plants the fullest opportunities to thrive; in particular, this required organizing the spaces (planting beds, paths, terracing) and providing the necessary water supply (cisterns, wells, conduits). The gardener also needed to orientate his site so that it responded most efficiently to the different winds, the different times of day, the different seasons (which were often associated with the four elements), and the specific locality.

That these were the essential constraints and challenges for garden-making was acknowledged even within the gardens themselves. Thus locality could be signalled by abstracting representations of its topography – river gods, allegorical mountains, painted panoramas. At the Villa d'Este channels fed by a hundred fountains run across the whole width of the site, connecting the waterfalls of the Tivoli region that was celebrated at one end with the model of the city of Rome in its

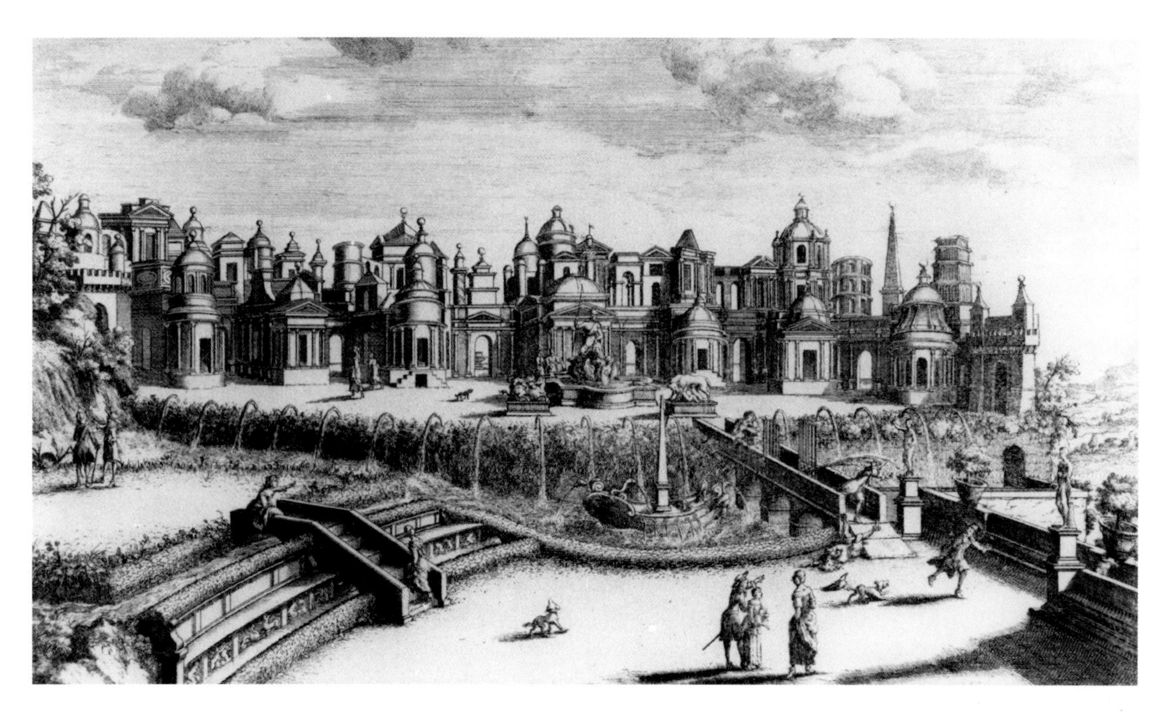

82 The Fountain of Rome, Villa d'Este, from G. B. Falda, *Le Fontane di Roma* (*c.* 1675). It represents the classical city, and shows on the 'stage' a sculpture of the Lion Attacking the Horse, a copy of which is at Rousham (see illus. 83).

fountain (Tiber) at the other extreme (illus. 82). Other examples acknowledged local conditions, like that proposed by Filarete for a garden with huge bronze figures of the four winds placed on pedestals on the sides of the garden square,[1] or statues of the four seasons at Castello to celebrate the beneficent rule of the Medici that enabled agriculture and gardens to flourish throughout the year. And just as the gardener might organize the garden to demonstrate the passing of the hours and seasons and to show off to best advantage the different elements (flower beds, aviaries, trees, shelters), so his resulting layouts determined how gardens were used and appreciated. Justus Lipsius urged his readers to observe the 'movement of the flowers, their growth and development' when visiting.

Pragmatic concerns determined a large part of a garden's layout, productivity and use. But their art could also be invoked to tell of matters beyond the garden, and alongside its delivery of an improved ('golden') nature of flowers, herbs, fruits and vegetables was often used to promote other associations and references. Just as the medieval garden's enclosure and fountains, required in the first place for reasons of good horticulture, could also come to represent allusions to Genesis, the *Song of Songs* and the symbolism that the Christian church extracted from those writings (see chapter Five), so Renaissance gardens extended their own range of reference. In short, as gardens came to be 'designed' as opposed to simply 'made', they assumed, like other arts, the obligation to represent many aspects of human nature and its concerns within and beyond the material world.

The Renaissance garden in Italy during the sixteenth century dedicated its considerable artistic and technical resources to displaying a variety of natures, and these displays grew in sophistication and intricacy. Botanical gardens (see chapter Eight) sought out and collected the plant world, and while the whole enterprise of a botanical garden may have had strong symbolic overtones, its presentation of the plants was largely designed to show them off as they truly were. In villa gardens, there was a growing attention to the effects that could be contrived: the agricultural practice of terracing hillsides to achieve level areas for growing crops was extended to make them more architectural, with porticoes and grottoes underneath and elaborate stairways connecting the levels. Trellis and other kinds of what was called 'carpenter's work' continued to have practical applications, but now also were formed into arbours, cabinets and tunnels. Water, brought into garden plots via irrigation channels and stored for later distribution in cisterns, was increasingly exploited for a whole variety of effects that far exceeded its utilitarian role. Indeed, the Renaissance hydraulic specialists had a whole vocabulary of terms by which to designate the different modes of manipulating and appreciating water.[2] Claudio Tolomei in 1543 explained how water brought into Rome by the Acqua Vergina aqueduct was 'directed, divided, turned, driven, intercepted, and made to rise or fall'. He then distinguished between its different effects, including currents (*correnti*), jets (*zampilli*), spattering (*acque spezzate*), bubbles (*bollori*) and ripples (*tremoli*); a later writer, Annibal Caro, saw a garden where the water came in trickles (*gemitii*), jets (*zampilli*) or imitated rain (*pioggia*), and so on. It was these exploitations of natural materials, extending them beyond the merely useful (irrigation, or cooling a garden), that constituted the art of gardening.

Water, one of the prime necessities for gardens, is also the natural element that lends itself most readily to control and manipulation: human ingenuity can be the more remarkable in that it forces water upwards, against the grain of its inherent character, which succumbs to gravity,

83 Lion Attacking the Horse (1740) by Peter Scheemakers, above the slope to the river Cherwell at Rousham.

84 Diana of the Ephesians at the Villa d'Este.

finds the way of least resistance and gradually erodes obstacles in its way. At the Villa Lante, near Viterbo, in the early 1580s Michel de Montaigne was rightly impressed by the repertoire of water effects. Indeed, the various treatments of water on its descent through this gradual hillside site put on display practically every aspect of water that, with engineering, could be devised; it also boasted the skills and ingenuities of those who contrived them. Montaigne celebrated its 'infinite distributions under diverse forms', its display of 'much more art, beauty and delight'.

Because a garden was, at its finest, considered an imitation of the whole natural world ('mirrors of the world'; the 'Great and Universal Plantation epitomiz'd'), such a collection of aquatic devices gave special credence to that claim. At Villa Lante (illus. 86), begun in 1568 for Cardinal Gambara,

85 The Walk of a Hundred Fountains at the Villa d'Este.

the top of the site was provided with reservoirs that stored water brought from surrounding hills; then, immediately within the garden proper, was the Grotto of the Deluge, which enacted in showers of rain the inundation of the world from which Noah escaped. The next Fountain, its octagonal basin decorated with dolphins, leads to the Water Chain: here from the jaws of a crayfish (Cardinal Gambara's insignia) issues water that rushes down a stone trough, its edges carved to look like waves; the water's flow over rough stone within the channel gives it, contrariwise, the abstract appearance of sculpted form. The different levels of the garden are joined by stairways, where jets spurt like flowers from vases or curve downwards between short pilasters as liquid 'handrails'. The waters flowing down the Water Chain reappear in an elaborate Fountain of River Gods, who pour

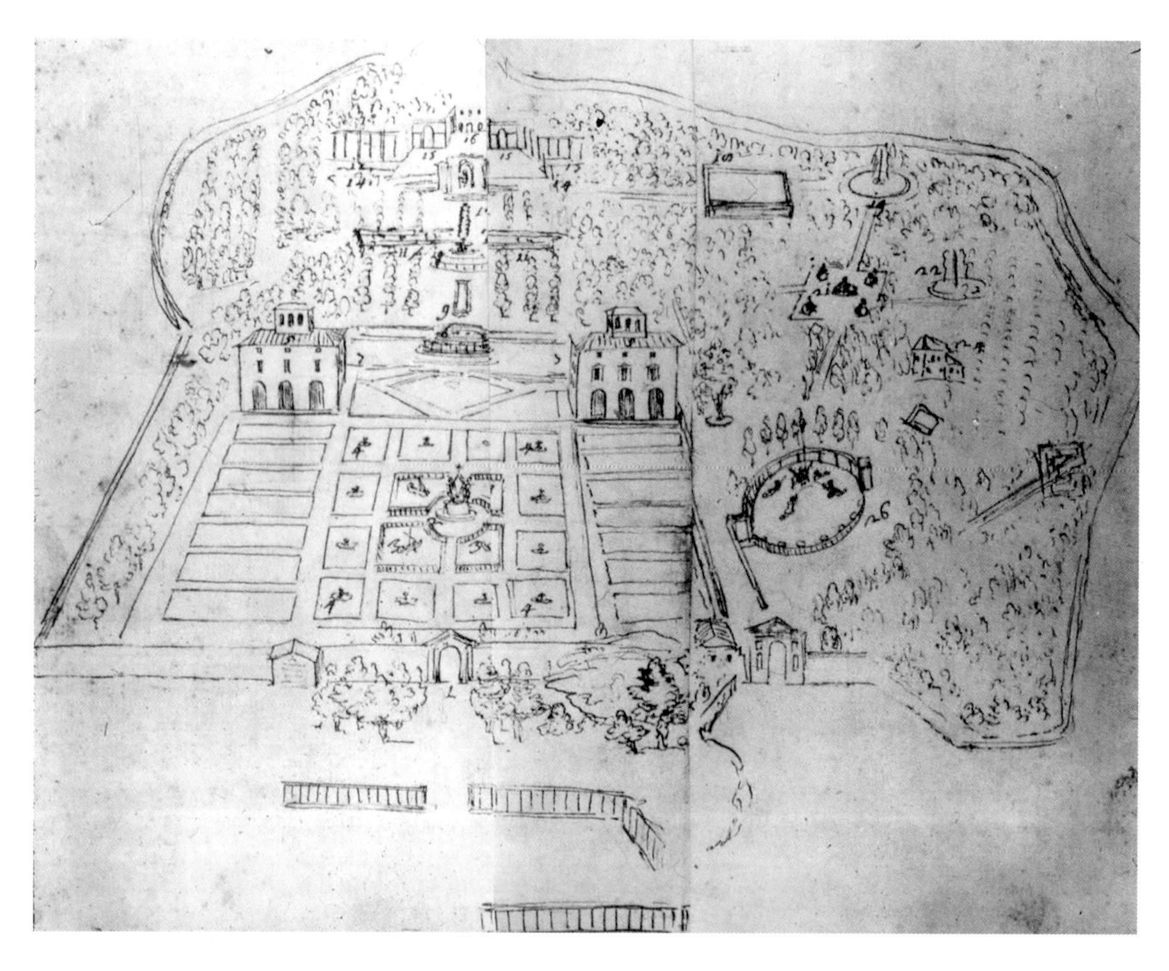

86 Giovanni Guerra, *The Garden and Park at Villa Lante*, 1604.

their waters from urns, and then emerge again to fill a channel cut into the long, stone dinning table, a device for cooling wine and floating dishes past assembled guests. The water is next enlarged into a conspicuous water theatre, comprising first a concave and then a convex series of stacked pools, along all the ledges or 'seats' of which are spouting jets. The last of the garden's three level terraces is the most spacious, with a square pool divided into quarters by walkways that lead to a central island crowned with a fountain. From the four pools soldiers in stone boats fire their water cannon towards the centre – an allusion perhaps to the ancient mock naval battles or *naumachia*; originally the central fountain had been a *meta sudans*, or pillar sweating water, another ancient allusion. The water finally passes from the Cardinal's gardens to supply the town of Bagnaia below, but not before it has been harnessed once again for the Fountain of Pegasus and the Muses, immediately outside in the parkland (illus. 87).[3]

There is much more that could be said about this still wonderful garden. Its playful and inventive manipulations of water, though, are its most conspicuous feature, rare contributions to the Renaissance delight in the collaborations of art and the celebrations of nature. To which dialogue or *paragon* within the central, main garden, however, it is necessary to add that its elaborate organization and technical engineering are themselves contrasted with the large hunting park out of which the garden was formed and within which, as a public park, it still remains: this provides a visible contrast between two ways of treating nature – the naturally graded parkland, dotted with fountains and through which avenues have been cut, though without any formal pattern, is juxtaposed to the carefully plotted garden. And linking them by its position in the park but immediately adjacent to the garden is the hemi-cyclical pool with the Muses gazing at Pegasus as he strikes his hoof against the rock, a reminder to visitors like Montaigne, if any would have been needed, of an ancient myth realized anew. For here in Renaissance Italy and at Bagnaia the winged horse releases the inspiring spring of the Greek Helicon, where the Muses imbibe and lend their skills and visions to the creators of this extraordinary garden. The complete arts of such a garden are dedicated here to a full and complex rendition of both natural facts and cultural narratives.

Montaigne visited several other fine gardens during his travels in Italy during 1580–81, including the Medici villas of Pratolino and Castello and the Villa d'Este. His *Journal* explains two aspects of the Renaissance *paragone* – he takes special

87 The Pegasus Fountain, Villa Lante, detail with four of the Muses behind Pegasus. This fountain is represented in Guerra's sketch (opposite) on the right in the parkland.

88 Giusto Utens, *The Medici Villa at Pratolino*, 1599, oil on canvas.

interest in how natural effects were enhanced and presented by art and technology, and he contrasts and compares the different achievements of these four gardens in their manipulation of art and nature. At Pratolino (illus. 88), he observes how the 'Duke of Florence . . . exercises all his natural five senses in embellishing' a site that was 'inconvenient, sterile and hilly' and apparently chosen to draw out the skills and craft of the garden-making there. He is constantly alert to how natural materials and effects are achieved by art: a grotto is encrusted with stone 'from some particular mountains, and they have joined it together with invisible nails'. Art here is used to enhance the natural cave. But elsewhere an aviary exhibits some exotic birds, where nature by herself is remarkable enough and acknowledged. Water, brought from a considerable distance to Pratolino, is used to activate machinery that produces 'music and harmony' and the movements of 'a number of statues and doors with different actions, which the water sets going, numerous animals that dive to drink, and the like things'. One fountain in particular shows a woman doing her washing: 'She is wringing a table-cloth of white marble from which water drops'. A long, broad allée was arched over, but its 'pergola' consisted of water jets springing from both sides of the avenue, under which the visitors walked dry. Montaigne also noted the start of construction on a huge statue of 'a giant': when completed, this representation of the Apennines – a giant metamorphosing out of the rock, or perhaps changing back into rock – would contain a chamber in the giant's head where music would be played.

Similar themes attract Montaigne on other sites. At Castello, just north of Florence, he finds that 'imagination exceeds reality' – that is, it delivers a more finished or perfect version of things; he observes 'a natural-looking artificial rock', 'a number of arbours, very thickly covered with interwoven branches of all kinds of odiferous trees', 'avenues of cypress . . . disposed in close order', a cabinet in the branches of a tree all walled with evergreen branches and supplied with fountains, and the Duke's escutcheon 'very well formed of some branches of trees, fostered and encouraged in their natural strength by fibres which cannot be easily delected'. At Castello, too, water makes musical instruments to play or is deployed in 'an excellent imitation of fine, trickling rain'. At the Villa d'Este Montaigne's attention was taken by the hydraulic effects: an organ plays, there is 'an imitation of the sound of trumpets', a rainbow is formed 'that in no way falls short of what we see in the sky', and birds chirp and sing until the advent of an owl (also driven by hydraulic machinery) arrives and they fall silent (see illus. 81). 'All these contrivances, or similar ones, [are] worked by the same natural causes'.

All these effects were the result of a growing fascination for what Alberti had earlier called the 'novelty of surprise'. Francisco di Giorgio explained how to enliven garden spaces with a variety of inventions:

> Lay out pools, fishponds, loggias and covered and open walk ways; set aside areas furnished with watercourses and green spaces, some open and some covered, where animals and birds may be kept. Paths and open areas must be set out in straight lines, running parallel and at right angles to each other; there must be lawns and glades, with a variety of trees that keep their leaves for most of the year, and other areas where grass and trees are interspersed with temples, labyrinths, loggias, seats and other delights. And the more varied they are, the more pleasing to the eye will they be.[4]

Such advice sustained many gardens during the sixteenth century, just as it was what attracted visitors. One result of this increased activity in the creation of pleasing gardens was that it moved thoughtful people to a more conceptual interest in defining just what was the particular character of the garden.

Some years before Montaigne's travels, a couple of Italian humanists were sufficiently attracted to developments in the art of Italian garden-making that they sought to understand conceptually how it was that gardens were constituted. Independently of each other, Jacopo Bonfadio in 1541 and Bartolomeo Taegio in 1559 each explained that gardens were a 'third nature'. This extended and amplified such activities as plantations, irrigation, roads, bridges, ports and other forms of civic constructions, all of which infrastructure the Roman author Cicero had termed a 'second nature', fashioned out of the antecedent 'first' primitive world of the gods. This new, third nature of gardens was described in precisely the terms of the *paragone*: 'the industry of the local people has been such that nature incorporated with art is made an artificer and naturally equal with art, and from them both together is made a third nature, that I would not know now to name'. Taegio's rather clumsy definition struggles with explaining the paradox of how nature, becoming an artificer (or artist), thereby achieves an equal status to art itself, while still retaining its natural character. What we have seen in Montaigne's reactions to gardens some decades later is a more acute and sophisticated understanding of these joint endeavours of art and nature and how they are experienced.[5]

The experience of a garden's manipulation by art (technology, engineering, horticultural expertise) and its raw materials (plants, earth, water, air) varied considerably over time and in different places. During the sixteenth and seventeenth centuries it generally grew more complex and sophisticated as both designers and visitors learnt to devise and enjoy garden experiences. English visitors, who would have been unable to observe few comparable gardens back home, found seventeenth-century Italian examples utterly remarkable.[6] Yet locations and cultures responded differently, so that the Veneto, for instance, was less committed to elaborate devices than Florence or Rome; and generally there was also a scale of complexity. At its most simple, what was appreciated was a display of the fecundity and beneficence of natural elements, from which their presentation did not unduly distract. Then it was the techniques that had been called upon to create such plenitude that came to be noticed. At that juncture, the dialogue, contest, collaboration between nature and art came into question – sometimes the sheer ingenuity was dominant and acknowledged, sometimes the astonishing effects it produced in imitation of the world outside the garden, sometimes both together. It is interesting that many sixteenth- and seventeenth-century visitors to elaborate gardens would marvel, say, at a grotto's scenographic *tableau*, then visit behind the scenes and inspect the machinery that drove it (illus. 89), and finally return and admire the *ensemble*, without belittling the performance because they understood its mechanisms. Their simultaneous appraisal of both the natural elements and their technological presentation may be likened to Wittgenstein's later conundrum of seeing either a 'rabbit or duck' in the same diagram.

These declensions can readily be tracked in two remarkable gardens from the sixteenth and early seventeenth centuries. Bomarzo (dating from 1552) and the Hortus Palatinus at Heidelberg (late 1610s) help us to understand the excitements, pleasures and intellectual stimulations of what Tolomei called 'natural artifice' and 'artificial nature'. Bomarzo has been hailed as a site of 'extreme artifice' and 'pure fantasy',[7] with its collection of fabulous, carved beasts, fictional and mythological creatures, bizarre structures (a

89 Giovanni Guerra, *Dining Grotto at the Villa Medici, Pratolino*, 1604.

90 The Giantess at Bomarzo.

leaning house), and many admonitory inscriptions (illus. 90). Yet these often disturbing and perplexing appearances are carved from the natural rock of the wooded site, which, though occasionally terraced and with paths descending the slope, is not at all as regimented as were many other contemporary gardens. The miscellany of items – elephants, giants, Pegasus, an orc, sphinxes, huge pine cones and acorns, a hell mouth – seem to spring from the very ground itself and be endemic to the place; so that the creation of the park by Vicino Orsini is hardly a 'denial of nature'. It was more probably contrived and understood as an expression – sometimes arcane, always startling – of its patron's human nature and of the Etruscan locality to which he belonged. For, like all garden-making in the Renaissance (and often since), Bomarzo was a means of declaring its creator's status, person and virtue: 'As is the Gardner', wrote Thomas Fuller in 1732, 'so is the Garden'.

Heidelberg was created by Salomon de Caus (illus. 91) also to assert the political, social and religious authority of the newly married Elector Palatine and his English bride (the 'Winter Queen'). The Elector was everywhere throughout the garden – in a 15-foot-high effigy accompanied by an inscription that lauded his mastery over all realms of nature in the creation of the Hortus, in a further inscription woven into the ground of one of the parterres, and in other conspicuous symbols of his role as a Protestant leader. The gardens were also a celebration of his marriage to Elizabeth, alluding to Vertumnus's courtship of Pomona

91 Matthias Merian, *View of the Hortus Palatinus, Heidelberg*, c. 1620, coloured print.

and to the transformative powers of natural fertility, with several invocations of the number eight (associated with marriage). Throughout the gardens, then, the evidence of art is palpable and emphatic, most obviously in the elaborate ordering of parterres, the manipulation of waters and in the grottoes, which are 'as close as architecture ever gets to the direct simulation of nature'.[8] But however much 'architecture' went into the fabrication of grottoes, it was their references to the natural world that were of prime importance: as the French gardenist Jacques Boyceau explained, grottoes were devised to represent caves (translated by John Evelyn as 'Grottoes are invented to represent Dens and caves').

De Caus invoked many ideas and images from earlier gardens, yet fashioned them afresh, including their 'inherent ambiguity', an ambiguity that was the means to a more perfect understanding of the world. De Caus wrote, for instance, that perspective, creating the illusion of depth on flat surfaces, was a way of imitating 'true nature'. So the Heidelberg Orangery – a prime site for the collaboration of art and nature in its production of what were often called the 'golden apples' of the legendary Hesperides – had stone columns carved to represent tree trunks, on which real ivy probably was trained, while inside were the actual orange trees maintained during the German winters through the intervention of the expert gardener. Ingenious, certainly; but probably designed with a view to enlarging visitors' understanding rather than overawing them with the cleverness of artifice. And come the summer months, a multitude of orange trees positioned throughout the parterres testified to the wonders of a golden nature.

The increasing elaboration of both technical effects and the invention of more and more complex configurations of garden imagery were pushed to extremes in the baroque period. Still sustained by the prevailing acknowledgement of a collaboration or perhaps rivalry between artifice and nature, and still accepting that art's business was to present purified and abstract versions of the natural world, the insistence upon invention, artifice and ingenuity increasingly saw to it that gardens were appreciated above all for their art; the concern to express and explain the workings of the physical universe was overwhelmed by the sheer virtuosity of the contrivances. Meanwhile, visitors, becoming attuned to such elaborate effects, learnt the satisfactions of fully responding to their achievements, thereby enhancing the whole experience of a garden. Today we may grasp some of the conceptual issues, but have lost touch with this rich and often demanding receptiveness.

Given a garden's essential elements and construction, the *paragone* of art and nature can never go away, even if it ceased to be a central concern of designers and visitors. A late, but essentially retrospective, formulation of landscapes considered in this regard is the frontispiece to the Abbé Le Lorrain's *Curiosities of Nature and Art*, published first in French in 1705 and later in other languages (illus. 92). Its very schematic visualization makes it a useful guide: the figures of art or science and technology (holding a globe) and the many-breasted figure of nature sit on either side of the view and jointly introduce the viewer to its landscape, which consists of three zones: first is a garden, with topiary and a central fountain, then a field with workers ploughing and sowing seed, then a lumpish hillside, its ground not unlike that on which the two foreground figures are resting. When the eye follows the upward surge of the fountain along the main axis of the garden it sees, emerging from the hillside, a natural spring of water. Simultaneously perhaps, it may suddenly notice that the hillside is peopled with figures playing on instruments – the number is blurred, but they are surely the Muses on a hill that we should associate with Parnassus and Helicon, the legendary mountains of ancient Greece where

92 Frontispiece to the Abbé Pierre Le Lorrain de Vallemont, *Curiositez de la nature et de l'art* (1705), re-issued in other languages, including English.

Pegasus struck his hoof and released the spring of Helicon from which the Muses drank and were inspired. As we saw at the Villa Lante, this scene was enacted in many designed landscapes of the Renaissance and later.

It hardly matters whether Le Lorrain drew specifically upon the earlier concept of the three natures, for the idea of such declensions of intervention in the land was widely established. But he has clearly diagrammed their sequence and relationship: the third nature of the garden is crafted from and exceeds in complexity the agricultural world, which is itself, as a 'second nature', distinguished from the unmediated world of the gods on the far hillside. The collaboration of art and nature are responsible for gardens and agriculture, while the Muses remind us of those who meditate in song or music upon what De Caus would have called 'true nature'. More sophisticated engravings of seventeenth- and eighteenth-century villa estates reveal the same tripartite zoning of the land, attesting to both the practices of estate management and conventions of responding to the different handling of land (illus. 93 and see illus. 119).

That the contrivances of a garden could still endorse lessons about the larger landscape is clear from commentary by the third Earl of Shaftesbury in the early eighteenth century. He found philosophical instruction in the shaped forms of nature within a garden; they were designed to draw out or abstract the true nature of the world so that, having absorbed its lessons, the lover of nature could perceive more clearly the shapes and structures encountered in the imperfect and messy environment outside gardens. For instance, Shaftesbury insisted that an old yew tree just outside his garden should not be removed, because the eye, tutored within by perfect geometric forms,

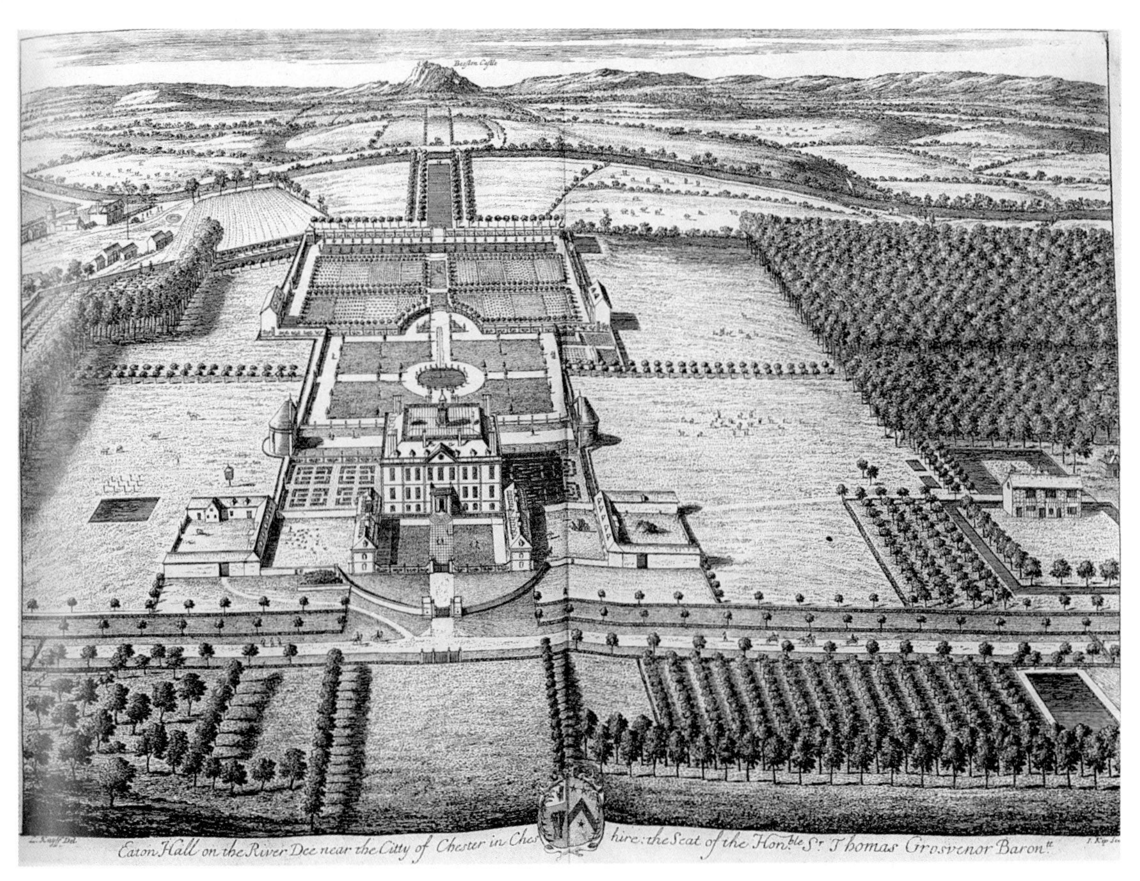

93 J. Kip and L. Knyff, engraved view of Eaton Hall, near Chester, from *Britannia Illustrata* (1707).

would nonetheless appreciate its natural character. As with Taegio and Bonfadio 150 years before, Shaftesbury was concerned with a conceptual understanding of the natural world and how it may be used to respond to gardens. Today's design students often deplore how seventeenth-century gardens distort natural forms, imposing unnecessary shapes upon their materials; yet it was their way of educating the contemporary eye and mind in the formal entities of the natural world. Thus, when André le Nôtre worked at Vaux-le-Vicomte, he channelled the wandering river Anqueil into a straight canal immediately as it entered the garden, but when it emerged at the other end of the grounds it resumed its meandering course through the meadows. He had 'taught' us to observe a river in both natural and mediated forms.

But the garden's 'lessons' or tutorials about the natural world gradually engineered some fundamental changes in the human attitudes to undesigned or unworked land. If, as Shaftesbury wrote, whoever 'studies and breaks through the shell' or exteriority of the world will 'see some way into the kernel' and appreciate its 'genuine order', it gradually became possible to admire the world outside the garden for its own sake. And for a variety of reasons – at once philosophical, social and cultural – a taste for larger landscapes grew exponentially. Their appeal called for fresh negotiations of the relationship between art and nature. Horace Walpole welcomed the elimination of 'canals, circular basons and cascades' and he thought (wrongly) that William Kent had eliminated the 'forced elevation of cataracts' at Rousham. But none of this rejection of conspicuous artifice meant that naturalism took its place. Indeed, if the new landscape art was to realize in its own medium 'the compositions of the greatest masters in painting' (Walpole's celebrated claim for the new, modern art of gardening), then designers nonetheless were asked to make nature over again as a painterly organization of space ('perspective, and light and shade') while employing a palette of 'none but the colours of nature'.

Here was a new challenge, or rather an old challenge re-applied to the art of landscape gardening. I take this up in chapter Twelve. Walpole noted how 'the modern gardener exerts his talents to conceal his art' – an ancient cry, since the Roman poet Ovid had enunciated the same *cri de guerre*. But it did indeed require real talent in the landscape designer, because, while concealing art, art is nevertheless employed to reveal and represent nature. So now it was the detailed, elaborate and unmediated world of nature that was used to disguise how much science, technology and art had gone into their presentation. The difficulty, of course, was that if nature completely camouflaged design, then design would go unremarked and designers would seem redundant. In fact, 'Capability' Brown seems to have been working within a very similar and subtle *paragone*, which the modern garden artist Ian Hamilton Finlay described succinctly as 'Brown made water appear as Water, and lawn as Lawn'. That aphorism is shrewd: Brown used his art to make conspicuous what was there – so water was encouraged to behave like water, trees to grow to their fullest arboreal potential, topography answered to its inherent ups and downs, and obvious artifacts (temples or statues) were banished from the landscape. And nature had successfully been used to camouflage the landscaper's art.

What was lost after the great eighteenth-century enthusiasm for naturalism (however mediated by appeals to pictures during the picturesque heyday) was any philosophical interest in the manipulations of art. There continued to

94 Bernard Lassus, garden for the COLAS HQ, Paris.

be those who wanted gardens to be clearly structured, like the proponents of the architectural garden at the beginning of the twentieth century. Contrariwise, their opponents argued the need for gardens to be natural and even 'wild'. It was, of course, a foolish debate, since, as already noted, the essence of any garden is its mixture of given natural elements with human intervention, however much one element or the other is downplayed. But the resolution of this debate in the arts and crafts movement was creative as it was sensible; it allowed both explicit formalism and evident structure to interact with a passionate attention to the life and shapes of plants (see chapter Seventeen). If there remained a hint of the *paragone* it served only to draw attention to the innovative mix of explicit construction and massed plantings. William Robinson's 'wild garden' was not utterly without human intervention, for his importations of 'wild' plants, unusual in conventional horticulture, could still draw attention to their selection and massing.

It is rare in today's landscape architecture for issues of art and nature to be either critically discussed or deliberately featured in built work. Some designers, nevertheless, certainly love to play with their visitors by offering scenes in which their own ingenuity is palpable alongside an equally naturalist palette. In many of Martha Schwartz's projects naturalistic elements are deliberately confronted with conspicuous, even flamboyant, artifice, and sometimes completely substituted for them. From her early Bagel Garden (late 1970s) and the parterre of Necco wafers at the Massachusetts Institute of Technology (1980) to the Broward County Civic Arena in Fort Lauderdale, Florida (1998), non-garden materials, distinctive forms and colours have served as foils to the surrounding natural sites, sometimes to be merely provocative without any apparent 'point'. But occasionally – as at Fort Lauderdale – the canopies of steel and coloured vinyl were a rebuke to the loss of trees once growing on this Everglades site as well as responding slyly to the pressure of budget restraints.

Recent work by the French designer Bernard Lassus also plays with the juxtaposition of real with artificial representations. In both private gardens (illus. 94) and corporate buildings he has taken to contrasting natural growth with steel cut-out patterns of trees and leaves; the resulting gardens wittily acknowledge both garden traditions and their necessary modern revisions, not least for purposes of easier maintenance. But throughout his career he has played energetically with artificial imagery. On the facades of apartment buildings in housing estates burdened with disaffection and dereliction in the town of Uckange, he devised landscapes painted in broad colours, so that the dreary, blank walls came alive to an astonishing degree, with 'hills' and 'dales' wrapped around corners and 'trees' rising into the eaves (illus. 95). Now its inhabitants could identify (literally) with where they lived by the location of their windows in these *faux paysages*, the estate bounced back, and the demand for the apartments rose dramatically. In the long tradition of dialogues between the various materials of nature or ordinary circumstance and the various arts that contribute to garden-making, Uckange occupied a distinctive position, at once modern and familiar.

95 Bernard Lassus, the painted facades of an apartment block at Uckange, France.

8 The Botanical Garden, the Arboretum and the Cabinet of Curiosities

Two different events raise the issue of 'where do plants come from?' In the sixth century AD the Chinese Emperor, Sui Yang Ti, was building his Western Park. As it approached completion, a proclamation broadcast an appeal for all those living nearby to supply plants, trees, birdlife and all manner of beasts (in particular, fishes and frogs were provided) for this enormous and extensive parkland. Eleven hundred years later Sir Richard Temple (Lord Cobham) was creating his large garden at Stowe; so he sent estate workers out into the surrounding landscape to collect trees for planting in the garden, very much as he might have sent his soldiers to forage when he was one of Marlborough's generals. True, Cobham could count on the resources of Brompton Park nursery, the best known at the time, although other provincial nurserymen were available and cheaper; but the Brompton accounts show only small bills. Instead, the search for around 60,000 to 70,000 trees for Stowe must have looked to other sources. Much of it was conducted on the large family estate – sowing acorns of holm-oaks and 'conkers' in a walled garden, taking cuttings, digging up seedlings or suckers in the woods; but friends and relations also supplied materials, often in astonishing quantities (37 cartloads of yew, in one instance).[1]

While today we would rely on our local garden centre or search online for supplies from nurseries, those events in China and at Stowe suggest that no decent garden could be created without substantial quantities of plants, and, moreover, that both Chinese imperial gardens and the rise of considerable eighteenth-century landscapes in England required not only quantity but quality, and quality of the finest calibre and curiosity.

The medieval world produced herbals and botanical manuscripts, and their authors had access to botanical gardens. Monastic institutions and private doctors maintained gardens dedicated to pharmaceutical and other medicinal plants; the famous plan of the St Gall monastery in Switzerland, dated AD 820, shows a variety of gardens – a cemetery with fruit trees marked, an orchard with neat, rectangular beds, and a physic garden next to a house for the doctor. Pharmaceutical imagery from around 1500 shows gardens for apothecaries, with herbs growing and their essences being distilled, or a physic garden where doctors are debating and distilling, with a patient in a nearby hospital bed.[2] Some places were known for their production and expertise. Venice was famed for its pharmacists, as John Evelyn duly noted in his diary, and owing to its extensive trading ('the emporium of the whole world') it supplied the pharmaceutical industry with products and medicine; its earliest known private botanical garden dated from the thirteenth century, but the Republic had already formally recognized the *speziali* (herbalist-cum-apothecaries, or druggists). Among the earliest books off its famed printing presses was a 1471 medical encyclopaedia translated from Arabic.[3] And it was the Venetian Republic that established its own *hortus medicus* at Padua in 1545. With this, and the creation of that at Pisa in the preceding year, the Renaissance marks the rise and spread of botanical gardens, and rather later arboreta.[4]

The main impulses behind these gardens were, firstly, prestige collection of plants for princes or regimes, drawn from worlds newly opened to exploration, notably across the Atlantic. But renewed scientific curiosity about the propagation of rare species, the rise of new medical faculties in universities and not least the creation of many new, and often much enlarged, pleasure gardens all contributed to the need for botanical collections. Thus, within a relatively short time, the importation of foreign plants into Europe began to overwhelm the capacities of even the biggest botanical collections, in part because it was no longer simply a practical collection of medical or pharmaceutical plants, but a hugely expanded interest in botany for its own sake.[5] That in its turn allowed some sites to focus their energies: medicinal plants for research institutions, or commercial collections, sometimes specializing in a particular species (America trees, alpine plants, rhododendra). Collections of trees, or arboreta, followed by the eighteenth century, again with emphasis upon materials from different climates, or on special and (at the time) often unknown shrubs and trees; arboreta themselves often grew out of earlier botanical gardens. Furthermore, some of the earliest botanical gardens were also linked to cabinets of curiosities or scientific museums, displaying *naturalia* and *artificialia*, while later plant collections were annexed, if awkwardly and briefly, to the foundation of garden cemeteries.

Four scholarly botanical gardens were established in the sixteenth century – Pisa, Padua, Leiden, Montpellier – and three more – at Oxford, Uppsala and the royal Jardin des Plantes in Paris – in the seventeenth century. Those at Leiden and at Uppsala were the work of distinguished botanists, Clusius and Linnaeus respectively. Their collections could be arranged both symbolically and practically: symbolically in that the circular garden at Padua was dived into quadrants (illus. 96) that suggested the four corners of the known world (Europe, Asia, Africa and America, also emblematized on the title page of John Parkinson's *Theatrum Botanicum* of 1640 – see illus. 104);

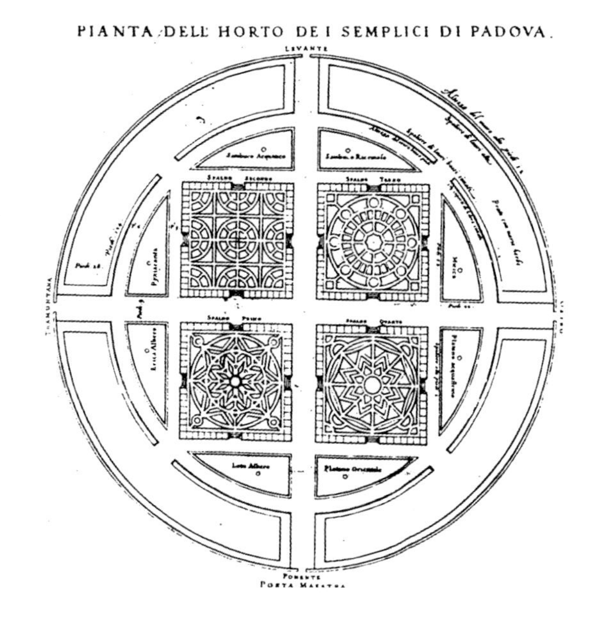

96 Plan of the Botanical Garden, Padua, founded 1545.

97 Leiden botanical garden, 1601 (detail), with a group of students and their instructor in front of the central archway of the Ambulacrum.

practically, in that, as at Leiden, the long narrow beds were laid out so that both professors and students could study the plants easily, and the gardeners could work on them (illus. 97). Each segment would contain a particular family of plants, further sub-divided and usually numbered for easy identification. The parterres at Leiden were austere, but others were laid out in elaborate patterns of geometrical shapes, and much ingenuity was expended on these designs for botanical as well as pleasure gardens: Agostino del Riccio loved 'beautiful patterns' and '*fantasie*' in his gardens, yet the Sienese writer Giovanni Battista Ferrari found shaped beds 'quite unsuitable both for sowing seeds and for growing them'. The plan of that for Oxford in David Loggan's *Oxonia Illustrata* of 1675 shows a mixture both of straight, oblong beds and of elaborate patterns (illus. 98).[6] The layout at Pisa in 1723 also showed a mixture, with four pairs of beds gradually becoming more simplified, so that the final pair used a version of the narrow planting beds like Leiden's, while the first was a dizzy embroidery of Maltese crosses and other geometrical forms; today, Pisa has a sequence of the very practical and rectangular plantings beds juxtaposed to a more 'English' space of curving paths and irregular lawns. After the vogue for 'English' gardens in the late eighteenth century, many botanical gardens changed their layout – like the Leiden one in the early nineteenth century – and arboreta, usually established later, generally used a meandering sequence of paths through what were in effect small parklands for the display of trees and shrubs.

The early world of botanical gardens was often part of a much more extensive scientific enquiry into nature and human structures. John Aubrey, an early member of the Royal Society in England, may be a somewhat extreme example, but he engaged in a whole host of learned activities: he devised, but never built, gardens on the Italian model, he studied and wrote about Stonehenge and Avebury, wrote a *Natural History of Wiltshire*, reflected on the geology of the world with its earthquakes and subterranean fires, and for his *Monumenta Britannica* he studied everything from coins to churches and architecture to handwriting, and he also wrote biographies in his *Brief Lives*. Given the range and limitless curiosity of these endeavours, it was inevitable that nothing got published during his own lifetime.

The role of a botanical garden was therefore more various than we think of it today. Beyond the collection and maintenance of plants, their identification and study, and the establishment of pharmaceutical uses (including the explanation of 'signatures', namely the idea that the physical properties of a plant were a clue or signature to its purported curative properties), botanical gardens gathered a range of other elements from the natural world, along with curious fabrications that humans had made from them.[7] These facilities were housed in adjoining buildings.

Aubrey's contemporary, John Evelyn, also worked on a monumental history of gardens and gardening, which also never saw publication until its transcription (of one surviving segment) was eventually published in 2000. Given Evelyn's primary focus, it was not surprising that he would include in the manuscript a design for a botanical or 'philosophical' garden (illus. 99). His sketch

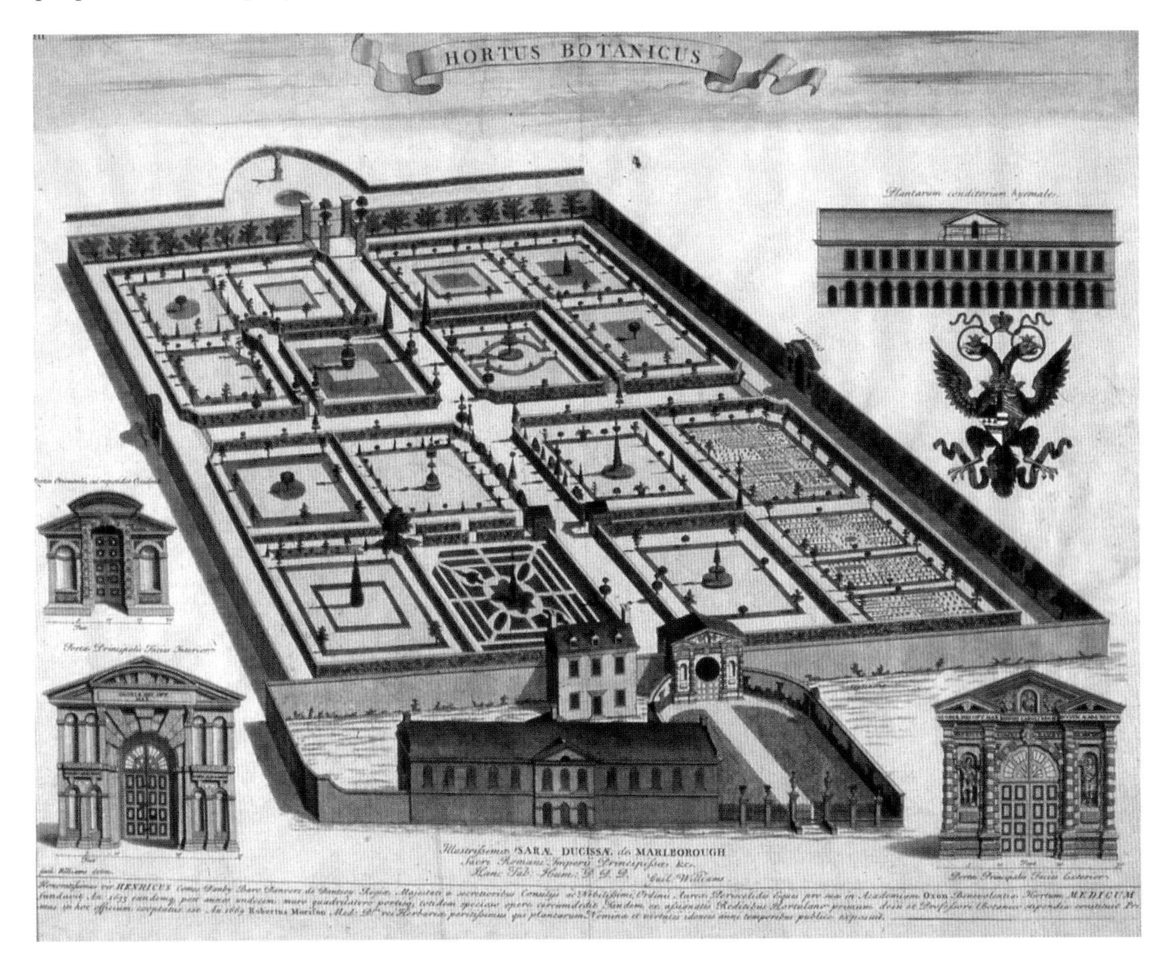

98 The Oxford Botanical Garden, from David Loggan, *Oxonia Illustrata* (1675).

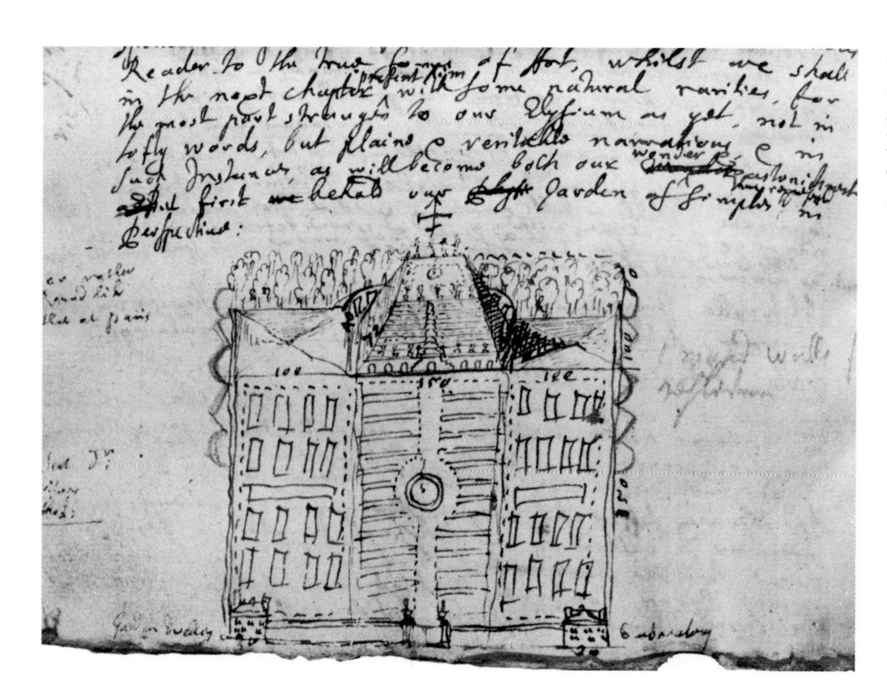

99 John Evelyn, sketch for a 'garden of simples' or medicinal plants, in his 'Elysium Britannicum, or the Royal Gardens' (c. 1650–1700).

draws upon both the botanical garden at Leiden – the narrow planting beds – and the Jardin du Roi in Paris, with its 72-foot mount that faced (in Evelyn's proposal) all four points of the compass, and established different ecological zones in each. Its forms and repertoire for all manner of plants would have made it, as Evelyn's text asserts, 'a rich and noble Compendium of what the whole Globe of her Earth has flourishing upon her bosome'.[8]

Besides their plants, botanical gardens often housed in adjoining galleries collections of *naturalia* and *artificialia* that would further have augmented the riches of the earth and its human occupants. When John Evelyn visited the house of Sir Thomas Browne in Norwich he described it as 'a paradise [that is, garden] and cabinet of rarities . . . especially medals, books, plants and natural things'.[9] These compendia, or early museums – an 'epitome' of the whole known world – were sought as a key to early modern knowledge (what Evelyn, writing of all the collections of the city of Rome, called 'the *World's* sole *Cabinet*'); but they also proved a threat to their very completeness. For the stuffing of these collections threatened to burst their banks, overwhelming a cabinet with the very multitude of its objects. Some early museums, like the Ashmolean Museum at Oxford or the first phases of the British Museum, still continued to include all kinds of material, but even these began to hive off parts of their collections to other institutions dedicated to science (zoology or geology) or the practical and decorative arts (the Victoria and Albert Museum, for example). It was the botanical collection that most obviously needed a separate and different, outdoor, space for cultivation, study and display. But even pleasure gardens continued to share both space and attention with cabinets of curiosity. In Mantua, Isabella d'Este located her sculptures, paintings and precious stones in a courtyard that included a grotto and a *giardino segreto.* Garden grottoes, along with garden loggias, became something of a museum in themselves, showing statuary and classical remains; different stones and geological devices were embedded for decoration of a grotto and also served to reveal a whole world of other sites from elsewhere. After Alexander Pope's death in the 1740s his gardener published a guide to his Twickenham grotto: its encrusted stones and rocks were identified by both their provenance and their association with the late poet ('Fine sparry Marble from Lord Edgcumb's Quarry, with different sorts of Moss . . . a fine piece of Marble from the Grotto of Egeria near Rome ... several fine Brain-stones from Mr Millan of Chelsea').

Many botanical gardens still clung to galleries where related elements of the natural world would be reviewed alongside the world of plants, all gathered within this one container. The inventories of the Botanical Garden at Pisa show that they collected what the Bolognese naturalist Aldrovandi called 'le cose sotterranee e le altre sopraterranee' (things below and above the earth). Evelyn remarked upon its 'Gallery . . . furnish'd with natural rarities, stones, minerals, shells, dryed Animals'; it also gathered pictures, engravings, drawings and portraits of botanists including Clusius. English travellers continued to be attracted to the Pisa garden and its collections: 'very commodious for Medicinall things ... abounding with all curiosities of Nature'; John Raymond thought the gardens were 'more for use than delight ... Cover'd with simples, outlandish Plants and the like', but he also appreciated the 'good walks, & water-works that well washt us' (water jokes).

At Leiden a similar gallery, the Ambulacrum, was built in 1599 to house similar items (illus. 100). Other means of relating plants to a wider cultural milieu included the system at the Orti Oricellari in Florence, where classical sculpture

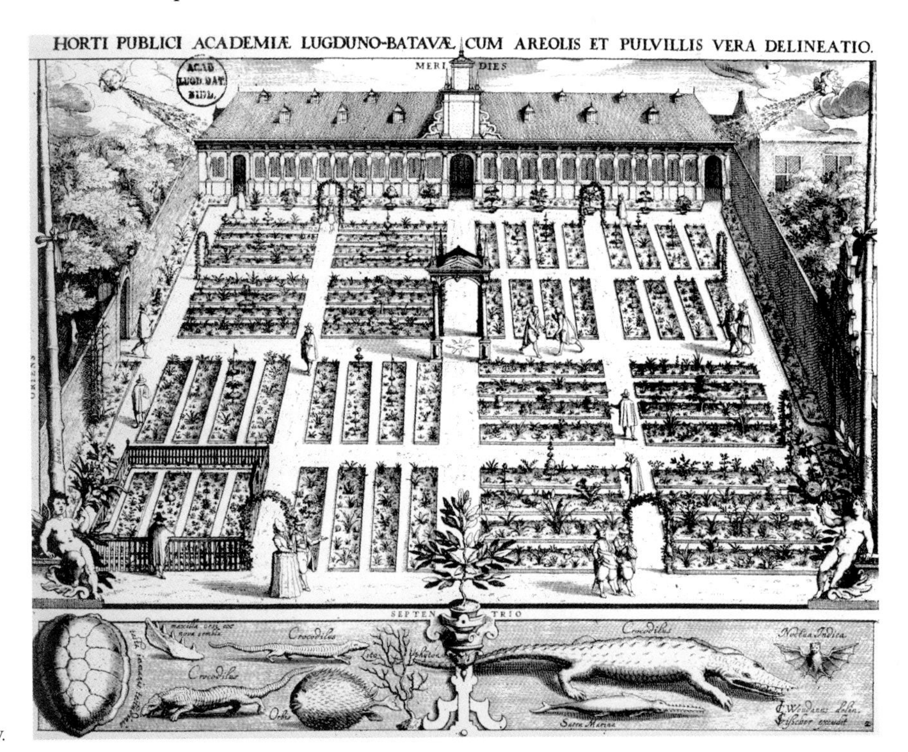

100 Ground plan of the Leiden garden with the Ambulacrum at the back, and a sample of its curiosities engraved below.

was displayed alongside every plant mentioned in classical literature. Gradually the term 'cabinet' became separated from a museum-like display, but continued its life in gardens, what Philip Miller's *Gardeners Dictionary* of 1737 called 'a summer-house or bower in a garden'; summer houses can still be the repository of miscellaneous findings in their grounds. But the scientific conjunction of botanical gardens and gardens began to lose its appeal, so that in his *History of the Royal Society* Thomas Sprat thought that 'Collections of Curiosities to adorne Cabinets and Gardens' was no match for the 'solidity of Philosophical Discoveries'.

One of the most interesting botanical gardeners, as well as a student of botany and a supplier of plants to Great Britain, was the American Quaker John Bartram (1699–1777), who was followed by two of his sons, John and William (two others became apothecaries, two others, farmers). The Bartram garden in the south-west of Philadelphia (illus. 101) still exists, its spaces reasonably well preserved; no longer used as a botanical collection, it is nevertheless a great teaching resource for local schoolchildren. The site of 102 acres was purchased in 1728, and Bartram began filling it with the result of his journeys along the Schuylkill River and in the Pine Barrens of New Jersey; but he soon extended his exploration northwards into the Hudson Valley and the Catskill Mountains and southwards into Delaware, Virginia, the Carolinas, Georgia and eventually Florida; his son William was also an extensive traveller and published a book of his southern botanical *Travels* in 1791. As a founding member of the American Philosophical Society in 1743, John Bartram's reputation was extensive, with visits to his garden by the Swede Pehr Kalm and the French-born J. Hector St John de Crèvecoeur; he also corresponded extensively with Peter Collinson, a wool and silk merchant and an amateur plant collector in London, who publicized Bartram's botanical work in papers read to the Royal Society in London, which published them in its *Philosophical Transaction*. Bartram received a royal commission as botanist in 1765 for travel to Florida, was honoured by the Royal Society of Sciences of Stockholm, and received a medal from a group of Scots in Edinburgh.[10]

Bartram's importations into England were widely distributed to other gardeners and estate owners at Bulstrode, Woburn, Worksop Manor, Thorndon Hall and Whitton. Bartram and Collinson corresponded frequently between the 1730s and 1760s, exchanging presents and seeds sent from England (including brown paper for dispatching branches with their seeds on them), considerable parcels of seeds and plants from America, and advice and intelligence from Collinson upon how Bartram's plants were being used in various landscapes: thus he explains how planting a 'mixture of Trees and their various Shades of Green' at Thornton was like 'painting with Living pencils', an early indication of how designers would draw upon a painterly picturesque for their landscapes. The scale and commercial success of Bartram's imports into England was astonishing (notwithstanding losses in transit and thefts from gardens in England), Collinson's care and generosity were exemplary for his many friends and botanical colleagues, and his European correspondence was extensive. Some specimens both delighted their owners and puzzled others who need to differentiate clearly between, for example, red and white cedars (*Juniperus virginiana* and *Chamaecyparis thyoides*). Another important benefit of this importation of American plants was that they could be carefully illustrated, even before they had prospered in

101 Bartram's garden, Philadelphia, showing the 1730s house and the upper garden, which, with the rest of the property, is well tended and used.

England. The flower painter Georg Dionysius Ehret watched a specimen of *Magnolia grandiflora* flowering in 1737, and painted every stage of its flowering from bud to button to 'full unfolding' so as to make available a 'perfect botanical study'. Bartram's plants sent to England were also collected, among others, in Lord Petre's Hortus Siccus at Thorndon Hall, Essex (now held in the Sutro Library, San Francisco); on many occasions the dried plants bear Bartram's own annotations (illus. 102). Many more important specimens were probably received from Philadelphia, but evidence from their provenance is rarely explicit.

One important function that botanical gardens, both institutional and private, performed was to commission and collect drawings of plants as well as animals and birds: this could be especially useful in that (i) plants didn't travel easily, (ii) plants could be captured in their fullest glory (often accompanied by details of different parts or stages of development), (iii) plants could be identified, studied and debated even when the specimen was no more, and (iv) these often wonderful and elegant images could be engraved for wider distribution; today they have become an especially interesting study of early botanical

102 'Our mountain laurel' identified in John Bartram's hand, in the *Hortus Siccus*.

history.[11] Some of the fruit trees that the elder John Tradescant found on his European travels for the Earl of Salisbury were planted in the gardens at Hatfield and were recorded in a delightful, rather naive series of watercolours called 'Tradescant's Orchard' (now in the Bodleian Library, part of the collection which was acquired for Ashmole's collection at Oxford after the death of the younger Tradescant, illus. 103). Between father and son, who himself visited America, the Tradescants doubled their collections over the next twenty years from some 750 species in 1634 (a measure of the influx of materials into such gardens), not to mention the accumulation of other trophies, including shells, that were crammed into what they called the 'ark' in Lambeth.[12] The owners of this 'little garden with divers outlandish herbes and flowers', along with its *Musaeum Tradescantianum*, were aptly identified on their tomb at Lambeth as possessors of 'A World of wonders in one closet shut'.

Another term for these botanical gardens and cabinets was 'theatre', in the dual sense of a 'compleat theatre', a show or presentation of items. So Jean-Jacques Boissard and Theodor de Bry called their emblem book *Theatrum vitae humanae* in 1596 and Henry Peacham described his collection of emblems, *Minerva Britannia* (1612), as 'a Garden of Heroicall Devises'; Parkinson entitled his compendium of plants a *Theatrum Botanicum, or the Theater of Plantes*, and on his title-page-cum preface he shows Adam (the first gardener) and Solomon (the archetypal scientist) gesturing on a stage between two pillars (illus. 104). These columns are the Pillars of Hercules, legendary guardians at the western end of the Mediterranean, but in early modern times they were also invoked to signal the ambition to voyage beyond them into an unknown world. But such pillars also supported the canopy of London's Globe Theatre, which Parkinson's stage also shows, a canopy decorated with imagery of the celestial heavens. On this stage ('all the world's a stage, and all its men and women merely players') the actors performed their roles in representations of human affairs. In a parallel move, botanical gardens presented the plants of the whole world for the delight and instruction of its visitors, to whom their curators displayed – like Adam and Solomon on Parkinson's theatre – their expertise and knowledge. We have today lost any colloquial sense of 'theatre' as a collection or performance of gardens, yet many gardens still aim to represent a whole repertoire of different garden styles and plantings, like Longwood Gardens in Delaware or the Parc de Bercy in Paris.

An early arboretum in England was that of Henry Compton, the Bishop of London, at his garden in Fulham in the late seventeenth century. His speciality was North American trees, and Stephen Switzer reports that he possessed over one thousand species of 'exotic plants' in his 'stoves and gardens', which allowed him to accommodate them for a colder climate. These collections had a dual purpose: cultivation and study, certainly, but also as suppliers of trees for other estates and gardens, and for their aesthetic value. There were many historical reasons for the establishment of tree collections – for show, of course, but also in later cultures simply to supply estates with plants for the production of timber and ship-building, even also as shelters for game.

John Evelyn's book on trees of 1664 was entitled *Sylva, or A Discourse of Forest-Trees, and the Propagation of Timber in his Majesties Dominions*, commissioned first by the Royal Society, and it was dedicated largely to the technicalities of forestry; its later editions (1670, 1679 and 1706) were augmented with lengthy essays on the Society and Evelyn's rumination on both tree cultivation and the lore of trees since antiquity. He himself claimed, rightly, that it was the 'occasion of propagating many millions of useful timber-trees throughout this nation'.[13] Indeed, the influx of new trees from North America and their distribution throughout the nation radically changed both the organization of estates and the design of their landscape gardens.

Given the increasing call for trees in a nation dependent upon ships for both commerce and

103 The May Cherry, in 'Tradescant's Orchard', *c.* 1611.

defence, England clearly devoted itself to their cultivation. The dedication of Evelyn's *Sylva* to Charles II hailed the glory of England's 'navigation', which cannot be impeached if the king and his nation 'continue thus careful of your *Woods* and *Forests*'. Thus the national attention to forest trees was an early, if private, dedication to arboriculture; countless engravings show the vast spread of woodlands – at Cassiobury in *Britannia Illustrata* (1707), or long avenues also lined with forest trees, as at Badminton in the *Nouveau Théâtre de la Grande Bretagne* (1708), or in Wray Wood at the third Earl of Carlisle's Castle Howard, whose father-in-law was the Earl of Essex at Cassiobury

104 John Parkinson, title page, *Theatrum Botanicum: The Theater of Plantes* (1640).

(see also illus. 119–24). Evelyn was particularly attentive, though, to the aesthetics and the historical significance of what he called the 'Sacrednesse and Use of Standing Groves', an essay which he added to the second edition of his book in 1670 and expanded yet again for its fourth in 1706.[14]

The modulations of English landscaping taste during the eighteenth century may have subdued the incidence of avenues and even vast stretches of woodland, but this newer interest in less formal landscapes and the greater emphasis on parkland that lay beyond the confines of the immediate garden simply took other forms. Indeed, a lesser regard for elaborate or extensive flower beds was matched by a huge increase in the underplanting of shrubs in woodlands. Douglas Chambers reports that between 1701 and 1750, 61 new trees and 91 shrubs were introduced into England, and these were distributed around a cluster of some of the most important landscapes in the country. Painshill in Surrey became, for Horace Walpole, one of the finest 'forest or savage gardens' in the land. It still organized itself with geometrical features, like the seven concentric circles of different species – beech, pine and beech mixed, and Scots pine – of the 'Keyhole' (aligned on the owner's house); but it assumed a more natural aspect within the other plantations of exotic trees (cedar, larch, juniper). It was, then, the accumulated riches and developments of woods and forests throughout England that gave to later designers like 'Capability' Brown and subsequently Humphry Repton one of their most usable materials and allowed some of their most exciting effects.

Arboreta were in the first instance, as the previous paragraphs suggested, largely promoted by private patrons. An early American collection was started in 1798 with an astonishing collection of species like the empress trees (*Paulownia imperialis*), and this was later purchased in 1907 by Pierre DuPont, privately developed and later opened to the public at Longwood Gardens in Pennsylvania. Even later at Westonbirt, near Gloucester, another famous arboretum was established by Robert Stayner Holford and continued by his son, George; they commissioned plant hunters to collect rare and unknown trees from all over the temperate world and indulged their zeal for propagation and hybridization throughout the property. It passed eventually to the Forestry Commission.[15]

One of the earliest public arboreta was established at Derby. Designed by John Claudius Loudon, this was opened in 1840 as a public park, donated by a philanthropist, Joseph Strutt, for the pleasure of its gardens and 'to excite an interest in the subject of trees and shrubs in the minds of general observers' (illus. 105).[16] Trees and shrubs were easier to maintain in a public park than a profusion of herbaceous plants, though Loudon did recommend that vases be supplied with a weekly assortments of different flowers. The layout of the Derby Arboretum was consistent with Loudon's own general proposals for public gardens and promenades: an intersection of two avenues, with a deflection to the south-east and some subsidiary or 'small episodical walks' around the edges of the irregular site. He recommended a circle of stone seats and a place for a fountain or statue at its centre, believing that straight walks 'without a terminating object' were deficient in 'meaning'. Trees and shrubs, obtained from different nurseries, were numbered and labelled according to the system of Bernard de Jussieu, and a pamphlet was published that proposed a tour of the collections with notes and quotations. As with many of his landscapes, each specimen was given its own space, kept separate from the others, a deliberately artificial rather than picturesque crowding. Loudon even proposed that after twenty years the whole collection at Derby should be taken up and replanted (in part so that visitors could see and understand single specimens rather than an emergent 'forest'). This did not happen, and the site has been extensively replanted now with sycamores. Some American visitors, who were probably looking for Picturesque design, found the Arboretum disappointing: A. J. Downing felt that the park was not 'scenery', and his compatriot Charles Mason Hovey found the mounds were not 'natural'. But in England the Arboretum at Derby was generally hailed for its combination of scientific interest and social thinking. The *Literary Gazette* thought it was 'the very treasury and epitome of the wide world's natural wealth'. And even Downing and Hovey saw that such an arboretum was a much desired facility for crowded American cities.

Loudon had published his eight-volume *Arboretum and Fruticetum Britannicum, or, The Trees and Shrubs of Britain* serially between 1834 and 1837, at great and crippling expense to himself. His delight and knowledge of trees was matched by his concern to make collections generally available rather than leaving them in an exclusive, private domain. He devoted a considerable and detailed section of his book *On the Laying Out, Planting, and Managing of Cemeteries* (1843) to the planting and use of trees, referring his readers to the earlier multi-volumed project. It was this specific and focused attention to trees that shaped his call for 'Gardenesque' design

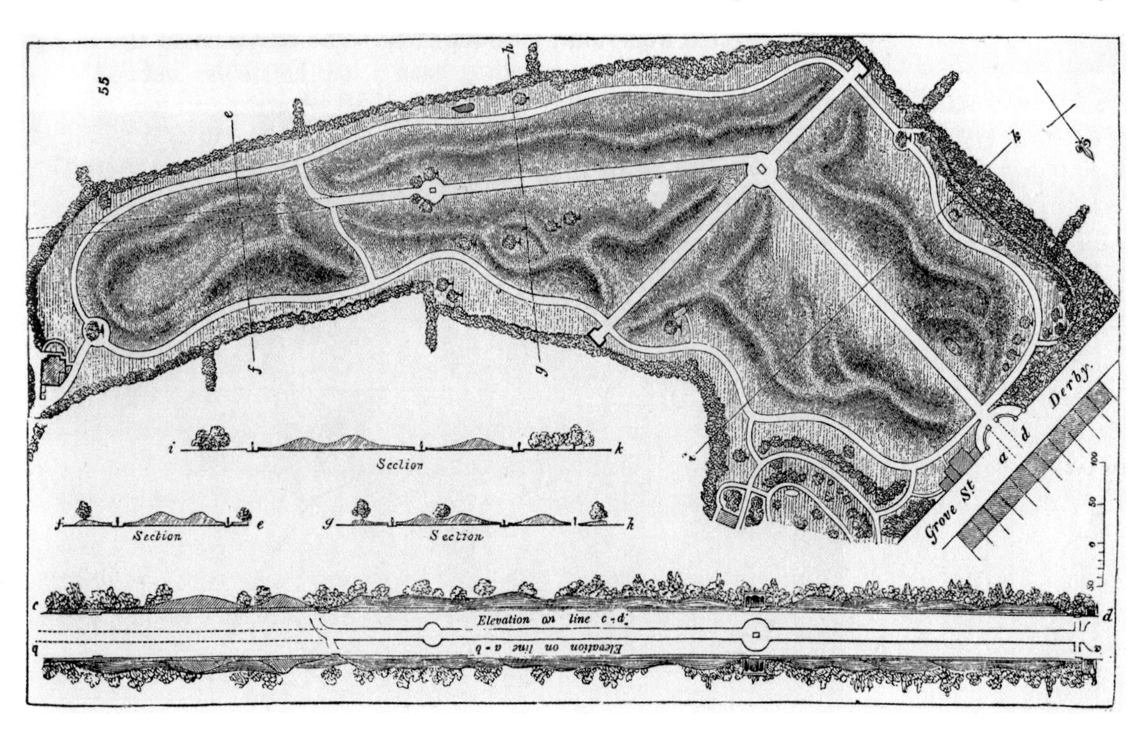

105 John Claudius Loudon, plan of the Derby Arboretum, 1839.

and determined his realization that planting should promote an 'ideal' nature (citing Sir Joshua Reynolds's call for gardeners to neglect accidental nature). As opposed to Picturesque scenery, where any promiscuous planting could run wild, he wished to promote the visible 'botany of trees and shrubs'; even flowers had to be used in such a way that they would lead to and show off their perfection in a given space. Yet he wished to augment the current fashion for Picturesque planting rather than simply reject it: he sought to bring his own botanical expertise (more substantial and informed, he claimed, than that possessed by earlier landscape gardeners) to bear upon a proper artifice of gardening. In his 'Remarks on Laying Out Public Gardens and Promenades' (1835) he urged that 'art should everywhere be avowed' and not mistaken for nature ('every garden is a work of art'). And when he obtained the earlier commission to design the Birmingham Botanical Garden in 1831 (never realized) he sought to group or connect all trees, shrubs and flowers rather than have them scattered randomly around. His use of the Jussieuean system, which grouped plants according to natural similarities in their forms, was in a sense the key to his how he wished to implement his vision, whether in a garden, park, botanical garden or arboretum.

Downing and Hovey's interest in public arboreta for the United States was answered in part by plans for Mount Auburn Cemetery, and by the foundation of the Arnold Arboretum in Cambridge, donated to Harvard in 1842. The evolution of these sites was involved. The Massachusetts Society for Promoting Agriculture, founded in 1792, had first established a professorship of natural history at Harvard and laid out a botanical garden (this survives only as marker on Garden Street near Linnaean Street in Cambridge); but Harvard wasn't interested and the Society later transferred its energies to establishing an agricultural college, what is now the University of Massachusetts.

Meanwhile, the Massachusetts Horticultural Society (founded in 1829) joined in an effort to create a rural cemetery with a group of suitably worthy Bostonians with a strong agricultural zeal; they included nursery-owners, the publisher of the *New England Farmer* and the people who grew all the apples available in the nation.[17] They were interested in cultivating flowers, ornamental shrubs and trees as well as food-producing plants. The Society held title to the land that was leased for Mount Auburn Cemetery and ran it for its first four years after its dedication in 1831. The Society had always wished for an experimental garden, and the first fruits of this enterprise were displayed at Mount Auburn in June 1833; they appointed a committee called the Garden and Cemetery Committee and the first superintendent was an experienced gardener and nurseryman. But the interests of the Society and the Cemetery steadily conflicted and eventually a plan was formed for separating them.

The Cemetery itself was concerned with a good layout of trees and flowers that would comfort the living and bless the dead; and this concern was, as Loudon would make clear in his 1843 book *On the Laying Out, Planting, and Managing of Cemeteries*, espoused by many other cemeteries of this kind (some were private stock companies, sometimes paying dividends, though Mount Auburn, did not). There was Kensal Green (1832) and Highgate (1839) in England, and Laurel Hill in Philadelphia (1836), being the first American cemetery to follow the example of Mount Auburn. Their layouts were various and irregular, following the contours of the land, with winding roads and

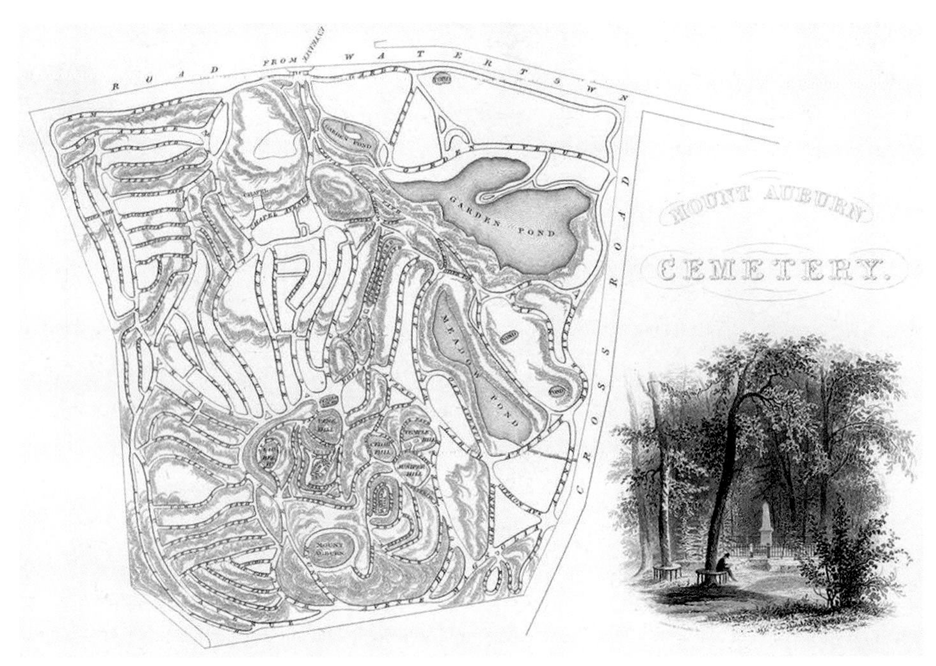

106 Mount Auburn Cemetery, by James Smillie after Cornelia W. Walter, 1847.

subsidiary paths (illus. 106), and they clearly subscribed to the vogue for Picturesque landscapes, but also – given the particular reason for their existence – to the opportunity for sublime experiences of walking among the dead. The cemeteries also catered to democratic visitors, which the majority of private parks and gardens did not, a point that Downing emphasized in his essay on 'Public Cemeteries and Gardens' in *The Horticulturist* of 1848. Soon guidebooks were produced for many cemeteries, identifying the sites of famous gravestones, but also identifying the planting: the *Guide to Laurel Hill Cemetery near Philadelphia* (1844), after narrating its history and some key monuments, provided lists of the 'botanical riches' that were found there and recommendations for the enhancement of graves for individual lot holders.

Inevitably, arboreta needed to establish their own sites and, more importantly, their own science and maintenance. Thus it was that the Arnold Arboretum in Cambridge, first started as a school of agriculture and horticultural, was then transformed itself into an arboretum in 1872 by Charles Sprague Sargent, the author of a *Manual of the Trees of America*. He summoned Olmsted to lay out the grounds and to link it with his larger project for the municipal parklands of the 'emerald necklace' around Boston. The aim was to collect and display 'as far as practicable all the trees, shrubs and herbaceous plants, either indigenous or exotic'.

Yet the establishment of other, modern arboreta and botanical gardens have not all maintained the exclusive and focused attention on their primary materials, and recent sites have also drawn out both new designs and other uses. Doubtless an interest in botany and trees will gain additional interest from those taken with other activities in the first instance. The Botanical Gardens at Wellington in New Zealand have been enhanced with a considerable display of modern sculpture, and the local State and University Morris Arboretum in Philadelphia involves its visitors with a variety of adventures in tree tops or with model trains as well as sculpture. The combination of these various objects of interest, or (as Loudon said) 'meaning', within sites formally dedicated to botanical study and pleasure have much enlarged the range of visitors that seek them out. Recent designs have also dramatically extended how we think of their forms and thus how we might view their contents: Peter Latz's master plan for the Arboretum and Green Areas on the Plateau de Kirchberg in Luxembourg, for the 1995 event 'European City', is a fresh version of the usual arboretum scenery. Catherine Mosbach's design for the new Botanical Garden at Bordeaux organizes its plants by their growing conditions rather than plant family, with sections for an Environmental Gallery (eleven different habitats), a Field of Crops (in elongated oblong strips, reminiscent of Leiden) and a water garden, where visitors can walk over the irregular grid of paths between the pools.[18]

107 The Greek Theatre at Epidaurus.

9 Garden as Theatre

We rightly think of the sixteenth and seventeenth centuries as times when the exchange between garden and theatre was most vital and inventive, when garden-makers were also theatrical designers and vice versa, when gardens became sites of performance before true theatres with prosceniums established themselves, when gardens and pastoral settings were a favourite decor for masques and operas, and where gardens themselves took the form of a mphitheatres and exedras. But long before the Renaissance, and certainly ever since, the liaison between theatre and garden (or landscape) was strong. Ancient Greek audiences watched their dramas played out against a vast landscape scenery (illus. 107, and see illus. 7). In ancient Rome public gardens, porticoes for instance, or in private villas, garden settings were both an occasion to be seen and to observe how owners or public figures behaved and spoke, how they performed themselves. Pliny saw the Tuscan surroundings of his villa as some amphitheatre, where the varied sceneries of wild and cultivated nature also seemed to present themselves to the viewer. Vauxhall Gardens in eighteenth-century London continued this theatrical role and was a prime, but by no means the only, example of a space specifically set aside for performances – for musicians and singers to be heard, fireworks and other spectacles to be observed, famous visitors to be spotted in the supper booths and the passing crowds to be scanned by the more fortunate diners. Today we still go to parks and gardens to observe and be observed, to be – consciously or unconsciously – actors and spectators in a civic and intrapersonal world; here each of us may perform and be observed, like T. S. Eliot's J. Arthur Prufrock choosing to 'prepare a face to meet the faces that you meet', a strategy that suggests both a question of theatrical costume and an assumption of an apt public persona.

Both religious and secular dramas had never lacked sites for performance: Greek and Roman temples, Christian churches, great halls of manor houses and castles, banqueting halls, cities where guilds could present their various narratives on stages or on carts drawn through the streets. But once Renaissance gardens, often aping classical prototypes (see chapter Six), grew more spacious, they acquired sites that were suitable and adapted for dramatic performance. The Belvedere Courtyard at the Vatican was devised partly to receive welcoming visiting dignities and ambassadors with the appropriate pomp; this was a familiar ceremony in the politics of the time, acted out on city streets, town gateways and other improvised stages, but increasingly presented in the setting of some grand garden or courtyard. The lower Belvedere courtyard was also conceived as an early site that could be flooded for presentations of sea battles (the Roman *naumachia*); a tournament took place there during carnival time in 1565. The highest area of its long courtyard probably took the form of a small stage, modelled on the Temple of Fortune at Praeneste where Roman rituals had been mounted on a platform, flanked by ascending and descending staircases. This elegant 'double staircase', as it became known and as it was codified by Serlio (see illus. 77), continues to suggest a performance space, with its audience or visitors on the steps below the central circle and the 'action' taking place on the higher range. Such a small but eloquent gesture welcomes visitors at the entrance to the Turner Gallery at Tate Britain, inviting them to view the exhibitions within, and the structure was much invoked in Arts and Crafts houses around 1900, gestures that invited visitors into whatever excitements the garden would reveal (see illus. 218 and 222). Furthermore, the Vatican's sculpture collection at the rear of the central courtyard was a showcase of recent antique excavations, one that offers to the visitor a display of papal prestige.

The whole development of places for theatrical events – what the French call a 'lieu théatral', or the Italian a 'luogo teatrale' – has been a major scholarly activity.[1] Yet while its search for and identification of the forms that we now accept as the beginnings of the modern theatre pay some attention to garden locations as sites for performances, the consequence and implications of these developments for gardens themselves are largely ignored.[2] The Belvedere was a very early chance to exploit the theatrical possibilities of its courtyards and gardens, and those opportunities were much imitated throughout the sixteenth century; by the seventeenth century, not just garden sites that could be utilized for theatrical presentations, but actual stages with 'wings' and 'footlights' were included, coinciding with the first creation of modern indoor theatres, public now as well as private.

We need to look not only at the gradual invention of garden places that could be used for performances, but also at the very way in which the theatrical possibilities of gardens affected new assumptions about their visitation and reception. The physical elements that speak most to the potential of dramas or performances in the garden are stage-like platforms or maybe terraces, arenas or hippodromes (an ancient Roman format), ranks of benches, often curved seating or *exedrae* (another Roman feature) from which visitors could admire events, and grottoes, or little theatres or 'peep-shows', where hydraulic machinery presented shows for the visitors as if upon a stage. Palladio would also suggest that the ground immediately surrounding a villa should be given an 'aspetto di grande teatro', and a glance at Utens's painting of the Medici villa at Castello (see illus. 18) shows exactly how that arena could look.

At the Pitti Place in Florence, the courtyard, and its elaborate ascent into the arena of the Boboli Gardens, offered a variety of theatrical settings. The hippodrome in particular, a natural declivity in the hillside, served as an amphitheatre for all kinds of displays. In this space, at first merely a grassed area before being provided from 1599 with wooden seats and then ranges of stone seats (still there), were performed pageants, ballets, displays of horsemanship (a 'horse dance' much admired by an English visitor) and musical dramas, of which a late example is the drawing

108 Stefano Della Bella, *Equestrian Ballet in the Boboli Amphitheatre*, 1661, etching.

by Stefano Della Bella, showing the festivities to celebrate the marriage of Ferdinand II in 1637 (illus. 108). So novel was the whole idea of theatrical spaces that English visitors not only made note of performances of 'musical Drammata' in Venetian theatres, and at an opera house in Siena with 'severall changes of Scenes', but admired gardens that enjoyed such spectacles, as well as other places as various as the piazza at Siena ('made after the manner of a theatre') or the port of Genoa. Burnet admired the Isola Bella on Lake Maggiore, the 'face of it [that] looks to the Parterre is made like a Theatre, full of Fountains and Statues' (illus. 109). John Evelyn visiting the Villa Mondragone at Frascati noted that it was a 'theatre of pastimes' (illus. 110): what he meant is unclear, except in his use of the word 'theatre' to mean a conspectus or complete collection. The courtyard there had a raised platform at one end, its apsidal back wall broken by seven niches that may have echoed the seven entrances in Alberti's reconstruction of antique theatres; so it certainly looked like a stage seen from the courtyard below, and the platform itself was enlivened by several fountains.

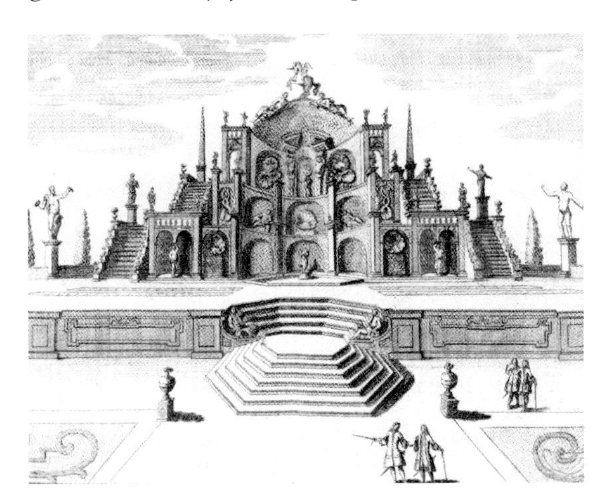

109 The Theatre at the Isola Bella, Lake Maggiore, from M.A. Dal Re, *Ville di Delizia* (1726).

Another, also Medici, family affair employed the Tuileries gardens in Paris for a performance with which Catherine de' Medici honoured the Polish ambassadors (illus. 111). Antoine Caron's drawing shows a *ballet de cour* at the point when ladies, representing the sixteen provinces of

110 Theatre at the Villa Mondragone, Frascati.

France, arrived at the specially constructed Mount Parnassus upon which Apollo and the muses are playing their instruments: courtiers here were both actors and spectators of the spectacle, and the presentation by the Muses displaying the riches of the French territories invoked this famous antecedent from classical mythology, a Mount Parnassus. But such ad hoc and mobile settings could also be realized as permanent features in princely gardens, often completed with hydraulically run performances. Such a Mount Parnassus was established at the Medici villa of Pratolino: a drawing by Giovanni Guerra, in the late sixteenth century, shows not only its Parnassus, but the small *exedral* seating that was used by visitors to listen to its musical performances (illus. 112).

Gardens not only invited suitable locations for major court performances or, as at Pratolino, small and local entertainments, but soon chose to supply stages specifically for theatrical events. There are several surviving open-air theatres in Italian gardens; a cluster in the area around Lucca suggests the incidence of these. At the Villa Garzoni at Collodi, a huge ranked garden that climbs the hillside and looks itself like some colossal theatre, with visitors at the lower entry level gazing upward to the spectacle, also contains its own small theatre to one side, with wings from which players would enter upon its stage. A similar, late seventeenth-century engraving (illus. 113) of the Villa Reale (today Pecci-Blunt) at Marlia reveals another theatre at the right, with 'wings' and a 'backdrop' of yew hedges, with 'footlights' in box and a 'prompt box' at the front of the stage, and a curved auditorium (still surviving, with statues of actors in three niches on stage). But the whole villa grounds also suggest how much

111 The reception of the Polish Ambassadors by Catherine de' Medici in the Tuileries, September 1573. Tapestry for the Fêtes des Valois; a Mount Parnassus with musician playing is at the right.

112 Giovanni Guerra, *The amphitheatre for watching the musicians on the replica 'Mount Parnassus', Pratolino*, 1604.

Italian gardens exploited similar theatrical spaces, even if they were not primarily used as stages – at Marlia there were two such areas: one behind the villa itself that looks also very much like another theatre, and to the villa's right is another curved alcove, also in yew, that fronts a square pool of water. Yet another Lucchese villa, Torrigiani, underwent some 'Englishing' of its parkland, like many in the eighteenth century, but it preserved, happily, a small sunken garden called the Garden of Flora: at one end is a terrace, with handsome divided staircases under which we can explore the different small grottoes (illus. 114); at the other end of the garden is a grotto of the winds, where – with encouragement from a gardener and the preservation of some of its hydraulic effects – unwary visitors can still be caught in a downpour inside the grotto and effectively prevented from exiting through a curtain of rising jets.

If Torrigiani included small rooms or grottoes underneath the small-scale terrace, Italian gardens generally accommodated larger and

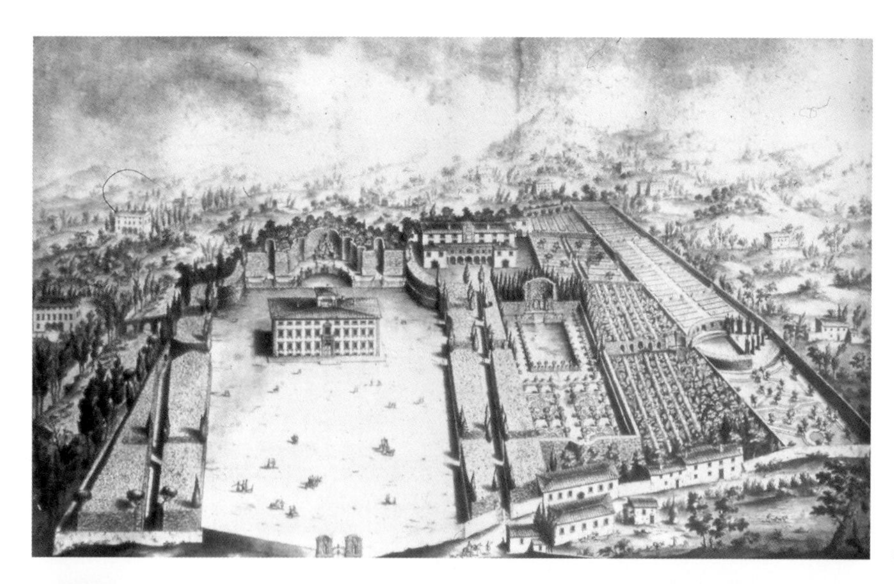

113 The Villa Reale di Marlia, near Lucca, late 17th century.

114 The Garden of Flora at Villa Torrigiani, near Lucca.

often complicated grottoes, most of which would either present some sculptural figures on the rear 'stage' or, more likely, utilize hydraulic devices to move and animate sculpture. These too would have been activated by some unseen gardener manipulating the machinery: John Evelyn enjoyed his visits behind the scenes as well as the shows themselves. One good example of this type was published by Salomon de Caus in *Les Raison des forces mouvantes* (1615): it shows both the sculptural figures – the nymph Galatea and the besotted, one-eyed giant, Polyphemus – and the accompanying mechanisms by which Galatea's conch shell would glide across the waves (the basin is shown here without water) and the 'off stage' winch that would propel her (illus. 115). Such devices rarely survive. What can still be seen are stationary equivalents: figures in niches and grottoes without any mechanic mechanisms, whose themes and activity visitors would need to imagine for themselves. Sometimes sculptures would be brought alive by the play of water around them – the roaring dragons at the Villa d'Este still make a plausible effect when jets gush from their mouths and the whole basin erupts in a frenzy of water. Others, like the niches that André Le Nôtre introduced into terraces at Vaux-le-Vicomte or Chantilly, might also be framed in water, but essentially were left to tell whatever story was apt for the different locations without the use – as Le Nôtre would himself have thought – of any 'childish' apparatus.

Italian and French gardens served to showcase gardens with theatres, not least because the climate made such locations propitious for outdoor display. And northern visitors enjoyed these encounters, and carried some of their culture back with them to England, for in England too gardens had become an essential part of royal progresses. This is not the place to trace these Italianate imitations and borrowings in England (I have done this elsewhere[3]); but the use of these much enhanced gardens became a crucial part of royal performances, as well as influencing the presentation of masques within theatres and court entertainments themselves that featured garden settings.

Royal visitations of country seats both paraded the monarchy and showed off the facilities of country houses – the monarch enjoyed some free hospitality, though the patron may have expended a fortune on both the actual hospitality and entertainments and upon the garden which featured them.[4] The Earl of Leicester welcomed Elizabeth in 1575 to his much refurbished garden: a contemporary, Robert Laneham, saw it 'beautified with many delectable, fresh, and umbrageous bowers, arbours, seats, and walks, that with great art, cost and diligence were very pleasantly appointed; which also the natural grace by the tall and fragrant trees and soil did so far command, as Diana herself might have devised there well

115 Salomon de Caus, *Les Raisons des forces mouvantes* (1615), showing the nymph Galatea and the giant Polyphemus, and the machinery that activated them.

enough too range [walk about] for her pastime'. And accordingly Elizabeth was herself represented there as the Virgin Queen, Diana the huntress. In 1591 Lord Hertford presented the Queen with an elaborate entertainment on a specially excavated lake, shaped as a crescent moon to represent the Queen as Cynthia.

Similar presentations marked the progresses of King James I and his Queen: in 1604 they were received by Sir William Cornwallis in his house and garden at Highgate, where both setting and dramatis personae contrived a seamless pageant in which actors, as if garden statues, greeted the royal couple and spoke to them, first as the god Mercury, and then as the goddesses May, Flora, Zephyrus and Aurora; after dining, the god Pan and a group of satyrs danced around a fountain running with wine. Nine years later the Queen was welcomed at Caversham with an entertainment devised by Thomas Campion, where the modest gardens were nonetheless exploited: as she approached the park limits, a rough figure emerged from the bushes to 'confront' her retinue, and when she reached the terraced gardens an elaborately dressed Gardiner and his Boy greeted her, serenaded by singers from a neighbouring arbour. Charles I and Henrietta Maria were welcomed at Bolsover Castle by the Earl of Newcastle in July 1634, with a celebration devised by Ben Jonson; his characters discoursed on the nature of love in a specially created garden, presided over by a fountain of Venus.

In masque designs, increasingly popular under James I and Charles I, the use of a garden setting was much employed, with considerable flair.[5] In Inigo Jones's *Oberon* in 1611, what appeared first as a rocky outcrop (with 'all the wilderness that could be presented') was then parted to reveal 'a bright and glorious palace'. It is as if a garden walker discovered new scenes as he explored its spaces, and it parallels exactly the terms in which a real visitor to an Italian garden would describe his experience. Sir Henry Wotton, writing of some unidentified Italian site, spoke of its various '*mountings* and *valings*' as if he had 'bin Magically transported into a new Garden'. Compare that with this stage direction in a masque scenario for *Coelum Britannicum*:

> the scene again is varied into a new and pleasant prospect clean differing from all the other, the newest part showing a delicious garden with several walks and parterras set around with low trees, and on all sides against these walks were fountains and grots, and in the furthest part a palace from whence went high walks upon arches, and above them open terraces planted with cypress trees, and all this was composed of such ornaments as might express a princely villa.

Edmund Warcupp in his *Italy in Its Originall Glory, Ruine and Revivial* (1660) translates an Italian text about Tivoli as if it was indeed perceived as a masque: 'In the descent into the first garden, shows itself the Colossus of Pegasus', or 'riseth an Island cut in the shape of a ship'.

Grottoes in gardens provided a key feature of these theatre sets. In *Tethys' Festival* of 1610, devised by Samuel Daniel and Inigo Jones, are described scene changes that involve 'nymphs in several caverns gloriously adorned', with friezes and sculptures depicting sea creatures and niches for the nymphs; a courtier, the Duke of York, goes in search of Tethys which he will find 'hard by within a grove', which itself suddenly appeared, as if the Duke had himself been wandering in some real garden. The theme of the sea goddess,

Tethys, was similarly displayed in the gardens at Villa d'Este. A court masque by Inigo Jones and William Davenant, *Luminalia* in 1638, plays with the *topos* of chaos and disorder being overcome through the metamorphosis of garden art: after its anti-masque, the 'scene is changed into a delicious prospect, wherein were rows of trees, fountains, statues, arbours, grottoes, walks, and all such things of delight as might express the beautiful garden of the Britanides'. While an actual garden stroller might proceed and discover a further garden, the masque audience, seated, standing, but stationary, must depend upon the illusions of the setting, and accordingly in *Luminalia* 'a further part of the garden' was disclosed itself to show the Queen herself. The unfolding of the garden, its association of royal presence with the arts of governmental and social harmony, the performance of actors by the courtiers themselves, above all the metamorphic metaphors of the masque, served to project an ideal world with which the realities of contemporary politics seemed often at odds. The garden was thus a place of grand illusions, but one which everybody was forced eventually to leave to confront the troubled worlds outside.

The incidence of garden settings in theatrical or operatic performances during the seventeenth century was considerable, and much of these designs obviously drew upon familiar experiences in actual gardens; Inigo Jones himself displayed his skills in both masque and garden work. Similarly, a prolific stage designer, like the Venetian Giacomo Torelli (1608–1678), produced many designs, where grotto facades and especially axial vistas could be excitingly represented on theatrical flats and backdrops.[6] One strange amalgamation of real garden effects with an illusionary stage setting occurred in the Barberini Palace in Rome, in 1656: a painted set of a garden was augmented across the font of the stage itself with a row of fountains for which real water was piped in from the gardens outside.

The specific exchange between actual gardens and illusionary theatre sets clearly established a rich response to how gardens were themselves received. Yet evidence of this reception is not always easy to come by. But almost every sixteenth and seventeenth-century engraving showed figures inside gardens, behaving and acting out their responses and reactions. People point and gesture, run towards events and behave in ways that often are drawn by artists in ways that mimic, or are even copied from, current technical manuals for actors to learn their gestures. There are such figures in Guerra's drawing of the amphitheatre in front of the Parnassus at Pratolino, or depicted in the views of Mondragone or the Isola Bella (see illus. 109, 110 and 112). The images may in many cases be using conventional *staffage*, but they do suggest that the experience of a garden was part of a visit, and its visitors are usually seen in front of some significant episode of a garden, to which they are reacting.

Too much modern photography of gardens ignores people altogether. We see the scenery, but not the actors or even the spectators. We admire the view, but see neither the viewers nor, though this is more difficult to represent, their responses. This seems an inevitable result of changes in landscape practice during the long eighteenth century, when the views of a 'natural' scenery were admired for their own sakes rather than as stimuli to people's behaviour. In its early years of the late sixteenth and seventeenth centuries, as a result of the considerable import of garden ideas from Italy and France, there were garden theatres in English gardens, but gradually

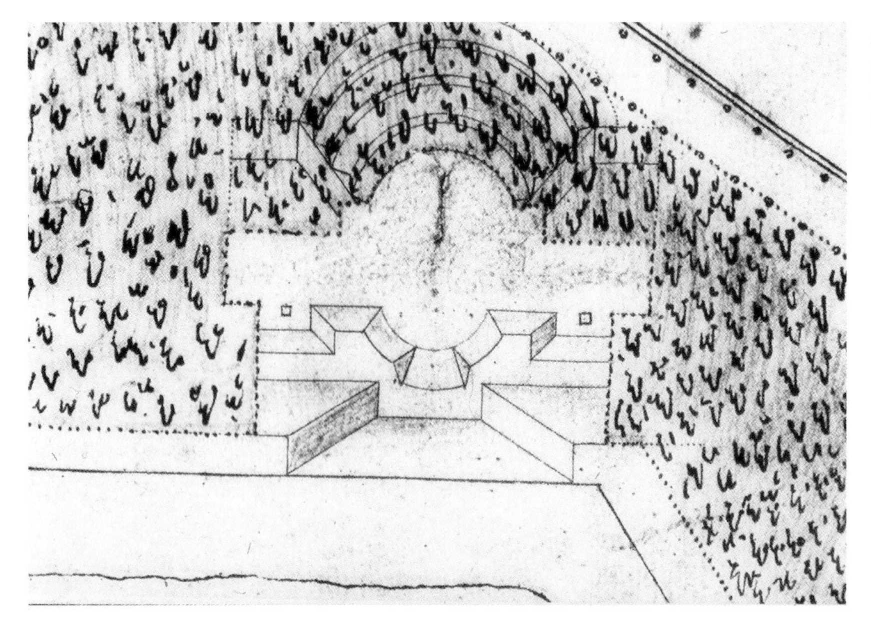

116 The theatre at Rousham, designed by Charles Bridgeman, detail from a plan attributed to him, 1720s.

117 The site today of Bridgeman's former theatre at Rousham, remodelled by William Kent, although the John van Nost lead figures remain.

such a formal shape was eliminated or absorbed into the topography.

Charles Bridgeman worked on many estates until his death in 1738, and one of his distinctive features was 'theatres'. Because they were most usually laid out in turf with grass ramps and settings, they were also liable to poor maintenance (sharp grass edges need constant attention) and became gradually absorbed into the surrounding landscape. There has been little explicit discussion of his theatres or their significance per se, but a glance at plans and engravings suggests how frequent they must be been on the ground and in his sketch plans.[7] It is unclear how much these spatial designs were used specifically as theatres, or even described as such (though we have one such at Rousham), but their depiction was always represented by a perspectival arrangement of hedges or ramps, narrowing like any set of theatrical wings and often located on a hillside at the far end of a garden design, its stage-like clearing surrounding by woodland, as at Amesbury. A slightly different format was provided for Tring and for Castle Hill, but the ascending hillside was still marked with five terraced platforms. At Eastbury, Dorset, as engraved for *Vitruvius Britannicus* volume III (1725), it is described as 'designed and finished by the ingenius Mr Charles Bridgeman', and the garden ends with a ramped theatre around a circular pool. Claremont in Surrey had a spectacular amphitheatre, what James Thomson called a 'terrassed height', with elaborate ramps and steps, adapting the Serlio engraving of the Belvedere double staircase, and this was engraved for Switzer's book, *An Introduction to a General System of Hydrostaticks and Hydraulicks* (1729), where again it was ascribed to Bridgeman. Whether a true or utilized theatre, or whether simply a way of marking a particular garden climax, with its narrowing groves drawing the visitor into its embrace, and sometimes associated with a mount, Bridgeman's proposals are found in sketches for Gunton and in his bird's-eye view of the gardens at Stowe; here the Queen's Theatre was so-called, its ramps leading upwards towards the column at its head. In the series of views, commissioned by Jacques Rigaud to document Bridgeman's work at Stowe shortly before the latter's death, we encounter, not so much views of conspicuous theatre layouts (the Queen's Theatre is shown in the far distance of one view), as a garden world where actors and spectators throng and respond in ways that bring wonderfully alive how these gardens both invited and accommodated the theatrical dimension of early eighteenth-century landscapes.

William Kent had perhaps different reasons for embracing theatrical episodes in his gardens and sketches. Bridgeman would have drawn upon the formal shapes of theatres and ramps in his landscapes from a variety of probably French models, while Kent had himself practised as a theatre designer in Italy and in England. Though information is scarce, Kent must have witnessed an extraordinary entertainment devised for the Academy of Virtuosi in Rome in 1711 by William Talman, and in England he provided both costumes and settings for a production at the Theatre Royal in Drury Lane in 1724, four years after his return from Italy. He designed costumes for a masquerade for the Prince of Wales in 1731 and is supposed to have designed sets for a semi-cantata opera, *Il Festa d'Imeneo*, five years later. However, we also have some sketches of actors and performances, including one in Sir John Soane's Museum, that shows actors/singers in a pastoral opera on a terraced stage, backed by a version of his own ruined hermitages at Stowe and Richmond, on the front of which is inscribed the word 'Arcadia'; this would be apt for the characters of a satyr and shepherdess that are shown on stage. And among the collection of designs at Chatsworth is a drawing of a forest pool, with a rising moon, that could easily be a backdrop: attempts to describe it as a project for Claremont seems far-fetched, as it is clearly indebted to one of Inigo Jones's masque designs. Also at Chatsworth, now bound in a folio of specifically theatre designs, is another proposal for what must be a backdrop design, showing pavilions in the woods, and a large lake with boats in a regatta, including a huge barge topped with a miniature version of Palladio's Rotunda.

Burlington of course purchased drawings of Inigo Jones's designs in 1720, which Kent would surely have known and which would have supported his own interest in both theatre designs and garden effects. At Chiswick, Kent devised two different versions of its exedra: one building with niches and a pyramid, which he later re-used for the Temple of British Worthies at Stowe; it 'performs' its coded message for those gathered to view its inscriptions and inspect its worthy British personages. He replaced this at Chiswick with a generous hemicycle of hedges, with full-length sculptural figures in niches. Its open gesture invites into the space those who wish to consult the statues, sit on its benches and observe, now from its 'stage', the gardens towards the house itself: it is quite literally a performance of Burlington's own programme for accommodating both Palladio and Italian garden cultures into the English scene.

In Kent's specifically garden and landscape drawings there is a host of spaces that signal some interest in similar experiences. There are openings in hillside groves, sometimes marked by an urn or by matching herms at the edges of an opening's 'stage'; platforms where his figures contemplate and gesture towards special views like cascades at Chatsworth, or arches and obelisks, even the large vista of Euston Hall in its parkland, or the agricultural countryside seen from the Rousham gardens. He proposed an elaborate exedral tunnel designed, but not used, for Holkham hillside. He sketched Alexander Pope's own grotto, now with a real and English occupant rather than the sculptures or mechanical devices that he must have found in Italy. All these instances, and their incidence is remarkable, point to Kent's interest in directing our attention in his sketches, to how the gardens would be used, directing viewers in actual gardens, to significant, composed moments of landscape experience. He took his specific theatrical experience and transferred it to exploit a new landscape aesthetic. And in one particular garden he seems to have done precisely what these suggestions can only imply. Bridgeman had designed a specific theatre for the Rousham gardens, alongside the river Cherwell; here it may have been a very specific stage for local entertainments, or perhaps was just an invitation for visitors to take their place on its stage and contemplate the Oxfordshire countryside beyond the river (illus. 116, 128). When Kent came to remodel the gardens in 1739 he eliminated Bridgeman's formal ramps and grassy platform shapes, but he left the space as some implied theatre, retaining three statues of pastoral figures at the three apexes, like actors waiting to play their parts. In the glade at Rousham we can still appreciate this theatrical moment and its possibilities for the modern visitor.

It is difficult to appreciate these garden moments, even on a site which retains these fossils of theatrical design, and even more difficult

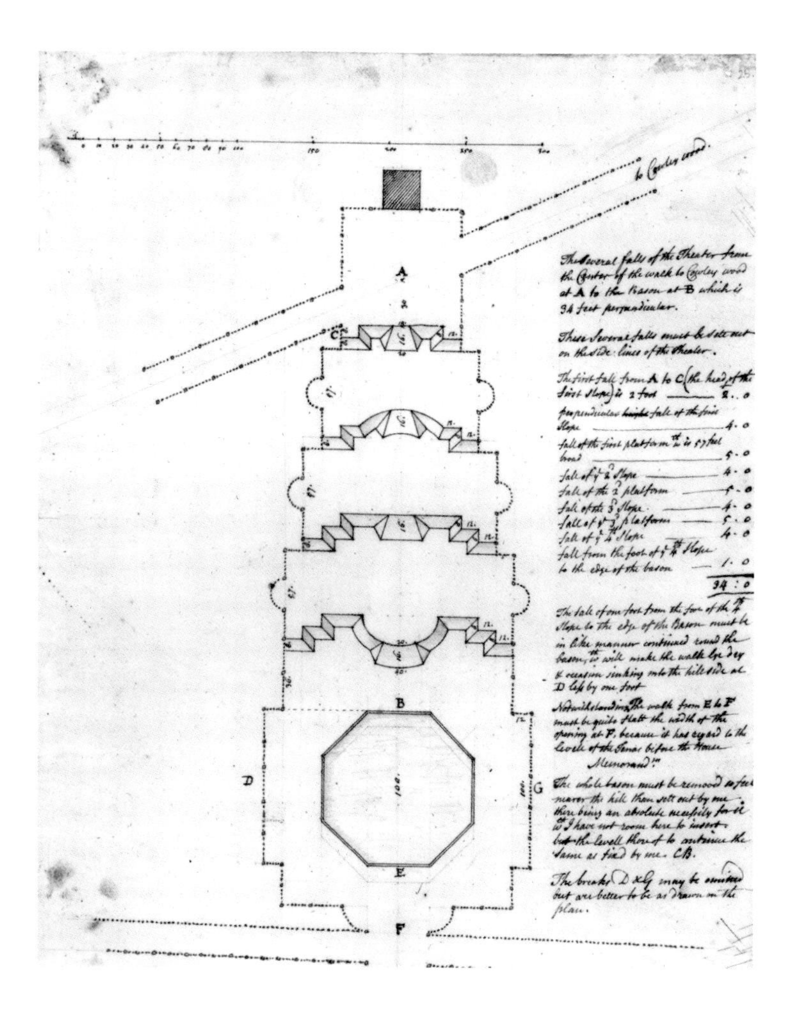

118 Charles Bridgeman (attrib.), design for a garden theatre at Tring, Hertfordshire, early 1720s.

to capture the experiences of them. Yet almost any visit to parks and gardens these days betrays their theatrical potential: we watch others do their thing, and we are watched in our own performances. And photographs of these activities are now Photoshopped into modern designs, where landscape architects can readily paste into their images all sorts of folk, roller-skating, talking on their mobile phones, chatting and walking along; this has become a basic tool of any design proposal (see illus. 243). Yet so many of these figures really provide no more than the old, much cliched *staffage*; for the most part they fail to show what earlier artists did by drawing attention to how landscapes would have been received. In an ironical way, we have rediscovered a new Picturesque, computer-generated ('computeresque'). Yet it largely enjoys little of what the original Picturesque promoted.

10 The Garden of 'Betweenity': Between André Le Nôtre and William Kent

English historiography in the eighteenth century sought to push theory and practice towards a climax that valued 'natural' and 'English' gardening above all else. Horace Walpole was the most vocal in acclaiming this inexorable and wonderful future:

> the reason why [English] taste in [natural] Gardening was never discovered before the beginning of the present Century, is that it was the result of all the happy combinations of an Empire of Freemen, an Empire formed by Trade, not by military & conquering Spirit, maintained by the valour of independent Property, enjoying long tranquillity after virtuous struggles, & employing its opulence & good sense on the reformation of rational Pleasure.

And the rhetoric of his claims continues to colour subsequent versions of this proleptic history, even if fresh materials and data have come to question his patriotism and design narrative. There were certainly some murmurs of protest, like those of the Abbé Delille, who decided (even though a Frenchman) that he would not subscribe to either the party of André Le Nôtre or that of William Kent ('Je ne décide point entre Kent et Le Nôtre'). And one later Englishman also resisted this either/or choice and could see the excitements of a new and wonderful gardening as it was developing, refusing the G-force of theory over actual historical practice: this was the architect John D. Sedding.

In the essays collected in his *Garden Craft Old and New*, published posthumously in 1891, Sedding championed the very beginnings of 'English' landscaping, and to some extent its pronouncements, during the seventeenth and very early eighteenth centuries. But he did so not to proclaim with hindsight the future of landscape architecture, but to honour a particular, historical moment. To call attention to the work that seemed to fall between Le Nôtrean practice and the English ideas of Kent and Brown, he termed it, somewhat awkwardly, 'betweenity – the garden of transition'.[1] Sedding is thus useful in the late nineteenth century for drawing attention to a period that should not be assimilated easily then into the annals of garden history, and he used his historical understanding to comment upon a new gardening that emerged by 1900 (see chapter Eighteen).

However, this chapter is primarily dedicated to the ideas and practice of the period around 1700 to which he pointed. So, what exactly was going on 'between' André Le Nôtre and William

Kent? There are two main answers, one that draws upon what was happening in France in the years after Le Nôtre's death in 1700 and the other that addresses theories and practice of early garden-makers in England, like Joseph Addison and Stephen Switzer, neither of whom could have known in advance where the 'English' gardening would eventually lead.

It was in fact Joseph Addison in *The Spectator*, no. 414 of 1712, who took issue with current gardening:

> Our English Gardens are not so entertaining to the Fancy as those of France and Italy, where we see a large Extent of Ground covered with an agreeable mixture of Garden and Forest, which represent every where an Artificial Rudeness, much more charming than that Neatness and Elegancy which we meet with in those of our own Country.

He may well be reacting against the niceties and smaller formalisms of Dutch gardening, much the English and Scottish thrones in 1688; but his emphasis suggests a less limited focus and a more interesting aesthetic. He values landscape that entertains the imagination, that mixes garden and woodland, which yields an 'artificial Rudeness' (that is, that contrives artificially to 'represent' a more complex topography). And he appeals to Italian and especially *French* gardening, which he knew from his travels on the Continent in the years 1701–3 (his *Remarks* on this topic were later published in 1721).

In this same issue of *The Spectator* he went on to ask why English estates could not be 'thrown into a kind of Garden by frequent Plantations' that would ensure the owner's profit as well as his pleasure: 'A Marsh overgrown with Willows, or a Mountain shaded with Oaks . . . [rather than lying] bare and unadorned' allows pollarding and forestry. Equally,

> Fields of Corn make a pleasant prospect, and if the Walks were a little taken care of that lie between them, if the natural Embroidery of the Meadows were helpt and improved by some small Additions of Art, and the several Rows of Hedges set off by Trees and Flowers, that the Soil was capable of receiving, a Man might make a pretty Landskip of his own Possessions.

Addison's appeal to a landscape painting is well know, but the terms of his proposal are more important. It is 'prospect' that concerns him, visions of 'a kind of Garden' that earns its keep as well as its admiration; what is needed in the meadows, for example, is not anything 'natural' but rather a 'natural embroidery', a term that translates the French gardening term, *parterre de broderie*, into a specifically English display of indigenous flora ('soil' that yields its own apt planting).

Addison is clearly acknowledging the experience he had gathered on his Continental travels, an experience of gardens that was able to register a range of territories, notably a diminution of landscape forms as the chateau or villa was left behind. This concept had been floated in the late sixteenth century by Italian humanists, but by the later seventeenth century its forms were visible on the ground and in many engraved views. The tripartite treatment of an owner's property and its prospects was organized in different ways depending on the site and the property in question, but the scheme was roughly like this: first, a much more regular treatment of land near the

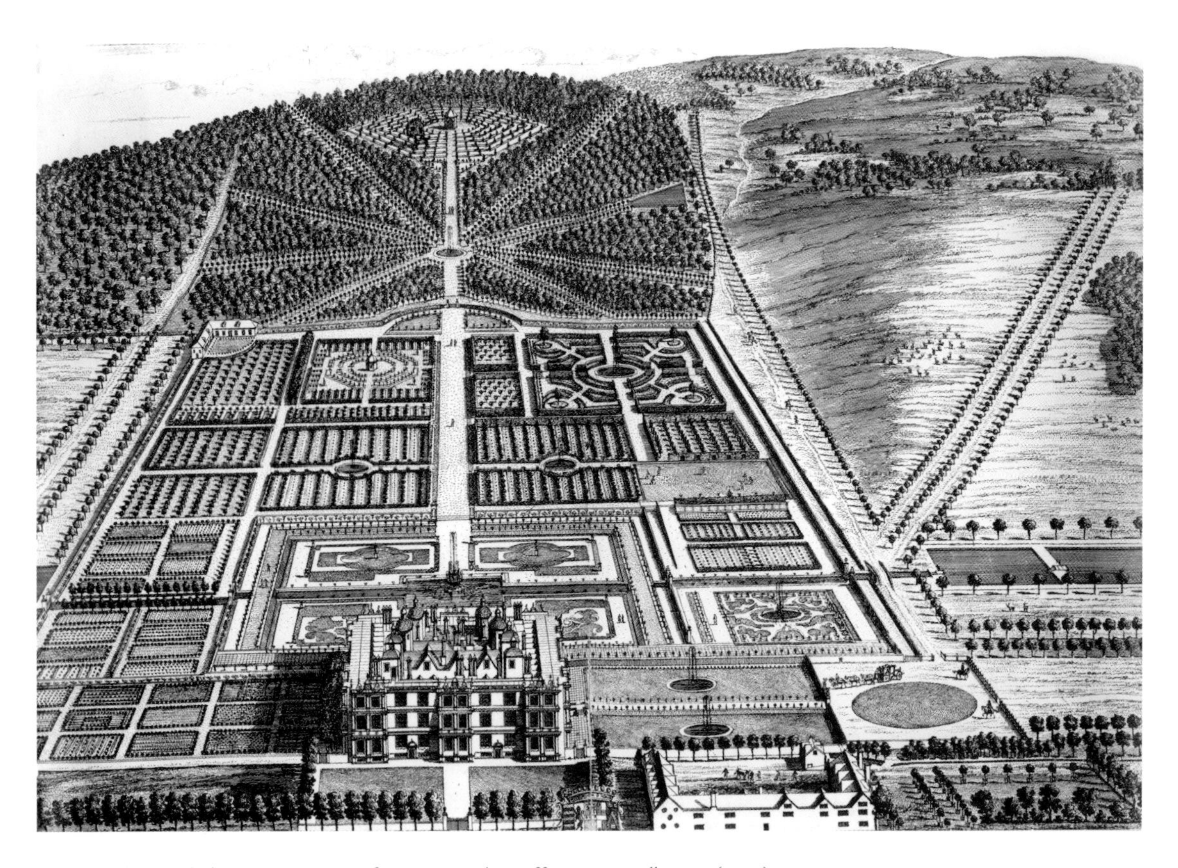

119 Longleat, Wiltshire, an engraving from Kip and Knyff's *Britannia Illustrata* (1707).

house, the garden *proprement dit*; then followed orchards or groves, often regularly laid out, maybe as a labyrinth; then finally fields or even waste land, which stretched to the near horizon and might have found distant termination points on hillsides or mountains. This format is clearly registered in engraved views that take their perspective from a high angle that reveals to the eye immediately this gradation of territory, an 'idea' or 'epitome' of the whole property. In one particular instance (illus. 119, but see also illus. 93), the 'long' view has been taken to the side, alongside the more organized territory, which nevertheless terminates in a labyrinth-like wood on the further hillside. When Sedding came to describe this very visible and accepted layout, he called it the 'garden of transition' (probably referring as much to the sequence of landscape modes within a given property as to his declaration of a transition between two historical modes of gardening).

Sedding describes this mode, with a proper touch of Victorian modesty, as 'a 'natural wildness' touching the hem of artificiality'; elsewhere, more aesthetically, as 'the attainment of this appearance of graduated formality':

> It is essential that the ground immediately about the house should be devoted to symmetrical planning, and to distinctly ornamental treatment; and the symmetry

> should break away by easy stages from the dressed to the undress parts, and so on to the open country, beginning with wilder effects upon the country-boundaries of the place, and more careful and intricate effects as the house is approached (p. 135).

He also repeats this vision of how the landscape should be worked, when discussing the seventeenth-century phase of English gardening in an earlier essay:

> You will note how the English garden stops, as it were, without ending. Around or near the house will be the ordered garden with terraces and architectural accessories, all trim and fit and nice. Then comes the smooth-shaven lawn, studded and belted around with fine trees, arranged as it seems with a divine carelessness; and beyond the lawn, the ferny heather-turf of the park, where the dappled deer browse and the rabbits run wild, and the sun-chequer'd glades go out to meet, and lose themselves by green degrees; in the approaching woodland, – past the river glen, the steep fields of grass and corn, the cottages and stackyards and grey church tower of the village; past the ridge of fir-land and the dark sweep of heath-country into the dim wavering lines of blue distance. (p. 69)

There is undoubtedly a strong flavour of late nineteenth-century English sentiment about this vision, suitable enough for the thinking of the 1890s and its revival of an indigenous Arts and Crafts garden-making; but the passages also reveal his keen sense of the landscaping that he encountered in his attention to early engravings and to various readings in Francis Bacon, the Earl of Shaftesbury, William Temple and John Evelyn, and Addison himself. It was, and still is, a somewhat neglected moment of garden history.

If we move forward just a few years beyond Addison, we can seize upon a recent English layout that seems to capture, almost diagrammatically, this three-fold sequence, this 'graduated' sequence of garden and forest spaces. It appears in Stephen Switzer's *Ichnographia Rustica*, first published with a different title, *The Nobleman, Gentleman, and Gardner's Recreation* in 1715, and expanded into three volumes in 1718 (a further edition came much later in 1742). Switzer illustrates an engraved view of the 'Manor of Paston divided and planted into Rural Gardens' (illus. 120). To describe what he terms as 'the extensive Way of Gard'ning', he opts for what the '*French* call *la Grand Manier*'[sic]. Perhaps, like Addison, this was to fend off a taste for Dutch garden-making; but his text also announces an English sensitivity to scale and proportion and to the range of landscape experience that the Paston map also suggests.

He divides the available ground into two kinds of layout, with a third suggested beyond the other two. First comes a symmetrical garden of groves, laid out below the house within a bastion-like garden reminiscent of fortifications; this contains a circular pool and a jet of water, and a series of straight pathways; this area is further extended down a long *allée* that ends in another circular bastion, now without any fountain; along the sides of this avenue are placed pairs of rectangular groves that answer each other across the avenue. Surrounding this layout, and engraved now in a lighter tone, is a highly artificial and curious landscape, with clearings carved out of its thickets, and a meandering set of paths that wind

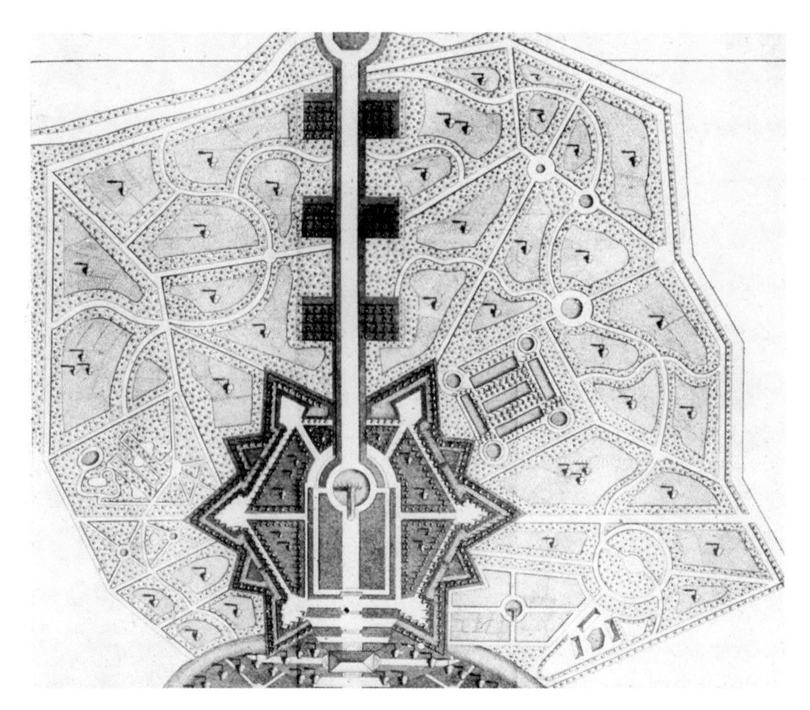

120 'Manor of Paston divided and planted into Rural Gardens', in Stephen Switzer's *Ichnographia Rustica* (1718).

their way through this woodland. In a few places there is some indication of a more regular feature: the squared plot to the right of the bastion garden might suggest a vegetable garden with fish ponds. And beyond this more irregularly shaped woodland is a blank territory that, deducing from the walkways along the edges of the woodland, is undifferentiated countryside over which views might be taken; as there is no indication now of bastioned walls, one may suppose a ditch or ha-ha separates the designed woodland from the larger landscape beyond.

In fact, this is exactly how this design appeared twenty or so years later in 1736, when Grimsthorpe was visited by the antiquary William Stukeley. His amateur but perfectly comprehensible sketches (illus. 121–124) show that the landscape of Grimsthorpe Castle in Lincolnshire, laid out by Switzer, presented exactly the same appearance as the diagrammatic plan that was given in *Ichnographia Rustica*. First came the fairly rigid parterre, then groves (well developed by 1736), and lastly view-points at the edges of the woodland, which indeed are shown as having ha-has across which figures are shown gazing into the open countryside.

For both Switzer and Stukeley there was undoubted pleasure in being able to experience a *variety* of landscapes, including a large prospect into the countryside. It was after all, a key concept in several aspects of contemporary thinking. The later vogue for English landscape gardening would be liable to opt largely in favour of the further, 'natural' prospect into the countryside, eliminating the rest. Yet a fundamental element of the Grimsthorpe landscape, and others like it, was that the visitor came to see the far landscape as a result of moving through or maybe gazing over the different gradations of design in the immediate garden. A person could appreciate all

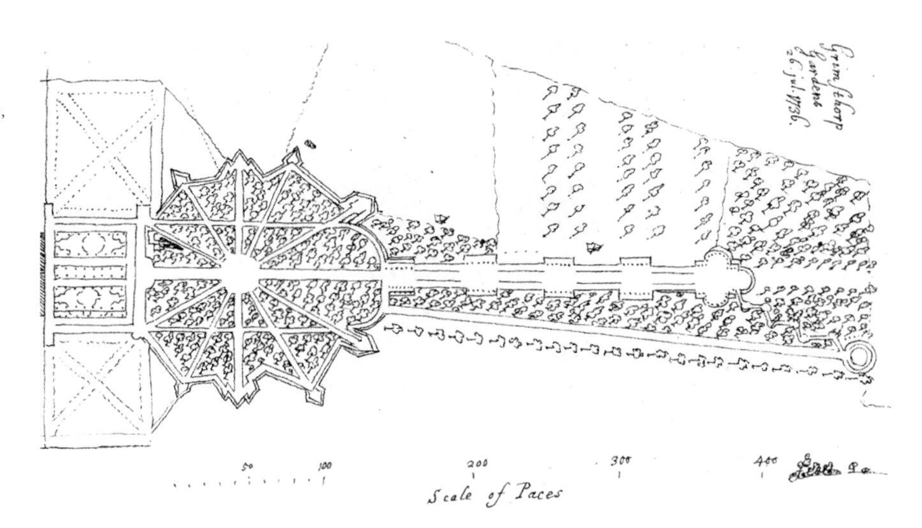

121 William Stukeley, Sketch-plan of the gardens at Grimsthorpe, Lincolnshire, 1736.

the different zones by virtue of experiencing them one by one and by comparing their various formats.

It was only a few years earlier, in 1707, that the third Earl of Shaftesbury had argued the same point, somewhat more philosophically.[2] He gave instructions for his estate of Wimborne St Giles that would enable a better 'Prospect from the House and Terrass'; this involved the manipulation or removal of trees and hedges so as not to hinder the views. Furthermore, he asked for sightlines to be established whereby large topiary bushes – 'larger Pyramids' and 'Tall Globe-Yews' – were succeeded by the 'intended middle rank' of Scotch firs and 'Cypresses [to] make a *Contrast*'. Finally, with 'no Prospect for them to hinder', the sight is taken further into the immediate countryside beyond the garden: 'The last row of Winter Greens to be continued by other plain Holly Trees of any sort . . . at large distances only for ye guiding of ye Eye up that Hill and so to ye end of ye reset Fields where ye great old Yew Trees stands'. The eye is

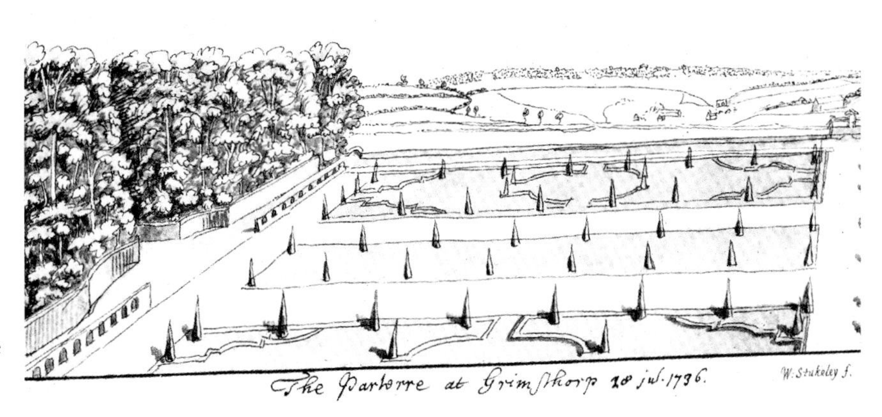

122 Stukeley's sketch of the parterre at Grimsthorpe, 1736.

123 Stukeley's view back to Grimsthorpe's bastion garden from 'Grimes walk', 1736.

124 'The Duchess's Bastion' at Grimsthorpe, drawn by Stukeley in 1736.

'guided' through the intermediate zones of the garden towards the ancient yew tree in the landscape beyond. But this is no merely visual composition, but an exercise in revealing the *characteristics* or 'characters' of the natural world.

For in 1714 Shaftesbury published his *Characteristics of Men, Morals, Opinions and Times*. Its frontispiece by Simon Gribelin engraved an earlier portrait of the author, placing him now on a terrace, beyond which are glimpsed neat lawns, fencing, a cluster of topiary trees, possibly a canal and in the furthest distance hills and woodlands (illus. 125). As David Leatherbarrow argues, this is unlikely to be a portrait of Shaftesbury's own Wimborne St Giles, so there is no 'ancient yew tree' for us to spot in the far distance of the engraving. The image is more abstract than such a local instance would be, and the emblematic portrait offers instead a

125 Detail by Simon Gribelin after John Closterman from the frontispiece to the second edition of the 3rd Earl of Shaftesbury's *Characteristics* . . . (1714).

neat epitome of Shaftesbury's general concept of 'characters'. He insists that all natural types be differentiated and classified and that, as in a garden, there should be no confusing or mixing of types and that the plants to be employed are clearly ranked. Thus proportion and measure are there to 'characterize' natural forms – in other words, to elicit their intrinsic character. In this fashion, once the eye has, so to speak, learnt the lessons of the garden's idealized forms, it may apply this 'tutorial' in the larger landscape beyond. Passing through the gradations of the garden world – both mentally and physically – its visitor can now grasp the ancient yew tree for its own sake and acknowledge its true character.

A central aspect of this kind of garden is that it not only observes proportions and types of garden-work, but utilizes and relishes the possibilities of perspective. The prevalence of avenues in early modern gardens is everywhere in old engravings (see illus. 119); they can still be detected as vestigial remains on the ground. But the destruction of many avenues in later eighteenth- and early nineteenth-century landscaping, not to mention the angry tirades it still elicits from modern designers who take particular pleasure in deploring 'axes', have lost for us the appreciation of seeing the different zones of landscape and comparing them. The modernist critic Christopher Tunnard complained, for example, of the 'deadening effect on the human mind' of such 'tricks of diminishing perspective on the main axis'; he clearly considered that the human mind was not invited to consider how the varied landscape might be grasped. It is not the axis or the avenue per se that we miss, but the opportunity it gave to a more nuanced understanding of landscape design.

This 'moment' in the narratives of garden history is interesting above all for the fact that is refuses to see gardens as ever-evolving towards a more perfect 'natural' perfection; for it incorporates all versions of 'nature'. I recall a remark by Frank Kermode from *The Genesis of Secrecy*, where he praises Hermes as the 'god of going-between: between the dead and the living, but also between the latent and the manifest . . .'. For later gardenists like Brown, when a world of 'natural' gardening was beginning to manifest itself in the mid-eighteenth century, they may indeed have looked back to the gardens of 1700 as some *ur*-landscape garden, struggling from the dead world of French gardening towards the living world of an English perfection, a movement towards something as yet unclear. But the very process of making the latent manifest was itself a distinct and even ideal moment, a moment when the multiple phases of garden-making were visible and palpable. Ranged in sequence, the different modes of organization brought out the fullest potential of each. I would even argue that these multiple gradations – this 'garden of transition' that seized and registered the complexity and layout between different garden moments – were one of landscape architecture's most accomplished forms. And it was no wonder that, after the acclaim that greeted natural landscaping in the late eighteenth and nineteenth centuries, the Arts and Crafts gardening returned to something like the garden of 'betweenity' as a version of garden-making. Indeed, to step out of a mansion straight onto the turf lawn is perplexing.

And it is still more plausible – even if garden-makers don't recognize it for what it is – to lay out even the smallest plot as a sequence from courtyard/patio, flower gardens, vegetable plot, to a 'wild' area, where dog woods and rhododendra grow in the undergrowth and shrubbery.

So it remains to look more closely at where the garden of 'betweenity' was *coming from* rather than where it was *going*. Just as the English landscape garden did not spring fully armed from the head of Kent, as Walpole and the Walpoleans occasionally imply, neither did the gardens around 1700 as envisaged by Switzer, Addison and Shaftesbury. There are multiple explanations of its provenance. Addison, as he says, praised the spreading and diversified landscapes of France; Shaftesbury also knew Continental gardens. Equally, an alert and intelligent eye cast over the English countryside could not have failed to appreciate the range and diversity of territory on estates and their multiple functions. Switzer himself applauds the 'Woods, Fields, and distant Inclosures [that] should have the care of the industrious and laborious Planter' and that could not be contained within 'the narrow limits of the greatest Garden'. He proposed that 'my Garden [be] open to all View to the unbounded Felicities of distant Prospect, and the expansive Volumes of Nature'. This clearly did not mean that there should be no small compartments and the usual amenities of country residences – 'Walks and Cabinets of Retirement, some select Places of Recess for Reading and Contemplation'. But it required that the varied possibilities of an estate be thoughtfully managed in consort with what Shaftesbury called the true characters of natural materials, respecting due difference and classification. One could argue that this is precisely what was accomplished at Castle Howard in Yorkshire, where Switzer's hand, along with his patron's, determined the handling of Wray Wood to the north-east of the house. Long praised, since Tudor times, for this 'fair young wood', Wray Wood would have surrendered with an ill grace to the contemporary and fashionable geometry of Baroque avenues and *rond-points*; instead, as Switzer described it in his 1715 book,

> 'Tis there [in Wray Wood] that Nature is truly imitated, if not excell'd, and from which the Ingenious may draw the best of their Schemes in Natural and Rural Gardening: 'Tis there that she is by a kind of fortuitous Conduct pursued through all her most intricate Mazes, and taught ever to exceed her own self in the Natura-Linear, and much more Natural and Promiscuous Disposition of all her Beauties.

Various proposals for designing the Wood were proposed, but the final version shows off 'its natural and promiscuous . . . beauties'.[3] The wood is drawn so that its 'schemes' are 'truly imitated' and find their proper relationship to the surrounding territory properly drawn.

It is almost certain that the rejected proposals for Wray Wood were designed or drawn by William Talman and the landscaper George London, who would (Switzer adds) 'have spoil'd the Wood'. Their landscaping debts and Baroque schemes were largely French-inspired, and give credence to those who would later reject French influences on the new gardening. But, in fact, a closer reading of French garden-making around 1700 suggests that Addison's appeal to French taste was more in line with his own and Switzer's ideas.

As is well known, André Le Nôtre himself never set out any systematic theory of garden practice, and it was left to A.-J. Dézallier d'Argenville in 1709 and to Jacques-François Blondel in 1737 to publish their detailed observations on the current state of French garden art. Dézallier d'Argenville set out what he saw as the main burden of recent French practice, a book which was Englished by John James in 1712 as *The Theory and Practice of Gardening*. Though D'Argenville was specifically concerned with pleasure-gardens, he did not deny the importance in a 'complete Garden' of a full complement of different activities and zones. While his book is mainly concerned with the ordering of French gardens, and does so with typically hierarchical plans, a reader would find many places in which he urges experiences that Addison and Switzer would also deploy. As one familiar with Versailles, it was no surprise that he valued level sites which rewarded him with prospects that allowed the most 'diverting and agreeable' aspect of a garden: namely, the 'pleasure of seeing, from the end of a walk, or off a Terrass, for four or five Leagues round a vast of Villages, Woods, Rivers, Hills and Meadows, with a thousand other Varieties that make a beautiful Landskip'; lines of walks or *allées* should be 'continued quite through the Woods and Fields'. He emphasizes what are termed 'natural features': that is, the use of turf for ramps and steps, hedges without trellis or architectural ornament. He discusses and illustrates theatre-like stages – grassy ramps and platforms – which provide ideal forums for viewing the countryside (see chapter Nine). In particular he noted the use of the ha-ha ('Ah Ah' in James's phrase), an English version of the French *saut de loup*, literally 'leap of the wolf'. The device had already been employed by a French gardener, Beaumont, at Levens Hall in the 1690s, and would become of considerable significance in England, being used on a small scale (Rousham) and around much larger territory (Switzer's 'Paston', or at Stowe). By 1737 Blondel's *De La distribution des maisons de plaisance* had taken note of English criticisms about French gardens, and he offered more various plans in both scope and scale. To contrast with a garden's symmetry, he valued both a 'countryside with little cultivation' (not quite what Addison had urged) and the use of 'picturesque negligence' to 'interpose' upon or enliven long, closed *allées*.

Despite such foreign hints, it was primarily local practice in the making of gardens and the consequences of cultural motives that gave to the English garden of 'betweenity' its niche in garden history. John Worlidge wished to 'excite or animate' the estates and country seats, and while his text advises on a range of Italianate items (grottoes, statues, fountains), the title-page of his *Systema Agriculturae* was happy to display a mingled rural prospect of fields, hills and woods beyond the garden (illus. 126). Other writers, like William Temple, praised the much earlier garden of Moor Park in Hertfordshire in the mid-seventeenth century (his remarks were published in 1692 in *Upon the Gardens of Epicurus; or, Of Gardening in the Year 1865*). The division of Moor Park into three levels brought the gradations of garden form right into the garden, with many steps linking the terraces; a further 'very Wild, Shady' segment, 'with rough Rock-work', had to be displaced elsewhere on the opposite side of the house, as a road intervened after the second level. Similarly, Timothy Nourse's essay 'Of a Country House', published in his *Campania Fœlix* of 1700, emphasizes variety and prospects, and again his frontispiece juxtaposes a rather *retardataire* parterre of taut planting squares with an expansive countryside with ploughing and sowing below its terrace (illus. 127).

126 Title page to the third edition of John Worlidge's *Systema Agriculturae*... (1681).

We tend to forget in the narratives of garden history how such pockets of garden-making survived throughout England. In many instances – more than later historiography likes to recall – estates could not implement the new range of fashionable landscaping; finances, inheritance issues or the variety of local traditions could preclude later interventions in the manner of Kent and especially Brown. Therefore the immediate context of older gardens continued to remind their owners of the varied *landskips* of their possessions.

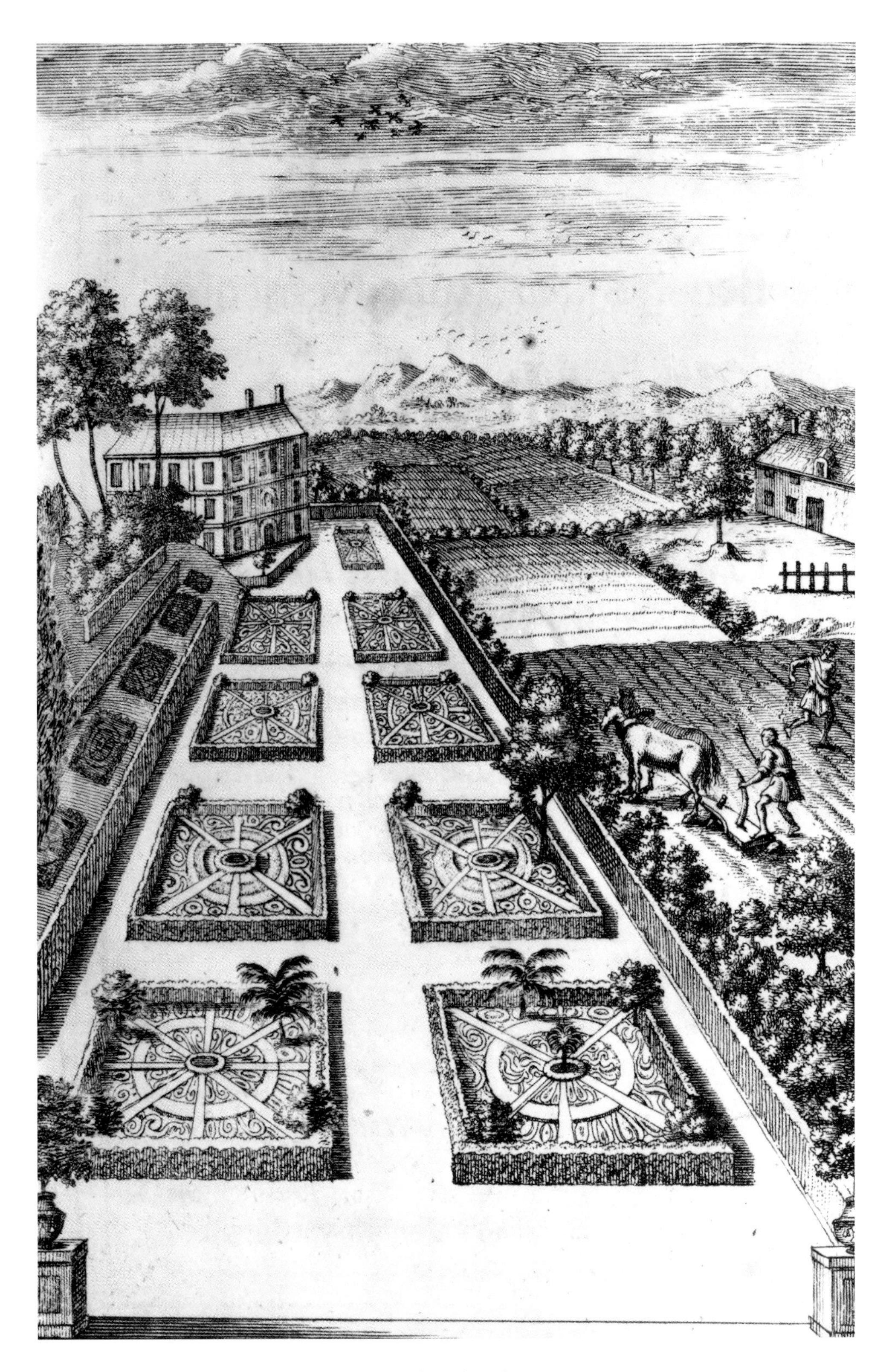

127 Frontispiece to Timothy Nourse, *Campania Fœlix . . .* (1700).

11 Leaping the Ha-ha; or, How the Larger Landscape Invaded the Garden

Horace Walpole famously defined what he saw as the major break-through in English eighteenth-century landscape architecture as William Kent's 'leap of the fence' to discover that 'all nature was a garden'.[1] We might, however, profitably look at this 'discovery' from the opposite direction, as the *invasion* of the confines of the gardens by the larger landscapes outside it. But either way Walpole's claim was not entirely as innovative as he wished to claim. Gardens before Kent had, in many different ways, looked outwards to territories beyond their enclosures, and elements of the larger landscape had even been drawn into the more immediate grounds. So that 'leap' was not startlingly innovative, even if Walpole wished to make it the key feature of the new landscape taste. Kent certainly exploited the immediate garden's juxtaposition to a wider territory. But just as interesting as his designer's move outward into a nature that is 'all garden' would have been a visitor's sense that this larger landscape had entered into and taken over the forms and character of the garden itself.

The two-way exchange between garden and landscape had been clearly articulated in 1748 by William Gilpin in his published *Dialogue* on the gardens of Stowe: one character argues for leaping the fence, the other for calling in the country.[2] The distinction seems quite sharp. Callophilus expounds the former, more conventional view: 'the Garden is extended beyond its Limits, and takes in every thing entertaining that is to be met with in the range of half a County, Villages, Works of Husbandry, Groups of Cattle, Herds of Deer, and a variety of other beautiful Objects are brought into the Garden, and make a part of the Plan' (52). Conversely, his companion, Polypthon, argues for bringing the countryside into the garden, when he opines that were he a nobleman 'I should endeavour to turn my Estate into a Garden, and make my Tenants my Gardiners: Instead of useless Temples, I would built Farm-houses; and instead of cutting out unmeaning Vistas, I would beautify and mend Highways' (45). He quotes lines from Milton's 'L'Allegro' in praise of the 'new Pleasures' of an extensive and un-designed countryside in support his, surely radical, version of landscaping; he may also be alluding to Addison's appeal in *The Spectator* for English estates to be 'thrown into a kind of Garden by frequent Plantations' that would profit the owner's profit as well as his pleasure (see previous chapter).

This two-way movement of 'natures' is perfectly visible at Rousham in Oxfordshire, Kent's surviving *chef d'oeuvre*, on which he worked from

128 William Kent, Drawing made in Rousham gardens showing the Temple of the Mill and the Eyecatcher to the north beyond the River Cherwell, *c.* 1739.

1739, only a few years after Gilpin's visit to Stowe. Kent's play with inside and outside is exceptional and exemplary, the result in part of his specific design, in part of his response to the site's contingencies. Visitors in the garden are given many opportunities to look out beyond the river Cherwell that flows along the site's lower edge; indeed, the topography enforces this, since its slope descends from the ridge at the back of the garden and ends at the river bank, facing the extensive water meadows and hillside beyond the river. Kent ensured that temples, terraces and lines of sight would all look out across the river and its water meadows: so the very first impression when leaving the house is of that extensive view, and its various seats and buildings, including the arcaded terrace of Praeneste, all face in that same direction. Kent also ensured that various improvements were to be made outside the gardens but in the sightlines from it. On the not too distant hillside he placed an appropriately termed 'eyecatcher' in the form of a Gothic, crenellated arch. Further, and to the left of visitors looking outwards from the garden, could be seen a mill, also Gothicized with buttresses and battlements (Kent's drawings for these survive, as do the structures themselves; illus. 128).[3] But these invitations for the eye to scan the middle and far distance are sustained by a deep sense that the exterior countryside is now not only colonized

129 Edge of the paddock near Cow Castle at Rousham with the ha-ha separating the garden from the paddock. To the right is the original garden gateway from the public road.

visually, but its forms have penetrated the lawns, groves and thickets of the garden itself.

A hint of this two-way exchange is rehearsed for the visitor immediately upon entering the garden. Outside the house and to the left of the bowling green – from which the distant eye-catcher is first seen – is a ha-ha, the sunken ditch that divides the garden from the adjacent paddock without any visual barriers. Looking then across the paddock, the visitor sees two structures, one a crenellated building and the other a classical gateway. They appear from this point to be outside the garden; but they will eventually be found to lie within it, for a path leads along the edge of the ha-ha from the end of the bowling green (illus. 129). The classical gateway and what turns out to be an alcove seat are part of the garden, and from that position, the visitor now looks back across the ha-ha to see the mansion rising from the green spread of the paddock. The illusion that it is all one space, however much the treatment of the paddock is rougher than the garden's, can of course be exposed by stepping to the edge of the ha-ha; but the overall sense of garden and farmland being uninterrupted and coextensive is still part of the experience, and the various buildings (house, gateway, alcove seat with a cattle shelter on its rear) are part and parcel of the ensemble.

130 Rousham: Peter Scheemakers's Dying Gladiator, *c.* 1743, with the grasslands of the garden and the paddock in the background.

In earlier gardens of the seventeenth century, different sorts of terrain could still have been visually juxtaposed, but with more palpable demarcations – fences, walls, even terraces and stairways or ramps leading from one part of the landscape to another. We have noticed this in chapter Seven, where the frontispiece of Le Lorrain's book on art and nature (see illus. 92) marks a tripartite division of land very clearly. Similarly, the early eighteenth-century philosopher the Earl of Shaftesbury used the tailored forms *inside* a garden to educate the eye about the 'true nature' of the world that would be encountered *outside*. So what the mid-eighteenth century begins to demonstrate is its ability to be at home in a landscape where the ideal forms of a designed landscape coalesce with the given forms of its materials. Where it was once the eye and the mind that made connections between different modes of landscape, now the very landscapes themselves seemed to merge and the feet may move between them.

Two changes that Kent made to the interior landscape of Rousham after 1739 make this clear. The first was the un-terracing of the slope that leads down from the end of the bowling green to the riverbank (Kent may have been responsible for this, though that is not clear). An old estate map

131 Walking south to Praeneste and the Vale of Venus at Rousham. To the left can be glimpsed a bend in the Cherwell.

indicates that early in the eighteenth century the slope was in some fashion cut into a succession of levels, perhaps for purposes of cultivation or to hold the bank in place; by Kent's time, those had gone and the ground dropped without interruption to the river, making the grassy slope a link between the bowling green and the water meadows across the river. Kent himself transformed another area, slightly upstream, from a sharp-edged and geometric outdoor theatre, created by his predecessor Charles Bridgeman, into a grassy amphitheatre, now marked only by the triangulation of a space between its statues. As we saw in chapter Nine, theatres and gardens had enjoyed an intimate and reciprocal relationship, of which Rousham's small theatre would have been a modest version; now its own theatrical moment has been considerably minimized, with much of the interest turned outwards across the water meadow and to the distant eyecatcher rather than on a Bridgemanick rostrum (illus. 116; contrast illus. 117).

So Walpole was both right and wrong about Kent's 'leap'. Right in that he clearly needed to demonstrate a decisive break with early design, although it is odd that he phrased Kent's achievement as only a movement outwards rather than recognizing how much Kent brought the country inwards; after all, Alexander Pope's advice for the landscaper in his *Epistle to Burlington* had been to 'call in the country', though he may well have implied simply a visual impulse. Walpole was wrong, or at least tendentious, in ignoring how much 'a fine view, and the Prospect of a noble Country' had been a criterion for earlier landscape design in both England and Europe. Even his metaphor of Kent's 'leap' may have been taken from the much earlier French equivalent of the 'ha-ha', the *saut de loup* or wolf's leap.

The phrase, 'a fine view, and the Prospect of a noble Country', is A.-J. Dézallier d'Argenville's, from his 1709 book, translated by John James in 1712 as *The Theory and Practice of Gardening*.[4] It reminds us that the admiration of larger landscapes seen from within confined and manicured gardens was nothing new. The expansive layouts of French gardens, about which d'Argenville was writing, not only took the eye far into the distance, as at Vaux-le-Vicomte (illus. 132) or Versailles, but did so by extending the immediate gardens and its parterres, first into bosquets or contrived groves, and then into the extensive woodlands and forests of royal and aristocratic estates. His insistence upon a 'fine view' or 'prospect' was as much a means of celebrating the owner's position and land holdings as satisfying a visual pleasure. This is exactly what Joseph Addison would be explaining to his English readers within a few years of the translated French text, when he argued in *Spectator* 414 for 'a whole Estate [to] be thrown into a kind of Garden by frequent Plantations', or that 'Fields of Corn make a pleasant Prospect'; or again, 'Hence it is we take Delight in a Prospect which is well laid out, and diversified with Fields, Meadows; Woods and Rivers'. His further plea that 'a Man might make a pretty Landskip [i.e. a landscape painting] of his own possessions' is effectively translated into practice by contemporary designers.

Stephen Switzer, for example, promoted what he termed a 'Rural and Extensive Gardening', allowing the 'Beauties of nature' to be 'uncorrupted by Art' and directing that 'all the adjacent Country be laid open to View and that the Eye should not be bounded with High Walls, Woods misplac'd, and several Obstructions . . . by which the Eye is as it were imprisoned'. In one particular estate with which he was involved, Grimsthorpe in Lincolnshire, Switzer practised what he preached (see previous chapter): in an engraving in his book *Ichnographia Rustica* (1718) Switzer presented a very abstract plan for 'The Manor of Paston divided and planted into Rural Gardens'; yet it is clear from sketches made there twenty years later that the place was less schematic and sensitive to the graded organization of land (see illus. 121–124). The very unnatural-looking woodland of the engraved plan in Switzer's book, with its mazy paths and contrived groves, none of which seems to modern eyes 'uncorrupted by art', may well exemplify the difficulty of finding adequate graphic formulae for the experience that Switzer was trying to convey.

An insistence upon 'calling in the country' was intrinsic to the vaunted *Englishness* of the new landscaping and was often seen as something specifically absent from French design (though this was, as already noted, not in fact the case, since of course it was the *Frenchness* of that countryside that was in question there). Proponents of 'rural and extensive gardening' implied that the special socio-political conditions and forms of the English countryside were both an inspiration and an integral part of their nationhood; this was indeed something that France could not match. Therefore their inclusion of that rural and extensive territories within a designed landscape was one among several aspects that showed English superiority to foreign styles (illus. 133).[5] And a dual contemporary meaning of the word 'country' should be noted:

implying both the nation and its topography or countryside, a doubleness that suited well with the patriotism of the new landscaping. No wonder Switzer also asserted that 'Gardening can speak proper English'.

Addison and Switzer may have, in retrospective, been in the van of the design of contemporary garden layouts; the presentation of Grimsthorpe in Switzer's engraved plan may certainly suggest that an implementation of such designs may not have matched what we now think of as the new mode of landscaping. Yet by 1736 Stukeley certainly saw the territory as less abstract and more responsive to the call of the countryside. People's experience of designed gardens or parklands, if not the actual forms of their layout, probably echoes how many would have already chosen to see 'rural and extensive' English landscape. As early as 1713 the Irishman Samuel Molyneux visited Petersham Lodge on the Thames at Twickenham: there was a parterre and then paths between hornbeam hedges about which he was somewhat cool, but they were overlooked by a wooded hillside 'interspers'd with Vistos & little innumerable private dark walks' that were lined with low hedges to allow 'unconfin'd prospects' of the countryside and river. Eleven years later, Lord Percival visited Hall Barn in Buckinghamshire, created in the French taste by the poet Edmund Waller from 1651 and 1687 but remodelled after 1715; he was particularly struck by the newer features, including the ha-has that terminated the woodland walks and 'over which you see a fine country, and variety of prospects'.

A more extensive meditation on the role of the 'country' in the new landscaping was implied by Pope's description of the landscape at Sherborne in Dorset, written in a letter to Martha Blount about 1724 ('my head', he told her, 'is so full of

gardens'). He hailed the estate as an exceptional example of 'those fine old Seats of which there are Numbers scatterd over England' and Pope immediately notes its special connection with Sir Walter Raleigh, for whom the house was built, though the mansion had been newly refurnished with 'beautiful Italian window-frames'. But it is the parkland that absorbs him, a wooded

132 Vaux-le-Vicomte, the vista down the garden towards the far hillside and woodland.

amphitheatre in which the irregular gardens fit perfectly:

> Their beauty rises from this Irregularity, for not only the Several parts of the Garden itself make the better Contraste by these sudden Rises, Falls, and Turns of ground; but the Views about it are lett in, & hang over the walls, in very different figures and aspects.

These views take in both the local town of Sherborne, with the ruins of a castle destroyed after the Civil Wars, and a richly varied countryside, to which his letter returns again and again. Focusing on the ruined castle, his letter explains how the higher parts of the landscape might be re-organized to lead visitors 'from one View to another'. The owner's cultivation of the ruins and their viewsheds would be, writes Pope, in the interests of both the owner (whose 'Goodness and Benevolence extend as far as his territories') and the landscape's historical significance (he urged the erection of an obelisk to signal this).

Pope was a prominent advocate in the first half of the eighteenth century of the new gardening, and his influence on his friend Kent was profound.[6] His own rented property at Twickenham was largely enclosed but – with Kent's help, who created a viewing platform on the river facade of his villa – he still enjoyed prospects over the

133 At Castle Howard: Vanbrugh's Temple of the Four Winds, the Palladian Bridge, and Hawksmoor's distant Mausoleum.

134 William Kent, design for Esher in the classical taste, with Bishop Waynflete's tower below at the banks of the river Mole, 1730s, pen and wash.

Thames, its river traffic and the country on the far bank. It is such negotiations, not just with the potentialities of a site (its *genius loci*) but with the reorganization of a site's relationship to its surroundings, that Kent was to explore whenever he found the opportunity.

His master work according to Walpole was Esher Place (here Kent was, he wrote, 'Kentissime'). Though the site, unlike Rousham, has virtually disappeared, it enjoyed the same topography, a sloping site that overlooked a stream at its base with views into the countryside beyond. In one of his unexecuted projections for a new villa there Kent places it on the top of the ridge, looking down upon the fifteenth-century tower where Cardinal Wolsey had been imprisoned and to the country beyond (illus. 134). Many of Kent's actual layouts have not survived sufficiently for us to judge them; perhaps ironically, many of them – like Esher itself – have succumbed to subsequent invasions by surrounding landscapes. However, his sketches reveal at least two essential elements of his design

135 William Kent, view in Esher Park, with temple and bridge, 1730s, pen and wash over pencil.

strategies: the invocation of non-designed materials and configurations of sites as the essential ingredients of his landscapes, and the placing of buildings, seats and such-like constructions from which views could be taken of the larger territories.

Sketch after sketch reveals Kent's signature move of breaking the strict exterior line of woodlands by carving glades into their edges on the one hand, and pulling out free-standing groups of trees on the other (illus. 135).[7] The double effect brings elements of the woodland into the meadow as well leading the eye back into its denser growth, and therefore blurring any division between them. In annotating a sketch for revising the terrace at Claremont, he proposes the removal of 'a level terrace' 'fronting the great room', so that the sight might be directed straight into the woodland, where again its edge is diversified. The contrivance is not disguised – a statue marks one opening in the trees and further views into the hillside are directed by screens of trees in the manner of theatrical wings – but the result is to stimulate our experience of the landscape beyond what strikes the eye in the immediate foreground. A later description of Kent's work at Claremont by Thomas Whately, though it neglects whatever specific insertions were made, is clear about the fashion in which the rearrangements of the woodland ensured an agreeable and various walk along the side of the hill

> and on the edge of a wood, that rises above it . . . The intervals winding here like a glade, and widening there into broader openings . . . The whole is a place wherein to tarry

with secure delight or saunter with perpetual amusement.[8]

Another of Kent's signature devices, his (in)famous clumps dotted across the landscapes in sketches for Euston Park and especially Holkham Hall, seem somewhat ridiculous now and were awkward then; Walpole in particular criticized them. But Kent was probably trying to move away from the abstract topiary of shrubs and bushes or from the rigid avenues of seventeenth-century estates and yet suggest a truer configuration of tree growth by isolating examples within the designed areas of the parkland.

Walpole's catalogue of Kent's characteristic moves responds poetically to an ineluctably British landscape: 'the delicious contrast of hill and valley changing imperceptibly into each other', 'distant views' glimpsed through the 'loose groves', and the intimate spaces where 'the rudest waste' could be cheek by jowl with 'the richest theatre'. To the givens of this indigenous land, Kent brought his inspired management of its capabilities: he broke up the uniformity of woodland or open ground, enlarged or obscured views, drew attention to free-standing specimen trees, and 'blended a chequer'd light with the thus lengthened shadows of the remaining columns'; they show his response to the larger landscape beyond the 'fence', connecting the garden with the park (illus. 136) and the garden's increased opportunities for 'variety and scenery', along with an apparent naturalness (for the 'modern gardener exerts his talents to conceal his art').

Walpole also noted how Kent inserted 'temples' to focus prospects and 'seats' from which to

136 William Kent, a drawing of Wimbledon (identified in Kent's hand on the verso), 1730s.

observe them. We tend to consider only how these pieces looked in their settings, partly because Kent's sketches show them from the outside rather than the prospects that they would face; whereas they were all designed to be places *from which* to look at the surrounding territory. All this, argues Walpole, was in the interests of presenting the 'richest scene' to the spectator or garden visitor ('scene', here, merging the connotations of topographical scenery, theatrical settings and painterly backgrounds). The rhetorical climax of Walpole's enthusiasm for the new landscaping for which Kent, at least in Walpole's narrative, had been responsible was his exhortation that 'prospect, animated prospect, is the theatre that will always be the most frequented'.

There were, as has been shown, English precedents for Kent's 'leap' into nature, and Walpole himself had acknowledged the 'leading step to all that followed was (I believe the first thought was [Charles] Bridgeman's) the destruction of walls for boundaries, and the invention of fosses'. There were also abundant examples from earlier times and cultures of the garden's exposure to larger landscapes. If the western medieval garden had been an enclosed and inward-looking world, many Islamic sites were not, and opportunities for taking extensive prospects from within the safely of palaces were typical. The early Islamic garden of Madinat al-Zahra, near Córdoba in Spain, was a huge complex in which the Caliph's palace dominated the site and gathered into its vision both lower terraced gardens and then adjacent and distant territories as an expression of the ruler's power. Viewing places or *miradors* would become features of both urban complexes like the Alhambra and villas like its neighbouring Generalife at Granada.

Of such examples, Kent would of course have little if any knowledge. But during his long sojourn in Italy he would have visited many villas in and around Rome whose gardens commanded similarly extensive prospects of landscapes rich with classical associations; Italianophile that he was, he was doubtless recalling those experiences in some of his English designs. Renaissance villas were established wherever possible on hillsides to benefit from the circulation of fresh air, but this pragmatic instinct was also informed by a wish to emulate ancient Roman examples which has also utilized Italian hillsides. The younger Pliny's letters describing his villas were well known – Kent's patron, Lord Burlington, would be involved with publishing an English version of them in 1728, Robert Castell's *The Villas of the Ancients Illustrated* (see illus. 33). And Pliny specifically emphasizes the views from his villas, giving, at the Tuscan site, a detailed description of a wonderful natural amphitheatre, allusions to which, or maybe literary emulations of its Latin text, seem to lurk behind early modern descriptions of other landscapes – like Pope's at Sherborne or Molyneaux's at Petersham Lodge. And beyond actual views, Kent would have noticed that villas incorporated many virtual landscapes, *trompe-l'oeil* visions of garden elements inside buildings on walls and vaults, including painted panoramas of villas depicted in their larger landscapes. On his journey across Europe, returning from Italy in 1719, he would have had many chances to observe the often large-scale spread of French country estates, as well as some 'perspectives', as painted landscapes were called, that were used to prolong the illusion of closed urban sites within Paris. This fashion for illusionist extensions of limited garden spaces was another antique mode that had been resumed with élan during the Renaissance, and while Kent never had occasion to invoke the same kind of device, he found opportunities in the actual English countryside to realize its artifice.

On Kent's development as a landscape designer, Walpole remarks that his 'last designs were in a higher style, as his ideas opened on success'. The metaphor is apt, if we consider Kent's opening of the garden's vocabulary and syntax to the surrounding countryside. Kent's landscape work, as Walpole so eagerly described its leap, did incorporate more and more direct quotations from the surrounding countryside. With hindsight we can see that Kent's successor as landscape designer, 'Capability' Brown, was particularly drawn to representing a much more 'natural' range of rural imagery; his work was cleverly explained by Ian Hamilton Finlay as making 'water appear as Water, and lawn as Lawn'. In other words, Brown drew out the essential features of whatever natural element he worked with rather than enhance a site with architectural or other additions. Towards that pure and abstracted landscape style, Kent had certainly pointed the way, showing in various manipulations of water, lawn and trees his recognition of their essentially natural forms. But it was left to 'Capability' Brown himself to extend Kent's display of English countryside. But he did so, not so much by opening the immediate garden outwards to the countryside, but by bringing that countryside into closer proximity with the house. Earlier illustrations of English estates had successfully displayed how gardens spread *outwards* from the mansion and diminished in artifice as they engaged with the larger landscape beyond; even if such bird's-eye views were not available at ground level, they did imply the notion that a garden and its landscape constituted one whole and modulated ensemble. Brown's work asked for different modes of imaging and experiencing his designs. Now the mansion was often depicted as sitting in an uncluttered landscape (on its lawn or 'Lawn'). And to experience the surrounding landscape, people now circulated on foot, horseback or carriage through its meandering rides and pathways, in its woods and over open grasslands, and along its bodies of water, which all unfolded along their route. This is a substantial development of garden design and use; so Brown's work and its afterlife in later times and places need to be taken up in another essay.

137 Fresh Kills, Staten Island, New York, two proposals for the new parkland, 2001.

12 The Role of the 'Natural' Garden from 'Capability' Brown to Dan Kiley

Many people today think that parklands are 'natural'. When they visit Hyde Park, New York's Central Park or Montsouris in Paris, they perceive what they'd call a piece of natural territory, even if sometimes they are aware that these spaces have been crafted and shaped in the past. And recent parks, not yet completed and still undergoing visible modification, like Fresh Kills on Staten Island in New York, are hailed for that same illusion of naturalism. Indeed, an article on Fresh Kills in the *New York Times* (25 January 2010) never mentioned that the landscape architecture firm of Field Operations was involved in implementing this new park; the text did, it is true, mention the Parks Department, but that for many that would have been read simply as a facility that looked after 'nature'. Accompanying pictures of birdwatchers overlooked a windswept grassland with an isolated, bare winter tree; despite the intervening high wire fencing, many readers would have seen just a natural habitat, and indeed Fields Operations' original design had stressed exactly this ecological paradise (illus. 137).

It is, of course, a challenge for some landscape designers to use their art to conceal its artifice. Yet the skill involved is surely not quite as simple as is implied by the enunciation of this famous principle (Ovid's 'ars est celare artem'). The obvious elements of artifice, to be sure, may be concealed in a landscape, but at the same time, is it not expected that visitors should be able to grasp this skilful *rendition* of an informed and mediated nature? In short, how do we know and appreciate a designed and at the same time natural-looking (or naturalized) landscape? One of the first landscape architects to be caught in that dilemma was 'Capability' Brown, both then and now seemingly praised for the artless fashion in which his landscapes were made to appear. His political adversary and rival, William Chambers, even argued that the work of Brown and his copyists was indistinguishable from 'common fields'.

A modern landscape architect like Dan Kiley is also often thought of as a minimalist in his interventions, notwithstanding published and illustrated books on his work that necessarily show the extent of his considerable artistry. At the garden he designed for Eero Saarinen's Miller House in Columbus, Indiana, despite the conspicuous interventions around the house itself and the views immediately from it, including the sculpture by Henry Moore, the land that lies beyond the allée of honey locusts seems a natural topography. So much is it deemed to be 'natural', that it

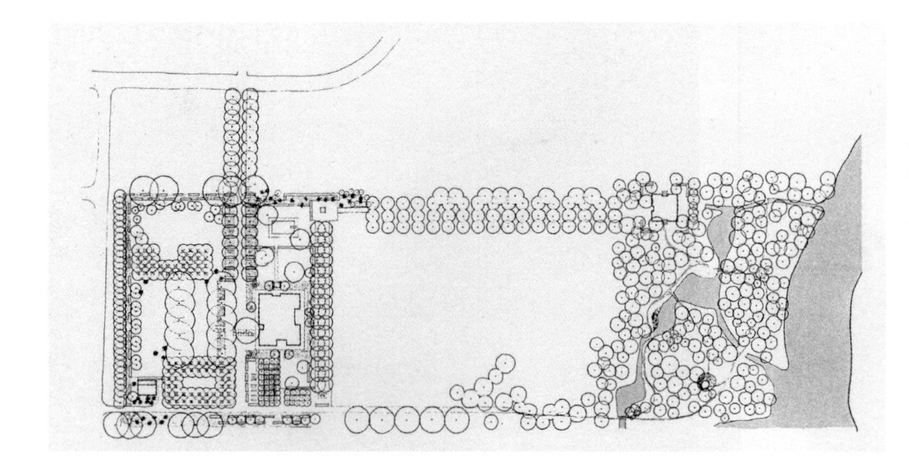

138 Dan Kiley, plan of the Miller House, Indiana, 1955. The house and its immediate gardens are on the left; an allée then separates that area from the open area that leads to the river on the right.

is not even shown in photographs of the other elements of Kiley's elegant design.[1] Yet when his meadowland reaches the floodplain, its lawn merges with the waters of a river, and the flooding and receding of the water makes this a seemingly unmediated landscape; yet the imagination which has allowed this effect is indeed Kiley's (illus. 138). Much of his work is extremely manipulated, far from being natural or even naturalistic (that is, actual or apparently natural); but the instinct to assume the contrary is often compelling.

This 'minimalism' was a recent development in landscape architecture in the later eighteenth century. Its causes are many and intertwined, but may be rehearsed as follows. Many who wished for their own landscape gardens, notably those aspiring to bourgeois status, has to settle for less land and/or less money to improve them; even the poet William Shenstone, for example, at The Leasowes during the years 1745 to 1763 found that he had to rely mainly on inscriptions and forego any visible artifice. Less acreage in itself probably meant that it was deemed unnecessary to stuff too many buildings onto a small piece of land, and therefore architects and other professionals were less called upon to design or to devise sculptures and inscriptions. Further, many in the later eighteenth century could not or did not travel abroad, what with wars on the Continent, or the lack of an education that looked to and profited from a 'Grand Tour'; so classical references were less needed and less regarded as essential in what was once a more learned gardening. Also English people started to be tourists in their own country. And if they wanted images of how to think about the countryside, they looked to Dutch and Flemish landscape pictures, rather than classical subjects by Poussin or Claude Lorrain, and these often displayed rural scenery. And finally, as the contents of Thomas Whately's book on *Observations on Modern Gardening* (1770) make very clear, the attention of landscapers was now also focused on *un*-designed landscapes, like the Derbyshire Dales, or the Wye Valley; in these picturesque places there was no insertion of buildings, statuary or any work that marked the place as the work of a landscape architect. All of these constraints and inclinations were noted by writers of landscape manuals, where their aesthetics and a taste for 'natural' landscapes could catch up with and help to justify new designs;

this was presumably an element of Whately's own theoretical profile.

Historically, older garden designs had always been deliberate, distinct interventions in the larger landscape, with which their formalisms were contrasted and their compositions made clear. Even the 'calling in' of the countryside into the garden itself (see previous chapter) was understood as part of an overall, *designed* landscape, and therefore part and parcel of a more expansive garden-making. But in the later eighteenth century many of what may be termed 'specific' gardens (parterres, stairways, flower beds) were eliminated, especially around the mansion (though this did not happen nearly as much as histories would have us believe). The far fields, the woodland groves and even agricultural land might now come right up the house. Terraces and other such conspicuous structures were erased, all except functional buildings were removed, and even these were hidden. Seats from which to see the view continued to be popular (illus. 139), but what were actually seen in those 'views' was the spread of the countryside with its indigenous hamlets, church spires and farm buildings.

What has become now a quite famous passage in Whately's *Observations on Modern Gardening* did much to foster this taste for uncluttered and 'naturalistic' landscapes. He opposed what he called 'emblematical' characters to his preferred 'expressive' designs. However, we should note that what he chose to prescribe was not necessarily taken up by every contemporary or even subsequent landscape gardener, especially those who professed to like 'expressive' responses but who did still rely to some extent upon older habits of mind, reminiscent of emblem books and iconographical motifs. However, Whately spoke for a new mode of responding to landscape design, when he wrote:

> Character is very reconcileable with beauty; and, even when independent of it, has attracted so much regard, as to occasion several frivolous attempts to produce it; statues,

139 Humphry Repton,from his Red Book for Babworth, Nottinghamshire, 1790.

> inscriptions, and even paintings, history and mythology, and a variety of devices, have been introduced for this purpose. The heathen deities and heroes have therefore had their several places assigned to them in the woods and the lawns of a garden; natural cascades have been disfigured with river gods, and columns erected only to receive quotations; the compartiments of a summer-house have been filled with pictures of gambols and revels, as significant of gaiety; the cypress, because it was once used in funerals, has been thought peculiarly adapted to melancholy; and the decorations, the furniture, and the environs of a building, have been crowned with puerilities, under the pretence of propriety. All these devices are rather *emblematical* than expressive; they may be ingenious contrivances, and recall absent ideas to the recollection; but they make no immediate impression, for they must be examined, compared, perhaps explained, before the whole design of them is well understood; and though an allusion to a favourite or well-known subject of history, of poetry, or of tradition, may now and then animate or dignify a scene, yet as the subject does not naturally belong to a garden, the allusion should not be principle; it should seem to have been suggested by the scene; a transitory image, which irrestibly [sic] occurred; not sought for, not laboured; and have the force of a metaphor, free from the detail of an allegory [pp. 150–51].

What is at stake in Whately's remarks is twofold: the designer must avoid any obviously artificial structures, triggers and prompts that require some specific or 'allegorical' response from a visitor; and now it is visitors who are responsible for adjudicating how a garden speaks to them rather than a designer, and they will be less interested in what is 'told' to them than how they can, from the natural arrangements presented to them in a scenery, learn their own responses.

Even in a site as peopled with emblematical characters as was Stowe, there is a hint of the coming changes in design. In 1747 a Temple of Concord and Victory (illus. 140) had been built to front a new landscape by 'Capability' Brown, implemented during 1746–7 and to be called the Grecian Valley (this area was enabled by the removal of a road along the eastern side of the garden). The Temple suited perfectly the character of Stowe's many-templed gardens, which some years later in 1762 Horace Walpole thought excessive: 'If Stowe had but half so many buildings as it has, there would be too many'.[2] But Brown's new valley was somewhat different, skilfully planted, with a few scattered sculptures in the woodland fringes to be sure, but the overall effect was of a 'naturally' wooded valley (illus. 141). Moreover, in 1764 a prominent building in the centre of the garden, once named Gibbs's Building after its architect and later rebuilt on the same spot and named a Temple of Diana, was removed and partly hidden in the glades at the very end of the Grecian Valley. Whether or not this was done to find a far less conspicuous position for a temple that was deemed less suited to the main gardens, its new situation was far less visible: in its new location it was renamed once again, this time as the Fane of Pastoral Poetry (one archaic term to indicate a somewhat old-fashioned theme and perhaps to signal its fresh role in a 'natural' scenery). Its new name and its far less visible location on the edge of the immediate garden were not, indeed, a big move in the overall conspectus of the Stowe

140 The Temple of Concord and Victory, Stowe.

scenery, but a hint nonetheless of how such more naturalistic designs were coming.

In 1770 Thomas Whately devoted a whole chapter to the Temple of Concord and Victory in this area of the Stowe grounds (pp. 243–5). But rather than articulating the building's message or meaning, he dwelt instead upon the lights and shadows of the sun, as it shone down the Grecian Valley. And he commented: 'Some species and situations of objects are *in themselves* [my italics] adapted to receive or to make the impressions which characterize the principal parts of the day . . . [communicating] the *spirit* of the morning, the *excess* of noon, or the *temperance* of evening.' His long disquisition, abbreviated here, combines items in a garden that are 'in themselves' meaningful, in part because they rely solely on the effects of a given situation, and in part because their associations were effected by their visitors: this is a major theme of Whately's garden aesthetic. Elsewhere he comments more generally upon how water scenery may raise ideas, stimulate imaginations via what he terms 'characters':

> So various are the characters which water can assume, that there is scarcely an idea in which it may not concur, or an imagination which it cannot enforce: a deep stagnated

141 John Claude Nattes, *The Grecian Valley, Stowe*, 1805, pencil and wash.

pool, dank and dark with shades which it dimly reflects, befits the seat of melancholy; even a river, if it be sunk between two dismal banks, and dull both in motion and colour, is like a hollow eye which deadens the countenance; and over a sluggard, silent stream, creeping heavily along all together, hands a gloom, which no art can dissipate, nor even the sun-shine disperse. A gently murmuring rill, clear and shallow, just gurgling, just dimpling, imposes silence, suits with solitude, and leads to meditation: a brisker current, which wantons in little eddies over a bright sandy bottom, or babbles among pebbles, spreads chearfulness [sic] all around: a greater rapidity, and more agitation, to a certain degree are animating; but in excess, instead of wakening, they alarm the senses; the roar and the rage of a torrent, its force, its violence, its impetuosity, tend to inspire terror; that terror, which, whether as cause or effect, is to nearly allied to sublimity [pp. 61–2].

These reactions to various landscape phenomena, whether or not programmed by their designers, are calculated to appeal to visitors and not to promote a specific meaning or message. Such explanations are very unspecific, which is inevitable, given Whately's need to explain a wide range of possible reactions. But the generalized effect makes even clearer the blander, less formulated structures within the landscape that he now prefers. How one meditates, how one responds to 'solitude', or how one becomes 'animated' are tendencies, but no longer imperatives, because they are linked only loosely to specific items. Indeed on a subsequent page (p. 63) Whately distinguishes various responses that no longer derive from any specific or a priori idea or character, but could be discovered anywhere.

Lancelot Brown would acquire his nickname from the 'capabilities' he chose to identify within a given landscape, as well as his own native capacity to draw them out. Much of his work does seem to argue that he elected to work on the natural elements of a site; indeed, the way in which his landscape designs were originally presented in plans or illustrated in contemporary engravings seem to demonstrate his emphasis upon the natural forms and elements of the natural world; in a site plan for Lowther Park (Cumbria) the sweep and rich clusters of woodland dwarf the representation of the footprint of the house itself (illus. 142). The Temple of Concord and Victory at Stowe made a less substantial impact at the head of the Grecian Valley, at least as we see it twenty years later through Whately's eyes, as much by its situation and its effects as by the building itself. Similarly, in enlarging the river Glyme at Blenheim into a larger body of water, Brown may have made it 'worthy of Vanbrugh's majestic bridge' (Dorothy Stroud), but at the same time it also masks half of the Brobdingnagian structure, thus diminishing something of its impact in the landscape (illus. 143). Brown was by no means above introducing buildings and other structures into his landscapes; he could propose 'A Place for the Flower Garden' at Lowther Hall, or a deliberately half-ruinous gatehouse as a prelude to that park (neither were implemented); he was, after all, an architect as much as a landscaper. Nonetheless, his distinction lay in earthmoving, planting and hydraulics, and the overall effect was to give a minimal profile to any structures: 'His characteristic style is that of open grassy expanses,

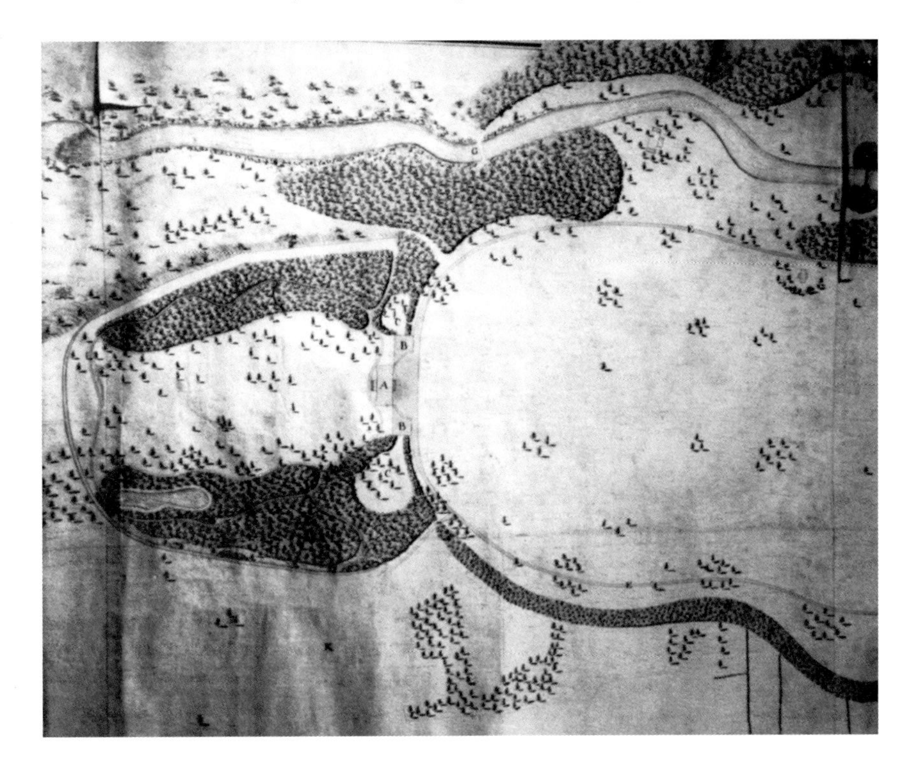

142 Brown's plan for Lowther, Westmoreland, 1763.

143 Vanbrugh's bridge in the park, which carries an approach road over the lake to Blenheim Palace, Oxfordshire.

punctuated by belts of mixed trees, using lakes and rivers to great effect and blending with the countryside beyond'.[3] And, despite reservations, this must be his proper claim to fame: to have shifted landscape architecture into a vernacular mode, where allusion and illusion was replaced by the empathetic affects that Whately promoted.

Further, Brown was in many ways a conservative aesthetician. He held to the longstanding belief that an artist must search for an ideal form of beauty, for a timeless, unchanging permanence of nature beyond and below the actual forms of the topography; so, while his place-making could not ignore the domestic and estate-related buildings for well-to-do clients, for whom he drew up considerable plans and designs, the larger picture was constituted by his eye for their place in an ideal and abstract landscape. Given, too, that so many of his designs were for large estates, the overall effect was also to give prominence to the territory as a whole, which his habit of providing elaborate plans further emphasized (this may be compared to William Kent, who rarely seems to have provided any detailed plan).

The change in the layout of gardens and parks can be grasped readily if we compare the Kip

144 Longleat House and park in Wiltshire, viewed from Heaven's Gate, planted out after 1757 by Brown.

and Knyff engraving of Longleat (illus. 119, contrast illus. 144) with the transformations that Brown implemented after 1757, or the elaborate footprint of Grimsthorpe with a later engraving showing the results of Brown's design of 1772, probably carried out by local workers. They record important changes in landscape design that took place during the eighteenth century. Yet they also hint at our modern and paradoxical view that, while specific sites may change (as they surely did), we like to think that 'nature' does not. If we are alert to Brown's historical and long-held traditional view that ideal nature must be seized by the artist, we may choose to see the natural world, too, as somehow a timeless and unchanging permanence (global warming notwithstanding). If we think of any ensemble of Brown's iconic designs (illus. 145), in their present state at Petworth, Harewood, Blenheim, Longleat, Heveningham, Moccas or Woburn Park, it is not too difficult to understand how his careful and deliberate art worked to conceal the very artifice by which his and others' work was actually produced. Yet this is undoubtedly more difficult if we register the physical materials of Brown's landscape designs that not only depended upon growing and changing forms, but have long since succumbed to inevitable outgrowth and even decay of woodlands, individual trees and vistas.

A different theoretical perspective was advanced by a French landscape architect, Jean-Marie Morel, even though the results of his designs would have seemed similar to the work of Brown. Morel (d. 1810) was the director of Le Jardin du Roi (later Jardin des Plantes) for close to 50 years, where he welcomed dedicated botanists like Rousseau and, most importantly, the great natural historian Georges-Louis Leclerc, Comte de Buffon, author of the monumental 44-volume *Histoire naturelle, générale et particulière*. Buffon saw the natural world as a system of organic and mutable phenomena, which Morel, trained in the engineering school of Ponts et Chaussées, realized was a much more effective concept of how landscape worked and how landscape could therefore be shaped. He also saw that the art of landscape has to be wrested from the domain of architecture, and dedicated to the proper understanding of landform or *terrain*. His engineering skills involved the

> identification of natural landforms, including surface and subsurface geology, topographic inclination (slopes), rock formations and outcroppings . . . and the natural landscapes features such as water bodies . . . flood plains, slopes, surfaces or subterranean springs, marches and wet areas, vegetation . . . predominant tree and plants species and their respective history.[4]

Few of his designed sites survive, but some plans suggest to what extent his work revealed the scope and processes of these natural, organic and mutable phenomena and how, as a keen observer of the physical landscape and its sceneries, their underlying structures could be employed in garden-making (illus. 146).

145 A glimpse of 'Capability' Brown's landscape at Petworth in Sussex.

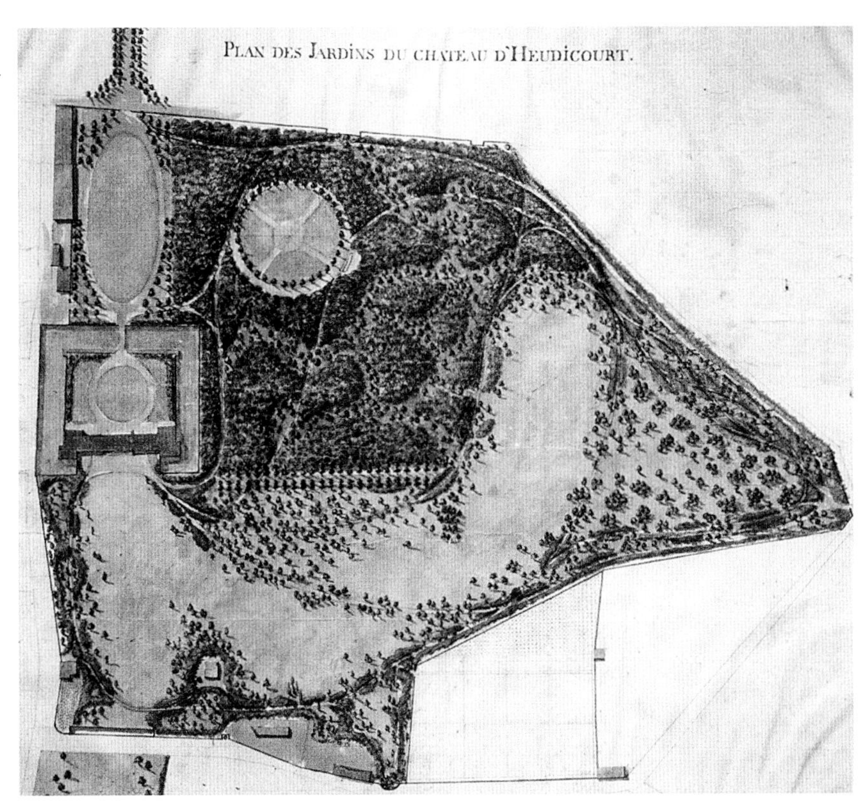

146 Jean-Marie Morel, plan for Heudicourt, *c.* 1805.

His theories were set out in his important volume (unhappily still without any English translation), *Théorie des jardins,* first published in 1776 (a second, expanded edition came later, in 1802). And in his careful explanation of how natural landforms could be shaped, he echoes the perspective of Thomas Whately, not only by stressing a necessary commitment to the forms and processes of different landforms, which he does with more care and detailed analysis than Whately, but by explaining the effect on visitors' responses of these natural, expressive, rather than emblematic characters. Perhaps with recollections of what Whately had written, or even relying upon the French translation of that work (published in 1771), Morel explains the different effects of water:

> Waters in landscapes are the body's soul; they animate a scene . . . if the waters are stagnant, if they flow slowly or advance rapidly, if they escape energetically or tumble noisily their impressions are neither equivocal or uncertain . . . The mirror of tranquil water reflects pictures that a spectator may vary at his pleasure. All these nuances, modifying their effects in a thousand forms, furnish an infinity of resources to enriching the pictures of Nature and vary her expression . . . [my translation from Morel's second edition, p. 126].

Much landscape architecture since Brown and Morel has invoked their naturalistic skills. Humphry Repton, who deliberately opted to establish himself in Brown's footsteps in 1788, often worked on similar sites or even at the same properties as Brown himself. Repton thought that his predecessor 'fancied himself as an architect', while he himself did not, though he often employed them. But his career, which began by practising the art of natural landscapes, shifted, for a variety of reasons, to respond to clients with smaller sites, who wished for a more evidently domestic scenery, or who owned suburban villas on the outskirts of towns; by the end of his career he was designing London squares. Ever since the fashionable 'high point' of natural gardening – as represented here by Brown, Morel and some followers – matters wavered between a deference to natural designs on the one hand, and place-making that did not disguise its interventions on the other.

Sometimes this wavering has been the result of alternating waves of taste and ideology. Partly as a result of the often misunderstood or poorly imitated lessons of 'Capability' Brown, nineteenth-century garden-making reacted against his 'bland' or 'natural' landscapes, and landscapes were soon dotted with a variety of *fabriques*, follies and other fabrications (see next chapter). These devices, in their turn, were challenged in the twentieth century with a fresh concern for the natural environment, either natural (unmediataed) or naturalized, and much landscape architecture since then has been characterized by a deep mistrust of over-manipulated garden designs (indeed, a wholly new dislike of landscape gardening per se). This minimal move also drew support from a peculiar tenet of modernist art, whereby each genre should honour and obey its own endemic materials: for painting, that might mean a rejection of story or narrative, a focus upon the forms and colours of the flat surface of the canvas; for landscape architecture, it appealed to the materials of the landscape itself, to the deployment only of its constituent parts of earth, air, fire (sunshine) and water.[5] However, it is also understandable that landscape architects wished to have their work recognized, to display their artistry and craft visibly to the eye, as well perhaps to contribute to individual sensibilities in the manner that Whately, Morel and other empirical philosophers valued. Even those designers, for whom the modern environmental imperative is strong, and who would favour much more nature designs, still want to display their work; recent work, for example, foregrounds how such practical items as storm drainage systems may become both an acceptable and visible feature of landscapes as well as educating the public to the work that they need to do.

Thus it is that landscape design for the last 200 years has wavered between claims for the 'natural' garden and the need to show off that such gardens are the work of professional designers. One can, I suspect, see some pattern that distinguishes private from public works. Public work needs to reveal itself, to talk eloquently to a public which does not readily notice landscape when it sees it; people may want landscapes that are part of, or adjacent to, urban centres to be natural or pastoral, while architects themselves may wish to make their own work more visible. Private properties need not insist upon such conspicuous publicity (though some certainly do), and for various reasons: owners will of course recognize that a professional has been responsible, however muted that work may be; size also suggests that a large estate can absorb more 'natural' elements. Some of these dilemmas, these competing values in landscape architecture, are clear if we compare public and private works by Dan Kiley or Laurie Olin; what they also suggest is that the adjudication of the 'natural' plays a different role in sites according to where they are and what other landscapes are in the vicinity, and in how individual responses are conditioned by personal experience or the quirks of photographic presentation.

Olin's public work has been far more prominent and visible than his private, a fact that his books and exhibitions endorse; indeed work for private clients rarely appears.[6] Yet the impact of his familiarity with private design (his own for private patrons, and his own study of others') has an undeniable impact on his public projects. None of these public designs, it must said, are inconspicuous, some are even downright busy, so one is left to tease out the place and importance of what is natural within spaces that present themselves decidedly and deliberately as artificial. So much depends on how one responds to a place, where one sits and what you look at – a proper element in public places. On the rooftop of the Church of Jesus Christ of Latter-Day Saints Conference Center in Salt Lake City, one can confront both the planting boxes that pile up as one approaches its summit, or take in a distant hillside and the prairie grassland, or admire both the city's profile and its planted rooftop. This versatility and the fashion in which the site affords one these very different apprehensions of 'nature' are aptly suggested by its photography; it isolates different moments, but it can also distort by virtue of its quirky or fanciful perspectives: overhead shots of the canopy of trees at Bryant Park in Manhattan give it a wholly different impression to being in the park itself; overheads of Wagner Park at the tip of Manhattan reveal the quirky disposition of geometric planting beds that on the ground are far less noticeable. Photographs of the Ray and Maria Stata Center at MIT declare a strong, distinctly urban milieu, whereas an Olin sketch foregrounds, literally, its treescape. Some sites require and receive a thoroughly naturalistic treatment in photographs and on the ground, like the foliage in the Midway Plaisance Winter Garden in Chicago, as do some of the firm's sculpture gardens, where sculpture often plays a less obtrusive role than the planting.

Some of these sites are conspicuously 'natural', like the planting on the roof at Salt Lake City or the Cactus Garden at the Getty; some are occasionally and ambiguously naturalistic, like early work of the Hanna/Olin firm for corporate headquarters and business campuses, where the main buildings were surrounded by a soothing and pastoral landscape. But one site in particular plays very convincingly with the possible roles of its 'nature'. The Wexner Center at Ohio State University was conceived as an elaborate play with the original site of its old armoury by Eisenman and by an abstract invocation of the surrounding prairie landscapes by Olin; the result is an intriguing manipulation of approaches, but what dominates are the massed grasses that waver over the pathways (illus. 147). The illusion, even for those in the know about the site's formulation, is to be tempted into accepting this landscape as a natural habitat.

Dan Kiley, previously invoked for his skill with drawing into the Miller gardens a sense of the natural, unmediated flood plane of the river, can mix and match his formal interventions with astonishing versatility. Private sites like his own house and office in Vermont house, Currier Farm (Vermont), Gregory House (Minnesota), Shapiro Residence (Westport, New York), Kimmel Residence (Connecticut), or Kusko Residence (Massachusetts) are by no means negligent of Kiley's own elegant and conspicuous formalisms; but they allow, in many cases because of the acreage involved, room for the 'natural' which can command as much attention as the remainder. His public works are equally versatile: one does not readily guess that the approaches to Dulles Airport outside Washington are his designs, and the mountain hillside that meets and blurs the edge of his extensive grounds at the Air Force Academy at Colorado Springs seems a conspicuous intrusion of the natural. But otherwise his public spaces are rich, busy and intricate in their inventions, like Fountain Place in Dallas, or Bank of America in Tampa. But the prominence of what one must perforce term its 'natural' elements, as in those two water gardens (illus. 148), or in the bushes encased in the platforms of the Oakland Museum or the lines of trees stepping down the Henry Moore Sculpture Garden in Kansas City, sometimes temper even his most organized formalisms with their own character.

The role of the natural in landscape designs is as varied as Morel knew that the effects of water would be. By the end of the eighteenth century there was a distinct sense that scenery had to be as natural as the topographical landscapes of the Wye Valley or the Derbyshire Dales, even if those landscapes were themselves moulded and changed by countless natural events and human activities. Such minimalism risked being mistaken for the 'common fields'. So landscape architects had once again to show their hand, perhaps by re-inserting architectural structures once again, or

147 Laurie Olin and Peter Eisenman, Wexner Center for the Visual Arts, Ohio State University, Columbus, Ohio, 1983–9.

148 Dan Kiley, Fountain Place, Dallas, Texas, 1985.

by displaying more conspicuously the resources of the natural world, reformed and reformulated in the designed landscapes. In the end it became as much the role of the spectator to detect and appreciate whatever manipulations of scenery had been introduced as for the designer himself to devise and dictate them.

13 The Chinese Garden and the Collaboration of the Arts

The French specialist Georges Métailié has complained that the 'terms "Chinese gardens" and "scholar gardens" are often used interchangeably', and that one must 'keep in mind that, if scholar gardens are indeed Chinese, not all Chinese gardens are scholar gardens'. The point is well taken. In fact, it is probably better to distinguish between private gardens and imperial gardens, though even these too may overlap in scope if not always in extent. Another distinction to be made is that between gardens and landscapes which are, or seem to be, remote, enjoying natural or naturalized surroundings, and those sites where a profusion of buildings is dominant even though they may be set in a considerable space of ground.

But what is particularly interesting, amid the quite considerable repertoire of different Chinese gardens, is that this preference for singling out scholarly gardens persists in discussions of Chinese gardens. The scholar garden, *tout pur*, exercises its fascination in part because it seems to subsume other landscape architectures and to epitomize the larger reaches of Chinese garden art: among other things, the gardens of the literati reveal a particular tendency for collaborations between the arts of poetry, painting and garden-making, and the role of calligraphy is in them all. These collaborations, a considerable feature of the eleventh century, continued to have a lasting influence on presentations of hermit and scholarly life, and of gatherings of friends and family. The Ming official Wang Xianchen spent sixteen years constructing the Artless Administrator's Garden 'to cultivate my garden and sell my vegetable crop', most likely a poetic gesture than something to be taken literally. And this garden was then depicted by Wen Zhengming in 1533 on 31 scrolls.

Yet a recent postcard from China arrived with a sequence of four stamps illustrating the Master of Nets Garden in Suzhou (illus. 151); although the Chinese postal services are hardly an adequate scholarly resource, the stamp surely implies that this image of 'a Chinese Garden' was indeed a scholar's garden, since one of the first buildings on this site was a library (the Thousand Volume Hall). Yet it is also a fact that not only did sites like the Master of Nets Garden change over time, but the 'Chinese' garden itself was never an unchanging entity throughout its history; furthermore many Chinese commentators themselves constantly noted the ephemerality of gardens.[1] But the focus of this chapter is upon the Chinese collaboration of the arts in scholar or literati gardens and how they were, and continue to be, viewed. One of its major themes is how the gardens, both real

149 A corridor along the side of the Humble (or Artless) Administrator's Garden, Suzhou.

and 'on the ground' and imaginary in poem or painting, were always mediated: texts were fundamental in designing and understanding gardens, and mediatization of cultural traditions was essential. Today much of this understanding is lost even to Chinese people.

Other cultures, too, have drawn upon a variety of arts – literary and painterly – for the establishment and description of their gardens, as for instance in the naming of temples in eighteenth-century English landscapes, where some skill was needed to read and interpret classical allusions, or the alliance of gardener and architect for the late nineteenth-century Arts and Crafts garden. But this collaboration is infinitely more intense and rich in Chinese scholar gardens, and this was especially true in the late Ming (1573–1644). The collaboration may also be true of Imperial creations, but they were less written about than those of the literati, whose popularity was marked by very personal links among artists, writers and their families. And beyond even this collaboration between calligraphy, paintings and poetry in the *making* of gardens is the fashion in which gardens were necessarily observed *through* the lens of these various and allied arts. The Chinese garden was essentially mediated. References even to paintings of actual sites may have been gleaned largely from texts on landscape painting rather than from studying the sites themselves, though this changed somewhat in the late

150 The Master of Nets Garden, Suzhou.

151 China postage stamp, showing the Master of Nets Garden in Suzhou.

imperial period with the publication of pictorial albums.[2]

Writing on Chinese gardens often relies upon a strangely essentialist discussion, where moving backwards and forwards between different periods with apparent ease gives a much more definite sense of unchanging practice.[3] Discussions, to-ing and fro-ing among historical materials and images, seem to present a undifferentiated garden culture. This is undoubtedly a function of the lack of extant early examples, so that later sites are discussed in lieu of earlier ones; but it also stems from the Chinese habit of celebrating earlier masters, whether poets, painters or even garden-makers: thus an image by Gao Fenghan (1727) shows and recalls the poet Tao Qian working in his garden, an incident that dates from over a century earlier (illus. 152); another painting by Qian Gu from 1560 depicts a famous occasion from ad 353 when poets gathered beside a stream in a grove by the Orchid Pavilion. Yet if the actual examples of the Chinese garden appears to be essentialist in many commentaries, even when its documentary sources and geographical locations are seen to be widely scattered, it nevertheless shares some characteristics with a special vogue for fictional gardens in the seventeenth century.

The Ming hermit Liu Yuhua called his imaginary garden 'The Garden that is Not Around'. It is not 'around', because it exists only on paper; for, he writes, 'only gardens on paper can be relied upon to be handed down'. As scenery is 'born in sentiments', he writes, and 'images are suspended on the tip of one's brush', so he prefers to invest in a 'composition [that] can be beyond limit', without restrictions of 'extent and arrangement'. He is clearly playing with the *topos* of ephemerality, a major concern throughout Chinese garden culture. Another work from 1674, the 'Make-do Garden' by Huang Zhouxing, also envisages a garden that invokes a plenitude of detailed forms and practices: 'I merely select the place where the landscape is finest under the Four Heavens.' It has rightly been noted that these imaginary gardens must not be mistaken for some ideal, quasi-Platonic, Western vision. The fictional descriptions are better seen perhaps as something in the manner of Western memory systems, as explored by Frances Yates: notably, the system for storing and utilizing ideas by placing items to be recalled in specific places in an imaginary house (here a garden), so that they may be called upon by an orator.[4] But a more useful way of understanding this Chinese account of imaginary landscapes is to see them as *topoi*, or commonplaces, clusters of landscape features, themes, ideas and associative features, drawn from an expert and often learned review of many centuries of gardenist writings. This may be a more plausible and relevant analogy for the Chinese situation, if we think of late seventeenth-century European landscapes, like the Hortus Palatinus at Heidelberg, that could also draw upon a whole cluster of garden themes and

152 Gao Fenghan (1683–1749), *Cultivating Chrysanthemums by a Thatched Lodge*, 1729, ink and wash on paper, showing a scholar working in his garden, a recollection of the poet Tao Qian who had died 1300 years earlier.

devices; except that it was undoubtedly built and was certainly 'around' (and still exists).

As Liu describes his imaginary garden, he does seem to be relying upon a complete recall of previous Chinese gardens (he specifically recalls Li Gegei's text on the 'various celebrated gardens of Luoyang'). He works in part by relating categories drawn from long traditions of Chinese garden-making. Huang Zhouxing, quoted earlier, also relies upon a topography recounted in an earlier story, Tao Qian's 'Record of the Peach Blossom Source', when he describes his 'garden'. Liu references the many famous instances of mountain scenery throughout China, where one may climb or gaze upward into the distance, and notes the 'piled-up angles' of garden's rockeries (this may well be a reference to brush strokes used in landscape painting). He is especially eloquent about the mountains and their rivers, because the latter would be excavated to make the former; the Chinese term for landscape (*shan shui*) means literally mountains and waters. Mountain streams, he says, are dredged for rivers and widened for ponds for fishing and for reflection, but can also be recirculated in many ways. Then he proceeds to trees, shrubs and flowers, where a taxonomy of their possible shapes and uses is expounded. Then come buildings and courtyards, and the versatility of their forms and associations: 'Flying galleries penetrate the sky, clouds nestle in the eaves, soaring towers spring from the ground, willows brush against carved balustrades'. Gibbons call in the night, cranes cry in 'a fragrant morning' and 'patterned fish jump across the waves' – 'things with an elegant charm and those worthy of appreciation'. Paths twist and turn to seek out secluded corners, caves, terraces and the glimpse of 'vegetables covering the fields' – all 'dangerous and wonderful spots in my garden'. And despite that frisson of danger (an 'alarming waterfall', a pavilion perched on a precipice), this wholly imaginary garden avoids all threats of illness and unpleasantness. These imaginary gardens have two main characteristics. Without any material elements, the garden will evade any ephemerality. And, since many actual gardens tried ever so hard and at great expense to achieve an inclusive and 'limitless' acreage and range of associations, Liu's and Huang's gardens can in the mind's eye accomplish this scope effortlessly and with no expense.

Another 'imaginary' garden from the mid-eighteenth century comes in the Chinese novel *The Story of the Stone* (also called the *Dream of the Red Chamber*). It is presented as if it were an actual, constructed site; and embedding the garden within a novel makes it fraught with the tensions and ambiguous perspectives that direct the characters and their interactions; the garden also succumbs to decay and ruin, as if it were a built work. Within the novel's dialectic, the meandering and intricate garden is contrasted with the formal layout, rituals and social proprieties of the central compound; to the less constricted world of the garden the young hero, Baoyu, turns as a relief from the male-dominated society and 'scholar-official' class of his family. Such a contrast between formal courtyards and 'naturalistic' gardens is indeed a widely observed layout (illus. 153).[5] Nonetheless, the *topoi* deployed in the narrative of the Grand-view (Prospect) Garden, or Daguan Yuan, and the discussions among its visitors, give or take some moments of irony or satire about prevalent notions, suggest a rich set of ideas about the formulation of such a site.

The narrative in chapter Seventeen concerns a sort of imperial garden within a scholar-official

153 A Chinese estate where living quarters mesh with different gardens and lead to a landscape beyond.

household. It revolves about a visitation to a newly 'finished' garden by a group of 'literary gentlemen' led by the patriarch Jia Zheng. Accompanying the party is the son of Jia Zheng, who is caught sneaking into the newly finished garden and whose father commands him to accompany the others. In fact, Baoyu's skill with names and the composition of couplets manages to make some very successful contributions, at least to the satisfaction of the literati. We are told that the physical structures of the newly established Grand-view Garden are all there, but it lacks a whole range of associations, including the naming of the various elements of the garden. As Jia remarks, 'All those prospects and pavilions – even the rocks and trees and flowers will seem somehow incomplete without that touch of poetry which only the written word can lend a scene.' And in advance of a visit by Her Grace the Imperial Consort or official concubine, Jia decides to review the garden and to propose 'provisional names and couplets to suit the places where inscriptions are required', so that subsequently the Lady may make her final adjudications.

The garden is enclosed, so they begin by closing the gate to 'see what the garden looks like from the outside'. After admiring its gatehouse, they proceed within (illus. 154). The assembled (and truly sycophantic) company is ecstatic at the steep hillside which faces them and screens everything else from their view, without which the 'mystery would be lost'. In the complete 'circuit' of the garden, that follows in the succeeding twenty pages, the party encounters rocks polished ready for inscriptions, a ravine full of trees and shrubs, streams and pools, gaily painted pavilions on their banks and the surrounding slopes, bridges with small kiosks or humped with a high platform, a whitewashed retreat with a cloister-like walkway hidden inside a grove of bamboos, a hut with rice thatching, apricot trees, a miniature agricultural tract, a cave from which a spring issues, and many other items.

At every point in this meandering and hilly garden, Jia Zheng demands of his son a suitable name and/or couplet; with much scoffing (as well as some reluctant admiration; convention would require him to belittle his own son in front of others) and alternative, if less elegant, contributions from the literati, this range of toponyms, literary

154 The Garden-view (or Prospect), an imaginary 19th-century reconstruction of the garden in Cao Xueqin's *The Story of the Stone*.

allusions and local associations are provided. The debates are leisurely and the give-and-take an evident part of what seems a somewhat pretentious gathering. Many names are forthcoming, sometimes twenty or thirty at a time, frequently banal or pretentious. But the boy reminds them that he had read in an old book that 'to recall old things is better than to invent new ones'; so when they encounter a mountain that is not the garden's principal feature he declines to emphasize anything more than a hint of later encounters and he advises 'Pathway to Mysteries', after a poem by Chang Jian.

Others now follow suit and think of poetic allusions, or, failing that, of ancient gardens from the past: 'there is nothing wrong with imitation provided it is done well'. This newly finished garden clearly needs the authority and the patina of classical references: 'naming' is a means of determining one's perspective and thus construing the world.[6] Besides a shrewd sense of location and an accompanying instinct for decorum (whether of the site or its proposed visitor), Baoyu has a keen ear for linguistic nuances, and he can quibble with epithets by other poets. Having succeeded with a satisfactory place name, he is often then forced to provide a couplet that speaks to the particular scene, of the sort that are still to be seen in the Couple's Retreat Garden in Suzhou. He knows his plants and can cite references to ancient writings, and he can produce couplets which allude more carefully to the surroundings. He dislikes obvious or too casual descriptions, and he engages in a very interesting debate about why one of the thatched buildings of a 'village' was termed 'natural' by his father:

> A farm set down in the middle of a place like this [says Bao-yu] is obviously the product of human artifice. There are no neighbouring villages, no distant prospects of city walls; the mountain at the back doesn't belong to any system; there is no pagoda rising from some tree-hid monastery in the hills above; there is no bridge below leading to a near-by market town. It sticks out of nowhere, in total isolation from everything else. It isn't even a particularly remarkable view – not nearly so 'natural' in either form or spirit as those other places we have seen [in the garden]. The bamboos in those other places may have been planted by human hand and the streams diverted out of their natural courses, but there was no *appearance* of artifice. That's why, when the ancients use the term 'natural' I have my doubts about what they really meant.

This exchange is not meant to prescribe any definite aesthetic, but in the context of the novel it raises interesting problems. Since a brand new garden does not enjoy any interesting environment nearby, it must rely upon creating a 'farm', still considered a necessary attribute of such an ensemble; yet it must also, Bao-yu implies, conform to the levels and perceptions of the 'natural' elsewhere in the garden. The narrator of *The Story of the Stone* himself is perhaps implying that the allusions to ancient poetry proposed by the group are themselves scarcely natural; indeed that they impose a particular perspective on the new garden. The garden is, however natural-looking, a garden that is read and responded to via its representations. Indeed, rather than any firm distinction between 'nature' and 'artifice' which goes back to early Daoist culture, what is involved is the lack in Chinese of any separation between 'man' and 'nature', which has an immediate impact on how we handle the treatment of artifice (very different from Western attitudes).

If there is one thing that is missing from these various exchanges in chapter Seventeen it is any explicit reference to paintings, though Baoyu invokes 'natural paintings' later in the same passage as an infelicitous example of what he calls 'forcible interference with the landscape'. But later in the novel there are references to paintings, when the inclusion of a concubine and maidservant in the newly formed 'poetry club' is permitted to figure in an imaginary painting of 'Scholars Enjoying Themselves in a Landscape': 'anyone who can write poetry gets into the picture'. Furthermore, we may also imagine that some of the recollected poets and ancient writings that are cited by the literati allude to or rely upon pictorial treatments with calligraphic annotations. Any garden, as eventually will be this one when the naming of places has been finalized, is a scenery in which words have been posted in descriptions and signs to match and enlarge what one sees. And the narrator himself has, as it were, anticipated this poetical and pictorial collaboration by inserting into his own narratives, lines that recite the simple elements of a scenery with distinct attention to how such landscapes may be envisaged:

> A marble bridge crossed it
> With triple span,
> And a marble lion's maw
> Crowned each of the arches

Or

> Past rose-crowned pergolas
> And rose-twined trellises . . .

In short, the imaginary gardens in both words and images are 'real' to the extent that they do not represent just a representation or reporting, but are an accurate account of how we might understand them on the ground. The Chinese *wenhua* means the 'transforming power of writing', and a modern commentator notes that a person 'ceases to be in the *garden* [when he] starts to live in a painting' (illus. 155).[7]

One dominant feature of existing Chinese gardens, whether real ones, painted ones or poetic representations, is that we view their scenes everywhere through a multitude of frames or lenses. These are often literal: seen through different

155 Drawing with annotations ('ancient pictures of garden design'), discovered by Chen Congzhou.

156 Osvald Sirén's photograph of the gourd-shaped garden gateway in the residence of Prince Cheng, Beijing.

frames in a garden: we sit on a terrace and see the landscape through its columns; we look through screens or patterned lattice work to see what is beyond; we glimpse distant bits of the garden through blossom-shaped windows or down the perspectives of their zig-zag or meandering verandas; we enter through framed and elaborately shaped openings, like moon gates, in garden walls (illus. 156). But also we see them through the lens of words: we encounter calligraphic inscriptions, by which we know the names or literary texts to associate with the houses and their separate parts, and so we arrive at an understanding of these places through their writings. The naming of both whole gardens or their individual fragments by references to some earlier image, to some notable

157 A moon door in a Suzhou garden.

poem, or even by recalling some notable predecessor is how we appreciate sites, a filter or lens for our imaginative perception of the garden. The invocation of a temporal or historical reference helps us understand the physical space of a garden, while the acknowledgement of past masters and owners ensures the intersection of time and space; it is rare to encounter the names of garden designers, but at Suzhou the rockeries in the Mountain Villa with Embracing Beauty, *Huanxiu shanzhuang*, are known to be the work of Ge Yuliang (and Suzhou is now a UNESCO world heritage site).

Paintings show remote hermitages and modest pavilions where inhabitants or occasional visitors are placed where the landscape is best grasped, where the experience of it will be the most felt and vital. In so many images, like the intricate and detailed Forty Scenes of the Yuanming Yuan, or the scroll of the Shao Garden by Wu Bin,[8] the landscape is clearly revealed as places where the surrounding landscapes could have been viewed or 'taken', as from fixed positions, what Chen Congzhou terms 'in-position viewing'. But there is also what Chen calls 'in-motion viewing', where the angle of observation changes as we walk through the scenery (illus. 158). Both in static views and in the moments seized while exploring, the full experience of gardens is made available.[9]

In the 1533 album of the 'Garden of the Unsuccessful Politician' by Wen Zhenming, this full experience is communicated in an astonishing and intricate presentation of a prose record and 31 separate paintings, each of which is glossed with a brief descriptive note and a rhyming poem.[10] So one way of grasping the distinctive character of Chinese gardens is to understand the *means* through which they are deliberately, even if subtly, communicated to us. The contributions to garden making from painting, poetry and calligraphy were what the eighth-century Chang'an emperor termed the 'three perfections'.[11] By the eighteenth century a critic could write that 'Both poetry and painting are scholars' occupations which help to express human moods and feelings', both of which in their turn might address, capture and design landscapes. The Tang painter Wang Wei painted a panorama of his country estate on which we may assume he copied out twenty poems about its perfections.[12]

The collaboration of the arts in the understanding of gardens (both real and imagined) became a feature of the northern Song period (960–1127), noted for its scholarship and such pursuits as poetry, calligraphy and painting, when garden-making and rock-collecting were of extreme interest.[13] The period saw an astonishing flurry of texts on gardens and landscape, like Li Gefei's record of the *Famous Garden of Luoyang*[14] and Zhu Changwen's *Further Records of the Geography of Wu Commandery*, which told of gardens built in Suzhou at this time. What is above all significant for the understanding of these gardens, beyond how they were described in poem or painting, was that these were worlds where owners could cultivate their own personality and character away from the duties of their public lives as government officials. This is not unique to Chinese gardens, as Western examples show, but the ability to draw upon the collective arts gave to their owners a special and enhanced meaning. Various scholars have shown how taking possession and composing of a garden gave the chance for a wide range of owners – merchants, imperial aristocrats, literati – to reveal economical and political as well as agricultural and aesthetic ideals, to reveal the concept of local consciousness or identity (vis-à-vis imperial and governmental politics), and to express

158 Figures on a bridge crossing a pond surrounded by galleries, an archetypal setting. Such routes surrounding central garden features like a pond or rock formation would enable visitors to see the garden as they move around or through it.

an owner's status, through social intercourse, often at elaborate lengths.[15] These involved detailed surveys of all an owner's or family's possessions, including every building, orchard, plant, tree, rockery and water feature.

One significant, earlier Song painting by Li Gonglin of his 'sprawling, unwalled garden in the Longmian Mountains' was entitled *Mountain Villa*: it was fairly rare for artists to portray their own properties at this time, but Li certainly knew of two even earlier and famous examples by Wang Wei and Lu Hong who depicted their *Wangchuan Villa* and *Thatched Hut* respectively in the early eighth century (illus. 159). It was painted when Li was absent from the villa, a kind of *guide-mémoire* to his properties and its surroundings; only six, scattered and individually incomplete copies of Li's own scroll survive, annotated with colophons or commentary by two brothers, Su Shi and Su Chen). These commentaries present *Mountain Villa* as a kind of mapping of the terrain, a 'recumbent travelling' for those who wanted a substitute for actual travel; it allowed them 'to trust their feet and walk about, naturally finding roads and paths', to know all the streams and rocks, grasses and trees', and to encounter the local fishermen and woodcutters. In particular, Su Chen penned his 'Preface and Twenty Poems Written on Li Gonglin's *Mountain Villa*', where he told how the artist had depicted a sequence of sixteen related sites, with further named sceneries, that were not seen in sequence on the scroll, but appended at the end. One version of the scroll, formerly in the collection of Bernard Berenson, shows the property in 'monoscenic units' interspersed now with Su Chen's texts.

Given that the scroll portrays how travellers are discovered moving through this scenery, it reveals the obligation and pleasures of viewing while in motion or imagining that exploration (illus. 160). Yet it also signals many named places where the visitors could have paused to review and visit significant sites, such as the Lodge of Establishing Virtue, the Hall of Ink Meditation or the Hut of Mysterious Completion mounted on a hillock of earth and stones with a surrounding landscape of waterfall and trees. Some, like the Necklace Cliff, seem inspired by the local topography rather than by the meditational possibilities of any one setting. The names themselves and the commentary by Su Chen juxtaposed to the images indicate not only how to respond to the representations of places,

159 Detail from one of the ten *Scenes of a Thatched Lodge* attributed to Lu Hong, 8th century, ink and wash on paper.

but suggests how the original sites would themselves have been received; indeed, Harrist suggests how much the scroll and its annotations constituted a dialogue between painter/owner and commentator, of the sort that might have taken place in situ. In the Hall of Ink Meditation, the art of 'ink meditation' includes both writing and painting: 'Every encounter with things [that] leads to meditation'; the 'single pellet of Ink' becomes mountains and streams, just as hills and rivers have themselves now become the subject matter.

The scroll of *Mountain Villa* allows us to appreciate that 'a Chinese garden in the broadest sense is a landscape for man's pleasure. Place a pavilion where he stops, pave a path where he roams, span a bridge where he crosses, erect a hut where he rests . . . and a garden is born'.[16] Physical remains of Northern Song gardens have long since disappeared, so we are dependent on our own historical and imagined gardens. Later gardens in Suzhou do exist, and may be used cautiously with hindsight to inform earlier gardens, though many have been reconstituted during the twentieth century. Yet for those who cannot read calligraphic inscriptions, who are merely pleased by the oddity of mysterious nomenclature, yet cannot divine their deeper meaning, who cannot relate, as Baoyu and the literati did, to the ancient references, and who are not able to judge whether a garden is or is not 'natural' or whether it conforms to original notions of simplicity, these are separated by an enormous cultural divide. Westerners are puzzled, too, that many gardens are not themselves seen as the work of a named designer whose work can thus be grasped, but rather are presented as the expression of an original owner. The owner may well have designed it himself, but it is to his character and personality that we are asked to respond: in writing of the makers of famous Luoyang gardens, Li Gefei said that 'the pavilions, platforms, flowers, and trees all emerged from the design of his own eyes and the craft of his own heart. Therefore the [places] that are winding or straight, spacious or dense, all [display] profound thought'.[17] This necessarily adds another distance between us and the garden itself, imposing yet another 'lens'

160 'Discovering Truth Embankment', 'Lodge of Fragrant Reeds', 'Necklace Cliff', scenes from *Mountain Villa*, a handscroll fomerly attributed to Li Gonglin (Song dynasty, 960–1279), ink on paper.

of personal response through which we view these gardens.

It is tempting for some Westerners to see Chinese gardens as in some way 'exotic' or 'Oriental'. Westerners also tend these days (mistakenly) to think of any garden as realistic or 'natural', without acknowledging any lens or mediation from the designer; this truly disables Chinese understanding (as it also does with much Western landscape architecture). But it is the Chinese culture, and not the garden-making itself, that is unfamiliar. The attention paid to Chinese garden-making from early parklands to mountain hermitages to urban or suburban sites, from imperial enclaves to literati retreats, is more sustained and intricate in its focus on garden-making than anything in the West. It responds to reception as much as to principles of design. In comparison to Western discussions of landscape architecture, it is the thoroughness and subtlety of how and why gardens were created that are at stake; references to geographical and historical circumstance and to mythological and religious belief, while indeed part of the culture, simply add another level of reference to the garden's meaning.

Something of this intensity and dedication to the making of gardens is clear from the analyses by the late Professor Chen Congzhou, gathered in five essays for publication in the 1980s.[18] Above all he insists upon the careful discrimination in the making and understanding of gardens: it is 'painstaking work', and there is 'much in the word composition'. This does not, however, mean that prescriptions are rigid (which would at least make things clear-cut!); there are 'no fixed formulas and what is important is the inventive application of these rules', yet there are, for example, 'certain rules' for rockeries. Gardens cannot 'arbitrarily' all look alike, since the arts and culture of a locality and period will determine their scope. Yet some of his pronouncements strain our comprehension as to the exact discrimination that a design would impart: 'a Chinese garden scenery should be sought where there is no scenery, sound in soundlessness, and motion in stillness rather than in motion'. Craftsmanship and above all 'mastery' are crucial in both design and appreciation, and mastery comes in part from a successful dialogue between opposing ideas: 'there is no motion without repose, and vice versa'. His pronouncements on the miniaturizing or epitomizing of gardens were premised on ancient enclosures that would create 'a sense of infinite space within a limited area'. It is sometimes difficult, when reading Li Gefei for example, to realize that despite the sense of fullness and variety, many sites were quite compact. So scale is vital. 'Spaciousness' and 'flexibility', contrast and interdependence of items, are the 'gists of garden designing'. However when he considers four famous gardens of Suzhou that can be customarily described by dynastic labels, he prefers to distinguish their characteristics or capabilities as 'in-position viewing', 'in-motion viewing', antiquity and 'great stateliness and magnificence'.

We may sometimes allude to the collaboration of arts in Western garden-making, its *Gesamtwerken*; yet Chen's attention to the 'lyrical and artistic values of a garden [that] depend on the poetic and pictorial conceptions' depends upon a much more fundamental embodiment of ideas in the design. Principles of painting also pertain to garden design and their compositions should be followed in garden dispositions; there is also 'something common between the composition of a poem and that of a garden'. Categories matter to him – 'There are things that properly belonged to the past and there are things of the present' (this doubtless signals some accommodation to

161 Rockery at The Mountain Villa with Embracing Beauty.

the times in which he wrote during the 1960s and '70s, when the revival of the tourist industry necessitated that old gardens be hastily reopened and restored). So how we view a distant mountain – whether in looking at a painting, or making a tour around it as when reading a scroll – involves emphasizing either the prominence of certain features, or the 'continuity and integration of scenery'. And the deciding factor is 'the presence of the self'. The subject of a painting will be a projection of the artist's mind or 'vital spirit' onto the exterior world, as also in garden-making. In responding to both landscape locations and garden sites, 'implication' is how one understands the 'quintessence' or what we might term 'genius loci'. Only when we attain a certain mood will it be possible to 'finish' off with an inscription, as is discussed in *The Story of the Stone* (to which Chen refers on several occasions). Finally, he is particularly attentive to the failures of even famous gardens when they are restored (it is something of a curiosity that much Western analysis, especially by modern practitioners with regard to their own work, resolutely neglects any discussion of failures[19]); so he repeats 'occurrences that ineluctably ruin the effect of good scenery' in 'The Miscellaneous Notes of Yi Shan', and accordingly he laments the poor renovation work of 'recent

162 The zig-zag walkway across the pond at Yipu, Suzhou.

years' (another oblique gesture to the political climate).

Clearly much is missing in this garden culture, not only for Westerners but for the Chinese themselves. John Minford feels how 'strange [it is] to walk round one of the marvellous gardens of Suzhou. In order to be there at all, one has to reassemble the entire universe of which it was a part, and which has been so thoroughly dismissed, in the name of progress or one sort or another'.[20] Yet, with respect, that is exactly part of this particular garden culture, and while it does require special insights, the place of one's imagination can be spacious enough to recover some of it. As discussed above in chapter One, the immanence of the sacred or even the play of significance has not entirely disappeared from gardens, nor has our sense of the political and economic foundations of their making. One may be neither Daoist, Buddist nor Confucian, but the former's dedication to compassion, humility and unselfish existence, and the latter's emphasis on moral conduct within worldly society, are not negligible qualities, for they allow 'the wise to take pleasure in water, the kind in happiness in a mountain' (Confucius).

14 Follies, *Fabriques* and Picturesque Play

We expect the cult for garden follies, or what the French call *fabriques*, to be most conspicuous during the Picturesque movement from the late eighteenth century into the nineteenth century, when that fashion spread from more elite designs to a myriad of smaller estates, suburban villas and public parks. This was accompanied and promoted by a considerable profusion of publications that showed how these structures might be devised and built and their significance or allusions understood. That said, it is also useful to note both the prevalence of garden buildings *before* the Picturesque flurry of pattern books and the *renewed* interest in follies in more recent times.

We may begin, however, by looking at the pattern books and other publications during the high point of the Picturesque.[1] These were clearly attempts to devise a whole range of items for different estates, including smallholdings and suburban gardens, mostly culled from aristocratic gardens; there was often the suggestion that these structures could be fabricated by local craftsmen or even by the diy gardener. By the early nineteenth century many of these books were published in more than one language, guaranteeing swift communication across linguistic borders. They contained suggestions that were both possible (swings, summerhouses, urns) and seemingly impossible (a floating Chinese garden, large replications of famous antique structures, elaborate gondolas manned by ten oarsmen). The range of cultural references was widespread, allowing landowners a chance to allude to a host of very different places, which few, if any, had actually visited: there were Greek and Roman structures, to be sure, but also temples in Turkish or Egyptian formats, boats and especially bridges in the Chinese taste, a nine-pin bowling alley in the Persian mode (whatever that was!), Tyrolean mills, Moorish tombs, Egyptian gateways, a Chinese umbrella under which revolving guests could whirl in their seats, or a Robinson Crusoe hut. While there was also a range of plausible, even necessary, items, they were often dressed up in fanciful or even symbolic dress: a vine-dresser's lodging was shaped like a huge wine barrel, or a hunting cabin dedicated to Diana, goddess of the hunt. They were wildly eclectic, exotic or quixotic, but clearly democratic in their ambition, and presumably eager to impart some 'meaning' to the selection and establishment of the different structures.

A very early example was Johann Bernard Fischer von Erlach's *Entwurff einer historischen Architektur* of 1721, one of the earliest European

Two illustrations from English and German pattern books:

163 Thomas Wright, *Six Original Designs of Arbors* (Book I of *Universal Architecture)* (1755).

164 A Chinese Floating Garden, from J. G. Grohmann & F. G. Baumgärtner, *Ideenmagazin für Liebhaber von Gärten, Englischen Anlagen und für Besitzer von Landgütern um Gärten . . .* (1796–1806).

texts to show off Eastern examples. But England saw a considerable flurry of design proposals, and an overview of these works and their titles suggests the richness and invention of the field (illus. 163): the hugely prolific author Batty Langley produced *New Principles of Gardening* in 1728 with suggestions for both buildings and for termination *trompes l'œil*; William and John Halfpenny, equally prolific, produced *New Designs for Chinese Temples* (1750), *Chinese and Gothick Architecture Properly Ornamented* (1752), *Rural Architecture in the Gothick Taste* (1752) and *The Country Gentleman's Pocket Companion* (1753), while William Wrighte put his *Grotesque Architecture* through three editions (1767, 1790 and 1802). Mention must also be made of Thomas Wright's wonderful but sadly uncommercial publication, *Universal Architecture*, which came out in two parts, *Six Original Designs of Arbours* in 1755 and *Six Original Designs of Grottoes* in 1758, of which there is a splendid modern edition.[2]

Between 1796 and 1806 J. G. Grohmann and F. G. Baumgartner published a five-volume *Ideenmagazin* (Storehouse of Ideas) in both French and German, and soon claimed that it circulated widely in Italy, Russia, Poland, Silesia and Holland; its various, more exotic proposals were interspersed with prints of actual contemporary German gardens like Machern and Laxenburg, practical devices like lawn rollers, sundials and a signalling system for communicating with neighbours, and a plethora of items that we now consider familiar in public fair grounds, such as carousels, dartboards, Ferris wheels and swings. Soon after the *Ideen-magazin* was started, a Dutch nurseyman, Gijsbert van Laar, started his own *Magazijn van Tuin-Sieraden* (Storehouse of Garden Ornaments) in 1802; he pillaged several designs from others, but offered his proposals in both plan and elevation. His materials were cheap and flimsy, but with items that clearly appealed to the bourgeois gardener: little hermitages, memorials for the dearly departed, cold-baths, birdcages and pleasure boats. The variety of his choices was as exemplary as it was absurd: farm buildings were offered in three styles – brand new, 'ruinated' or dilapated'!

More such publications soon followed. In 1809 came the two-volume publication by J.-C. Krafft and P.-F.-L. Dubois, its title self-explanatory: *Productions de plusieurs architectes français et étrangers, relatives aux jardins pittoresques et aux fabriques de divers genres qui peuvent entrer dans leur composition* (Works by several French and foreign architects, touching Picturesque gardens and the different kinds of fabriques that can be introduced into their design). Its acknowledgement of architectural competence assures that the picturesque vogue is now in the secure hands of professionals! Later in 1820 Gabriel Thouin published his *Plans raisonnés de toutes espèces de jardins*, with two further editions, that of 1828 with coloured plates; he combines ground plans for 'all sorts of gardens' with sectional views that show the different *fabriques* (illus. 165). Given that he distinguishes firmly, if not always distinctly, between different kinds of layouts and their decorations, the structures shown might be expected to be suitable or apt: there are Chinese, English and irregular landscapes designs that are distinguised from what he terms *paysages*, which consist of agrarian land, woodland, pastoral parklands and *carrières* (spaces set aside for riding). Other publications followed, some very carefully directed at bourgeois residences: Victor Petit lived up to his name and provided small-scale structures for homes, outhouses, chalets and kiosques (illus. 166) in his *Habitations champêtres* from around 1855.

It is both fun and amusing to review these suggestions. But more serious questions arise: where did these suggestions come from, and what exactly was their purpose? The origins can more easily be traced: for example, the Chinese umbrella with revolving seats was clearly modelled on the feature installed at the Petit Trianon by Marie Antoinette, and latter-day versions of this became a standard item at pleasure parks and fairs. Yet the very provenance of such ideas in elite and aristocratic sites raises the issue of what meanings could have been attached to them as the vogue for these items spread more widely and to less learned situations.

An earlier and very important publication, issued serially from 1776 onwards by Georges-Louis Le Rouge, comprised small paper-backed collections of engravings, or *cahiers*, variously know as *Jardins anglo-chinois* or *Détails des nouveaux jardins à la mode*.[3] These were not pattern books, but collections or records of various gardens throughout Europe and China, totalling nearly 500 engraved plates. It is an impressive record and repays careful study. Le Rouge borrowed, not always accurately, illustrations and plans of

165 Gabriel Thouin, *Plans raisonnés de toutes espèces de jardins* (1828).

famous gardens in England – Chiswick, Stowe, Esher, Claremont and Kew – and he pillaged texts and figures from William Chambers, George Bickham's *The Beauties of Stowe* and William Wrighte's *Grotesque Architecture*, among others. He also presented French gardens like Ermenonville, Monceau, the Désert de Retz, Rambouillet, Maupertuis and Montefontaine. He also devoted four whole *cahiers* to Chinese gardens (nos. XIV to XVII), which must in themselves have stimulated a whole lot of Chinese lookalikes in Western gardens.

166 Victor Petit, project for a belvedere, from *Habitations champêtres* (c. 1850).

Le Rouge was concerned, clearly, to broadcast the work of different up-to-date gardens *à la mode*, and they must in their turn have had a considerable impact on those with no first-hand knowledge of the cultures depicted. Yet it is difficult to know how well informed were those who perused his *cahiers*, or how, if at all, they interpreted the various *fabriques*. Indeed, the net effect of Le Rouge's many images is to celebrate the considerable range of *fabriques*, their eclectic styles, their often busy ground plans and their distribution throughout many countries, but to leave aside any larger explanations or any adjudication of meaning. The general result of his publication was focused (if at all) upon its future usefulness, rather than upon any thoughtful and discriminating commentary upon earlier sites. Two examples will serve.

Le Rouge was right to take notice of the extensive gardens at Stowe, already a well-known English landscape with a considerable number of temples and other monuments, on which published discussions had already appeared. Indeed, over a decade before Le Rouge featured engravings of Stowe in 1776, Horace Walpole noted both Stowe's profusion of structures and at the same time distanced himself from them. (Similar protests were voiced elsewhere: a visitor to the Folie Saint-James at Neuilly-sur-Seine in the late 1780s thought its *fabriques* would be 'very beautiful, if there was less of them'.) Yet when Le Rouge illustrated Stowe in the fourth *cahier* he noted only a scant collection of various columns with their dedications to specific persons – Lord Cobham, General Wolfe, Captain Grenville, George I, George II, Queen Caroline – he made no mention at all of the Gothic Temple (Temple of Liberty) or the Temples of Friendship, Ancient Virtue or the British Worthies. He indicates these by numbers on a plan, but makes no effort to explain or illustrate what they could have signified. Indeed, his general presentation of designs throughout his engraved views is focused largely on their location in such and such a garden, identifications (via captions) but no explanations, and (by implication, because we are looking at them) their formal styles.

Another site to which Le Rouge gives, rightly, much prominence in his thirteenth *cahier* was the Désert de Retz. This is provided with a plan, and in this case with a fairly complete repertoire of its *fabriques*. Like other expositions in the *cahiers*, Le Rouge is not concerned with why those particular items have been selected and gathered there by the owner, Monsieur de Monville. With hindsight, we can now respond to this wonderful place, understanding it in a variety of philosophical traditions, as Michel Baridon has done.[4] Monville arguably organized his retreat after 1774 to invoke a cluster of historical and cultural references: twenty items were devised and inserted, including a Chinese Pavilion built of teak (illus. 167), a pyramid (doubling as an ice house), a Temple of Pan, a Gothic church (a genuine ruin), a Turkish tent and the Broken Column (see illus. 2). The truncated enormity of this last feature suggested (not unlike Claes Oldenberg's gigantic half-buried tricycle would do later in the modern Parc de la Villette) that the parkland was long ago inhabited by a race of stupendous beings; indeed, the Scotsman Thomas Blaikie was put in mind of the Tower of Babel. Lovingly restored in recent years, the Désert de Retz is still a place of mysteries, 'at the antipodes of ordinary life', in the words of Yves Bonnefoy. It recalls cultures which haunt our consciousness, yet remains emphatically modern in its determination to characterize this particular place and time by a celebration of one man's wish to transcend them all. The broken column at the Désert de Retz is at once a colossal absurdity and visionary poetry, and no wonder it has attracted a range of enthusiasts from Thomas Jefferson to Gustavus III of Sweden, André Breton and the Surrealists.

The role of theatrical designers in devising and filling these parks and gardens is intriguing. Theatrical stage designers have been involved everywhere in creating several Picturesque gardens – for Venice's Public Gardens after the fall of its Republic,[5] or much later in Genoa, where the scenographer Michele Canzio designed the park of the Villa Pallavicini between 1840 and 1846. His invention was impressive, and much of it is still visible as a public garden: an Egyptian obelisk with hieroglyphs and crocodiles in the grotto, Chinese bridges and pavilions, Turkish, Indian

167 The Chinese House, Désert de Retz, from Alexandre de Laborde, *Description des nouveaux jardins de la France et de ses anciens châteaux* (1808).

168 The Naumachia at Parc Monceau, Paris.

and Hellenistic structures, Etruscan motifs and medieval castles and tombs, along with carousels, see-saws and a miniature Ferris wheel. Another earlier theatre designer, Louis Carrogis, known as Carmontelle, also contrived the Jardin Monceau in Paris for the Duc de Chartres, for whom he had previously worked to provide festival events: like the impulse behind the Désert de Retz, he said it mirrored 'all times and all places', a bewildering assemblages of sceneries each focused around suitable structures – Dutch, Turkish, Chinese, Egyptian, ancient Roman (the Naumachia still survives, illus. 168, but not the Temple of Mars), cheek by jowl with agricultural farms, vineyards and a chemical laboratory that led into greenhouses. The whole was so astonishing, ambiguous in its mixture of frivolity and zeal, that the need to claim it as a site designed for Masonic rituals has been irresistible (a similar claim for the Désert de Retz is far less convincing). Whatever subtext is plausible, the effort of the whole was to invent variations that would outdo any merely local and historical understanding of the site's *genius loci*. Today the Parc Monceau (as it is now called) evinces, if at all, a different sense of place: well-to-do Parisians walk their dogs and their children play amidst a few relics of this pre-Revolutionary fantasy, in a green space re-invented for them by Baron Haussmann in the nineteenth century.

For *homo ludens*, it is, of course, important to allow 'play' to be serious, as well as frivolous, or

169 An 18th-century print of the Temple of Modern Virtue and Walpole's headless statue at Stowe, both since demolished. This Temple and statue were designed to serve as a caustic counterpoint to the adjacent and noble Temple of Ancient Virtue.

plain silly. To the extent that sceno-graphic designers with experience in inventing settings for theatrical plays or devising festival events were involved in the creation of these gardens, some type of play was usually involved. Stowe or Désert de Retz had, beyond doubt, serious scenarios – political and philosophical – whereby the apparatus of their inventions contrived to make ideas come alive. But that does not mean we should pull a solemn face when confronting William Kent's Temple of British Worthies (with the memorial to a greyhound on the rear), or should not smile at the idea of a headless statue of Robert Walpole once placed next to the ruinized Temple of Modern Virtue (illus. 169). Yet when Le Rouge played down or even neglected meanings in the various landscapes he published, he was encouraging a display of less strenuous, fanciful or whimsical effects. That was even more the case for those who pillaged the pattern books to indulge fantasies or local attitudes on their home ground. And playful items, however facetiously offered, serve to betray as much as proclaim an owner's wishes and ideas. This applies as much to the design of monstrous mansions as to suburban gardens (like the snobbish one in Jacques Tati's film *Mon Oncle*, with its fish fountain turned on only for important visitors). They can be hugely disproportionate (large Chinese porcelain lions guarding a modest suburban driveway) or unintentionally revealing of their inhabitants (where I live, in a suburb proud of its garden appearances, is a front garden covered completely with black plastic, on the middle of which has been placed an urn containing a dead bush).

Much modern attention to follies usually assumes that they are 'big, Gothick, ostentatious, over-ambitious and useless . . . with a wildly improbable local legend attached'; their significance, if any, would lie in the 'eye of the beholder' (both citations from the National Trust's *Follies: A Guide*). These are not very appealing, let alone strenuous, assertions. And while a folly or *fabrique* in a garden could be, by turns, silly, vain, trivial, foolish or fragile, it is more interesting – and, yes, more fun – to review a few occasions in which these devices provide a magical *open sesame* to a permanent world of imaginative scenery.

The habitual rebuke of follies, as in the National Trust's guidebook, misses two important points: the identification and presentation of some *genius loci*, and what, in rhetorical usage, is called *prosopopoeia*. *Prosopopoeia* was a device by which a poet or orator imagined something or someone in the landscape as speaking directly to a privileged visitor or passer-by: a famous example would be when Moses, in the Bible, is told by a voice from heaven to put 'thy shoes from off thy feet, for the place whereon thou standest is holy ground'. The idea of some genius or spirit of place discovered in, or that could be attached to, many landscapes, allowed localities or landowners to mark certain places with significance and meaning. The communication of such significance in a particular site can be found, most notably, in memorials – as when the shepherds in Nicolas Poussin's *Et in Arcadia Ego* discover a tomb, or when Thomas Gray's poet muses upon the gravestones in a 'Country Churchyard'. In this way, many items devised and placed within a garden could advise the passer-by to attend to their 'message', deducing this from a scrutiny of their form, inscriptions or other labels, and the site itself.

Gardens associated with an important owner (Versailles) or a long tradition of ownership (Rousham) could mark such places with a variety of items that would speak to and identify their importance. Sometimes these sites yielded self-evident and local 'messages', prompts and triggers that seemed both natural and appropriate to the place. Other sites, without lineage or local significance, had however to contrive an apparatus of signification, through which visitors could respond to the place, or through which even owners could reassert and find reassurance in their ownership and location. This is primarily what follies, *fabriques* or other garden items were called upon to do. They were a lens through which the place was viewed and understood. They could be devised and carefully studied, and would profit the thoughtful visitors by what they were able to communicate. Some might well be boastful; others pedagogical. Even if contrived, they could also be thoughtful and usefully provocative. Others could just be fanciful, playgrounds for the curious.

The use of *fabriques* for a truly pedagogical purpose was outlined by Joseph Spence in his *Polymetis* of 1747. Spence was a professor of poetry and later of history at Oxford University, and a gardenist much involved in contemporary projects; his idea in *Polymetis* was to invent a landscape filled with temples and other edifices, all designed to educate a generation that he felt was exceedingly lacking in knowledge of Roman gods and their mythology. Each building would celebrate a given deity, and provide it with a whole reference collection of medals, statues and other descriptions:

> You see that Rotunda, with a Colonnade running around it, on the brow of the hill? Within that, are the great celestial deities . . . The statues are placed in niches made for them, and ornamented with copies of such antient relievo's or pictures as relate to them . . . [also] drawers, to put in the medals, gems, prints and drawings . . .

An equally moral if more sentimental mode of pedagogy was the valley landscape actually created

170 The monument dedicated to 'Gothic Friendship', from W. G. Becker, *Das Seifersdorfer Thal* (1792).

at the Seifersdorfer Tal, near Dresden, in the early 1780s (illus. 170). Here, amidst forest, meadows, streams and cliffs, were scattered temples, inscriptions and portrait busts, altars to Truth and to Virtue, a Monument to Gothic Friendship, and a rich anthology of references to Petrarch's passion for Laura, Goethe's young Werther, Edward Young's *Night Thoughts* and Lawrence Sterne's Parson Yorick. Seifersdorfer Tal was readily accessible to the sentimental souls who thronged it.

Other sites were more learned and/or obscure, and some hedged about, too, with Freemasonic mysteries. At Machern, west of Leipzig, a meandering trail visited both classical and gothic structures, a Ritterburg or Knight's Castle, and an Egyptian pyramid with a Neoclassical portico. But one of the last great landscapes dedicated to a muster of buildings and cultural sceneries may have been the Freundenhain (Grove of Fields), created in the late eighteenth century by the Cardinal Archbishop of Passau at his summer residence (it changed its name to Freudenhain, Grove of Joy, in 1889). This park was open to the public, part of the diocese's campaign of good works (roads, a hospital, a theatre). During the 1780s its ravine-bisected and thickly wooded landscape was filled with an assortment of sceneries. Linked by avenues, visitors could encounter at least 24 buildings and even villages, some the work of Dutch workers. Much of the terrain is now overgrown and spoilt, but a record of all the major *fabriques* was made in 1792 by J. F. Karl, the drawing-master at the Passau Academy (his drawings were reprinted in a pamphlet of 1921, illus. 171).

Every nation, every locality, could find its suitable, plausible or implausible way of identifying the place and reaching out to attract visitors' attentions. English gardens that focused upon local or national history could at the same time celebrate the events and personages recalled there. Actual buildings and ecclesiastical ruins could reference whole histories – 'ruined choirs where late the sweet birds sang' – like Shenstone's Priory at The Leasowes, or Fountains Abbey in Yorkshire, subsequently absorbed into the park and regular water gardens at Studley Royal (illus. 172). If authentic structures were lacking, then suitable inventions could be accommodated. Painshill in Surrey has its Gothick tower and an open-sided belvedere – elegant enough, but a touch incongruous (Walpole did not entirely approve). Pope's friend Lord Bathurst built King Alfred's Hall in his park at Cirencester between 1721 and 1734 so as to have a 'truly' old piece of architecture, but also to declare his alignment with ancient British traditions of liberty that mattered enormously after the Glorious Revolution of 1688. For another landowner, Henry Hoare at Stourhead, King Alfred had a special meaning because of his local affiliations: on the ridge above his gardens Hoare erected Alfred's Tower that displayed the king's statue in a niche and an inscription proclaiming 'The Founder of the English Monarchy . . . the Giver of the most excellent Laws, Jurys, the Bulwark of English Liberty'. (Thomas Hardy got the point exactly when his poem 'Channel Firing' of 1914 linked 'Stourton Tower' with 'Camelot, and starlit Stonehenge' as British sites to be protected in time of war). Nonetheless, Hoare added other Gothic elements that were not specifically local, like the genuinely medieval market cross from Bristol and another authentic item called St Peter's Pump, along with stained glass from Glastonbury Abbey, all of which joined the existing parish church in a declaration of long-established Englishness.

Around 1800, French landscapes seemed particularly invested in recounting the *longue durée* of cultural history, as at Retz, or the current state of national politics, which sustained the estate of Ermenonville. There the Comte de Girardin created scenes in the manner of different painters and with inscriptions that referenced a cluster of writers from Theocritus to Gessner and Rousseau. But his strong political convictions directed many other installations: a Temple of Modern Philosophy, left incomplete, with columns lying on the ground to be erected when philosophy had realized its full potential (illus. 173), unique among follies in calling attention to the future, not just the past. No less prospective and modern is the modest, vernacular cabin built for Jean-Jacques Rousseau to inhabit in the wilderness, and after the philosopher's death at Ermenonville the landscape turned elegiac with his tomb, designed by Hubert Robert and installed on the island of poplars. But Girardin used other insertions – the Prairie Arcadienne, the Liberty Tree, archery butts where local people could practice, examples of primitive architecture and prehistoric burial chambers, as well as memorials of Rousseau – to make of his estate a place where the peasants

1 Cabinet de Porcelaine Chinoise
2 Veauxhall
3 Gloriette a l'americaine.
4 Niche de Rocher
5 L'Erémitage.
6 Le Petit Sallonet le jardin de Rose en arriere de l'érémitage
7 Amphitéatre du Carousel
8 Terrasse in dem Parc nächst Freudenhain zu Passau
1 Le Pont volant à la suisse.
3 L'arcade du grand Pont et la Cascade
5 L'allée des Soupirans et la Grotte de Pluton.
7 La Triple allée qui Conduit à la Terasse du Belveder

171 A facsimile of the 1792 drawings by J. F. Karl, the drawing master at the Passau Academy, for the park of Freudenhain at Passau, 1921.

172 Studley Royal, with the ruins of Fountains Abbey in the distance as an 'eyecatcher'.

were well regarded and educated and where a propitious socio-political future could be projected.

The European fascination with garden structures, then, provided a serious purpose as much as it loved a display of invention and frivolity. Sometimes it is difficult to get one's mind beyond the amazing scenarios or, if that is appropriate, to see how to take a serious meaning from their structures. As we know, foolishness is never-ending and the fun of folly doesn't go away. So it is no surprise that modern architecture has played with the revival of the folly. The eccentric Lord Berners erected a Gothic Tower, 110 feet high, at Faringdon House in Oxfordshire in 1935. Garden festivals have witnessed a resurgence of the folly, its invention doubtless sustained by the knowledge that its life on such sites is short and the impact required must be big. For the Dutch biennale Floriade of 1989 a dozen young architects proposed a set of follies, while their publication also traced a concise and sympathetic history of the genre. But without even the opportunity of a temporary site some architects have enjoyed playing with eccentric forms that evince no palpable usefulness or plausible geography: in 1983 galleries in both New York and Los Angeles hosted an

173 The Temple of Modern Virtue at Ermenonville, to the northeast of Paris.

174 Rousseau's Tomb on the Island of Poplars at Ermenonville.

175 The Charcoal-Burner's Hut at Ermenonville.

exhibition on *Follies: Architecture for the Late-twentieth-century Landscape*, for which Rizzoli published a catalogue of proposals by nineteen architects. In the introductory essay Anthony Vidler championed the folly's now familiar ambiguity ('at best sublime and at worst frivolous'; p. 13), celebrated its perverse internal logic ('withdrawn from the world . . . in a sense pure') and argued for its relevance as providing an asylum for 'the forbidden, for the repressed, for the denied and the absolutely impossible'. That appeal at least undertakes to extend the genre as Freudian follies *de nos jours*.

Bernard Tschumi was among those exhibitors in 1983, but the follies he actually was able to erect at Parc de la Villette (conceived for the international design competition at the same time as the American exhibitions) are sited in a real place. The grid of bright red follies (illus. 176), a colourful but empty gesture to many people, strikes me as having a shrewd, if ironic motive: the grid itself is an ideally modern format, since it can be established and extended, according to Tschumi, anywhere; at La Villette, if the grid of red follies meets another structure it simply attaches a fragment of itself to that building and moves on to the next point; otherwise it ignores the existing and historical topography. Tschumi's use of the folly, then, is a reminder of how the folly itself and the whole notion of *genius loci* seem to have been emptied of meaning (most of the structures at La Villette have no function); if the structures have any reference at all, they are to Russian constructivism elsewhere, while the grid itself is placeless and limitless. Tshumi's design seems to celebrate the

176 A line of Follies at Parc de la Villette beside the canal.

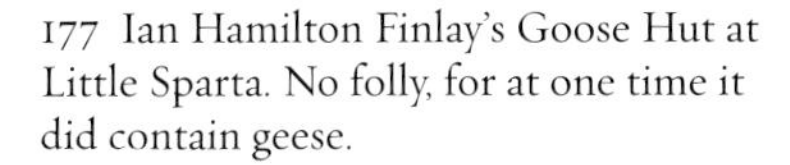
177 Ian Hamilton Finlay's Goose Hut at Little Sparta. No folly, for at one time it did contain geese.

end of a tradition, even while cynically relying upon it.

But that, fortunately, has not been the end of the matter. Recent books, like Udo Weilacher's *In Gardens*, or Louisa Jones's *The Garden Visitor's Companion*, reveal – without focusing themselves specifically on the theme of follies – the sudden florescence of garden works that rebuke, distantly and loudly, any relish for the 'natural garden' (see chapter Twelve). So, too, might other designers in these chapters, like Bernard Lassus' belvederes at Nîmes-Caissargues or Paolo Burgi's Geological Observatory (see chapter Nineteen). Just as Lassus, at Nîmes-Caissargues, incorporated an unwanted facade from the nineteenth-century opera house at Nîmes when Norman Foster redesigned it, so other derelict buildings find their role as follies in new landscapes. Gas Works Park in Seattle features the disused buildings there, while the steelworks at Duisburg-Nord has, contrariwise, become a much acclaimed and much touted folly, with the industrial remains now an exciting part of the new park (both as associative object and as facilities for outdoor activity; see illus. 187 and 235).

But here, to end, are three British sites which have exploited and relished the opportunity of inventing and utilizing structures for both their own visual forms and as bearers of meaning: Ian Hamilton Finlay's Little Sparta, Charles Jencks's Garden of Cosmic Speculation, and Niall Hobhouse's projections for the landscape at Hadspen.[6]

In Little Sparta a stone is inscribed with the words man a passerby. Visitors there pass by many other inscriptions, confront various items (gateways, the miniature superstructure of a nuclear submarine items, miniature battleships, stiles, stepping stones, a Goose Hut, illus. 177), large numbers of which are inscribed, though not everyone notices or reads them. Nonetheless it is a very readable garden, and its technique of *prosopopeia* holds out the opportunity to listen and respond to the

178 The Garden of Cosmic Speculation, Dumfriesshire, Scotland: the Black Hole Terrace and the distant 18th-century octagon, now converted to a study lab.

garden's injunctions. Finlay's insertion of his 'follies' (a dangerous word, as he very properly objected to the term as conventionally used) are small scale, since modest prompts and triggers best suit the site, and Finlay's own finances had largely militated against any more flamboyant items (a particularly large temple, planned for Little Sparta, was deemed too large and a home for it found elsewhere). Otherwise, this wonderful and provoking garden works by sly and even oblique admonitions.

A rather different presentation of *genius loci* and its rhetorical admonitions, also in Scotland, is the garden at Portrack House in Dumfriesshire, designed by Charles Jencks and the late Maggie Keswick, which plays to the full with startling installations; it is far more loudly spoken than Finlay's, and also in some ways more obscure. On the one hand, the site is filled with large-scale sculptural forms and garden layouts that delight the eye and are endlessly absorbing; the formal invention and the sheer unexpectedness of finding all this in a fairly large landscape are very exciting (illus. 178). The *locus*, the *genius*, which this garden presents, is in fact the whole world of cosmic creation; so, on the other hand, it takes Jencks's own rich and elegant book on the garden to unpack what we need to know about DNA, fractals, chaos theory, black holes, quarks, the *longue durée* and the 'jumps' of creational history that takes us back

179 The Belvedere in the woods at Hadspen, designed and erected by students at London Metropolitan University.

nearly 13 billion years; if you are not familiar (as I was not) with the new science of complexity and cannot grasp the meaning (beyond the beauty of their sensuous forms) of waves, twists, spirals or folds of the universe, or are unable to follow readily the forms and structures of the atom and the double helix, then the book is needed to expound and explain it all. Jencks makes early reference in his book to two Italian gardens – the creation narrative at the Villa Lante (the 'development of civilization' from the deluge fountain at its summit to a golden age of cultivation and design in the gardens below), and the 'esoteric and popular themes' of Bomarzo. In both those places we are confronted with a rich but also challenging repertoire of garden items, devices that stretch our skills and abilities to probe the local and universal meanings. So, as with the Enlightenment contexts out of which another garden like the Désert de Retz emerged, we may need to wait until our own age has sufficiently mastered its scientific and philosophical legacy to make full sense of the Garden of Cosmic Speculation.

Niall Hobhouse's family landscape at Hadspen, Somerset is a work in progress. It envisages re-examining fresh approaches to gardening as a horticultural pursuit and as part of rethinking what a new agricultural estate could be. A radical remake of a nineteenth-century walled garden (shaped like a parabola D, with the straight edge facing down the slope and into the agricultural area beyond) started the process, first under the auspices of foa, of inventing a truly conceptual and modern site (or *lieu*). The Hadspen project moves beyond designers and other professional specialists, even the patron/owner, beyond feeble historicism or the ruins of modernism and the pieties of even the most esteemed gardenists (Jekyll, Lutyens, Vita Sackville-West at Sissinghurst, et al.) and, above all, understands architecture as just a subset of gardening, rather than the other way round. That is where follies, broadly understood, truly become the prompts and triggers of a new vision, as perhaps they once were in the eighteenth century. For the landscape will involve *fabriques*: there is a wooden tower by Peter Smithson already there, and a belvedere in the woods by students of London Metropolitan University (illus. 179). Both Smithson's tower and two industrial containers (beautifully refurnished inside as libraries) are moveable anywhere on the estate, a distant recall of the 'dining house on wheels that could be rolled into the park and turned to face the sun' at Hesdin (see chapter Two). Now we have moveable follies, and the play continues!

15 The Invention of the Public Park

We have encountered parks for hunting and for amusement, twin ends of the spectrum that stretches from elite to popular. Situated between these is an alternative parkscape that gained a huge following in the nineteenth century and has continued ever since. These parks were firmly democratic, in two senses: they were used by a wide range of users, by all sorts and conditions of men, women and children; and they were used – consciously – as a means of inculcating the values of good citizenship, even before suffrage was extended to a complete population. This role seems as much, if not more, significant than the customarily held view that 'the urban park ... mitigates the debilitating effects of congested urban life'.[1]

Two early parks generally acknowledged in the history of public parks are the Englische Garten in Munich (1789) and the 'People's Park' at Birkenhead (1845–6), both created for a general public and unrelated to any mansion or castle. Yet long before these parks there were spaces in many cities that fed the need and were used for 'common pleasures'. Both these early versions of public parks and, more crucially, the widespread establishment of them in the nineteenth and twentieth centuries raise interesting issues about the world of gardens.

First, what *were* the debts of park-making to the idea of the garden? Garrett Eckbo remarked that 'garden design is the grassroots of landscape design ... Private garden work is really the only way to find out about relations between people and environment'.[2] Others have seen different connections that can be made between gardens and parks: the architect John Carrère, in a talk at the Johns Hopkins University in 1902, said that the experience of public gardens would lead the less privileged to 'reach for a beautiful garden of their own and eventually to a whole new demand for a city beautiful'.[3] Yet many contemporary landscape architects, while they may well design and provide gardens for private citizens, are often unwilling to see any connections between these and their call to invent public parks. As parks have also been described as 'public gardens' – an odd paradox, for we generally assume gardens to be essentially private – they still raise the issue of how the inspiration for parklands came from the world of garden-makers, who were called upon to create them in the first place: this influence is still detectable in the language if not in the forms of park-making today. Public spaces always define themselves vis-à-vis private spaces.

Second is the very issue of what constituted the public, and how its needs and requirements shaped the creation of parks; how much did locality, municipality, citizenship or nationhood impact their design? There has always been a 'public', a populace that in some form or other had to be both acknowledged and 'controlled', or to be given scope in which to pursue its own activities. Athens and Rome had gathering places of common resort, but private gardens in Rome were also opened by people like Julius Caesar for 'common pleasures' (see chapter Three). Medieval cities arranged a range of open spaces for their populations to use for different purposes: in a city like Venice, from its earliest urban formation each parish had its *campo* (literally a meadow, many of them being unpaved until the nineteenth century) and these were the focus of local citizens in that particular parish; even today in a far more secular culture, these *campi* are still the centre of local activities. But in Venice there was also the central piazza, the focus of its own government (which was the first to be paved in 1268). Typically the first truly public garden in Venice, the Giardini Pubblici, created by Napoleon in the 1810s, was an imposition of a modern, French culture upon a city that found it largely incompatible with its traditional social practices.[4]

Public parks have generally controlled visitors by both indicts and posted rules. Romans displayed a *lex hortorum* (garden laws), listing what could be done in gardens, and similar notices continue to be posted at park entrances; endless notices still tell youngsters what *not* to do – *no* ball games, *no* skateboards, *no* loud music, *no* riding bicycles. When sports were allowed in nineteenth-century parks, men and women were still segregated. Olmsted had clear ideas about what could be admitted in his parks – it took some years before a zoo and a metropolitan museum arrived in Central Park, but he did permit the provision of fresh milk from its dairy, or milk station. In 2004 the Central Park authorities refused permission to let demonstrators against the Republican Party Convention gather on its Central Lawn.

Clearly what constitutes 'common pleasures' in contemporary parks has changed fundamentally in the last hundred years, and it varies widely from place to place and is qualified in different circumstances. But whether the idea or formal concept of public parks has been affected by these changes, apart from catering for specific sporting activities, is less clear. Early commentary on spaces used, if not specially designed, for public consumption insists upon things that have hardly changed at all. When Colbert was persuaded in the late seventeenth century to open the royal Tuileries, it was on the basis of it being a place for people who had nowhere else to go: to 'recover from illness [in] the fresh air; others come to talk business, marriage and all those subjects more easily discussed in a garden than a church'. A Swiss visitor to London, Béat Luis de Muralt, noted in 1726 that 'People go there [St James's Park] to get rid of the Dirt, Confusion, and Noise of the Great City, and where Ladies in fine Weather display their ornaments'. He also described it as 'a large extent of Ground with Walks set with Trees all around' that, though within the city, 'we imagine 'tis in the Country'.

Third, if early parks were in some fashion or other a presentation of 'nature', how has that concept been sustained in gardens of the late twentieth and early twenty-first centuries? Do we still see them, and indeed cling to them, as the open, green areas for relaxation and recreation they began as – everything for which Colbert realized public gardens were for, as Luis de Muralt acknowledged in St James's – or do we assent to Adriaan Geuze's rhetorical claim that 'there is no need for parks anymore . . . The park and greenery have become worn-out clichés'?[5]

Especially, how does the park respond to the 'nature' of the city at a time when a fashionable 'landscape urbanism' is currently much promoted in the United States?[6] Since the public park is one of *the* defining marks of modern society, how have its traditions been challenged by innovation, and what are those innovations?

As in other nations, England's first modern parks were all royal: St James's Park, Regent's Park, Greenwich Park, Hyde Park and its adjoining Kensington Gardens; Hyde Park was the scene of fierce protests in support of the Reform Act of 1867, held precisely on land that was available to the public but was not yet in the public domain. These parks and garden were usually open to properly dressed and hopefully well-behaved visitors: in France, gentlemen could even hire swords at the Tuileries, thus granting them 'instant' gentility. And it was exactly these same gardens that were carefully preserved and fully opened to public visitation after the French Revolution.[7]

But other urban areas apart from royal gardens were set out for public use. An early prototype was the Mail near the Arsenal in Paris, where the game of *Jeu de Mail* was played and where spectators walked up and down to watch it; it was shaded with lines of trees. Such installations became very popular throughout France and, of course, London had its own *malls*, which survived even when the game lost much of its early popularity. But in 1616 Marie de Medici commissioned the Cours de Reine in Paris, 1,500 metres long and 50 wide, with separate lanes for pedestrians and carriages, divided by lines of trees; this had no function as a ball field. This separation of traffic and pedestrians was an early model for Paxton's Birkenhead Park and Olmsted's layout of Central Park (illus. 180). By the following century such promenades were hailed by the amateur architect Louis De Moondran as key 'embellishments' of a city; other French cities, Toulouse in 1752, Bordeaux in 1755, Rouen in 1757 and Nîmes in 1774, all incorporated wide avenues and promenades for walking.[8]

The essential design features of a promenade were that it was a green space and that it was

180 Central Park, Footpath Under Archway, for Bridle Road South of Playground, *Third Annual Report of the Board of Commissioners of the Central Park* (1860).

suburban. Those created (as many were) on disused battlements also enjoyed further views out into the countryside; the term for these, *boulevards*, eventually migrated to inner-city avenues, generally tree-lined, but without views of adjacent countryside. The linear form encouraged movement, and there was little else to distract the stroller (no sculpture, paintings or garden features). Such activity has no real equivalent in English for *passaggiare* in Italian, or the French term for describing the place where such activity takes place, *promenade*, or sometimes *promenoir*; even the English seaside promenade lacks a certain nuance – it is a place to walk, but walking is not promenading ('taking a turn', however, or 'strolling' may get us closer). Promenading is still deemed a particularly French activity: Alain Finkielkraut, a philosopher and supporter of President Sarkozy, opined that 'Western civilization, in its best sense' – by which of course he meant France – 'was born with the promenade' (reported in the *New York Times* in July 2008). The modern Promenade Plantée, created on an abandoned railway line in Paris, makes this meaning clearer, though the advent of jogging has radically revised the concept of the promenade; the High Line in New York is certainly a 'promenade' and at the moment too short to be plagued with joggers.

London, besides malls, had its squares, which also offered opportunities for parks, though they were largely open only to surrounding neighbours, and their land was anyway leased from aristocracy. But they were distinctive features, and much admired by Napoleon III when he was exiled in London. He introduced them to Paris as 'Squares' under the direction of Baron Haussmann in the 1860s (illus. 181); his architect, Alphonse Alphand, filled out the interstices of the boulevards, what have been called the 'leftovers' of the new urbanism, with these 'Squares' (still so called by the English term, and much frequented).[9] Napoleon III was, in part, perhaps repaying the debt that John Claudius Loudon acknowledged to 'our continental neighbours [who] have hitherto

181 Six squares throughout the city of Paris, from Alphonse Alphand, *Les Promenades de Paris* (1867–73).

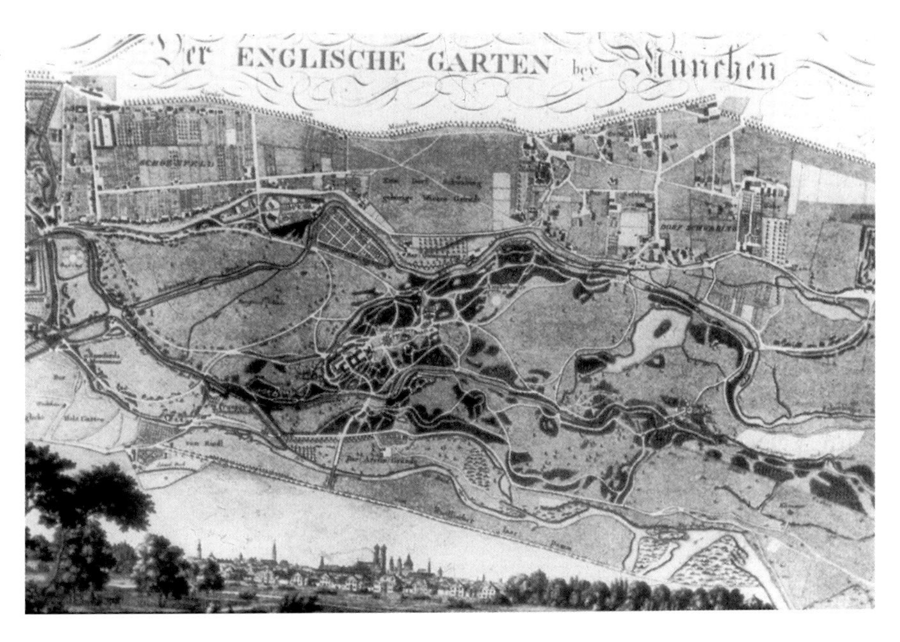

182 F. L. von Sckell's plan and view of the English Garden at Munich, 1806.

excelled' in the creation of public squares. In his histories of different national gardens in the *Encyclopaedia of Gardening* (1822, with further editions) Loudon noted the Prater and Augarten in Vienna and the Champs Elyseés, among many other public promenades, in which foreign repertoire he also included botanical grounds, cemeteries and even semi-public hunting parks throughout France. He decided that the English 'anti-social character' had hitherto prevented much attention to public parks, but hoped exposure to the Continent would 'rid' us of this handicap!

The creation of the Englische Garten in Munich in 1789, the year of revolution in France, was the work of Ludwig von Sckell and the American Benjamin Thompson (later Count Rumford, illus. 182). Its generous meadows, woodlands and water course make it an ideal pastoral country for walking, somewhat (as its name suggests) in the 'English' landscape garden mode; a marble tablet instructs its visitors to 'Roam carefree here and then return to all duties with renewed energy'. The durability of its design depends upon a combination of simplicity, robust forms and its openness to the demands of two centuries of public occupation. The park at Birkenhead is somewhat less relaxed in scope (illus. 184): surrounded by a wide bridle roadway, bisected by another road through the park, it has open lawns with small hidden lakes, the excavations from which formed mounds for the planting of mature trees (evergreens and exotics), with bridges over the water, rock gardens, formal bedding and shrubberies, all in the style of English nineteenth-century private gardens. A refreshment building was added in 1849. The park was the result of a cluster of parliamentary acts and commissions, like the Select Committee on Public Works (1833), and was a proprietary venture financed by the sale of private residential plots surrounding the park. Joseph Paxton, who was given the commission in 1843, designed various lodges at the park entrance and plans for the

183 In the English Garden at Munich.

surrounding villas, which would take their views over the parkland; he brought in others like Thomas Bayley, a gardener from Chatsworth, and Edward Kemp, who assisted in laying out the landscape and afterwards became superintendent there for 40 years. When Olmsted visited Birkenhead in 1850, as A. J. Downing did two years later, their admiration for the park inspired and promoted their plans for Central Park.

Olmsted explained his visions for public parks in a lecture of 1870: 'recollect that the[ir] object for the time being should be to see *congregated human life* under glorious and necessarily artificial conditions' (his italics).[10] But this was, as he realized, 'for the time being', and the designs were 'necessarily artificial'. These reminders, often forgotten by those who write about him, must direct our attention both to changes in cultural circumstance and to differing conceptions of how nature and artifice work together.

In contrast to what Olmsted and Vaux had established in Central Park and then in Prospect Park in Brooklyn, by the late nineteenth century New York was provided with a cluster of very un-Olmstedian small parks (see Iannacone), which by their size and their usage militated against anything as 'natural' as Central Park; they had 'greenery', but it was fenced off and to be observed from the pathways between them or from the nearby streets. They were designed to introduce the crowded occupants of tenements, where many new immigrants lived, to the benefits of the open air, swings and playgrounds (separately for boys and for girls), milk stations and shelters for nursing mothers, and opportunities for gardening and farm schools. But they were also designed, somewhat paradoxically, to give new citizens an identity as Americans, while (often) locating the parks in areas where their own immigrant cultures were predominant.

What developed, then, in New York City as a mixture of large, largely pastoral enclaves and the small pocket parks scattered in various localities, was replicated across the whole of the United States. This was not untypical of other countries, but the sheer size of the American continent and its need for an array of open spaces for common pleasures and informal business provided hundreds of towns and municipalities with small parks and public gardens; state houses and courthouses got their landscaped surroundings; academies, schools and hotels acquired established landscapes that invited semi-public use. The immense prestige and influence of the Olmsted firm and its thoughtful response to different topographies and localities has tended to put in the shade the work of many other more local park-makers, some professional, but some with local, vernacular talents (illus. 185). The range of these parks and their spread throughout the United States are astonishing, but they have rarely received much scholarly attention (apart from scattered local historians who can provide the best information of this multitude of public parks and gardens); they may not have shown much originality of design, and some were undoubtedly naive and short-lived, even if locally popular, lively and curious.

One use of these small public gardens and public spaces, wherever they were, was to celebrate local worthies (some of whom had paid for these facilities in the first place): to memorialize local celebrities, artists or craftsmen and eventually those members of the community fallen in times of war. In the aftermath of the French Revolution lessons of patriotism and citizenship were

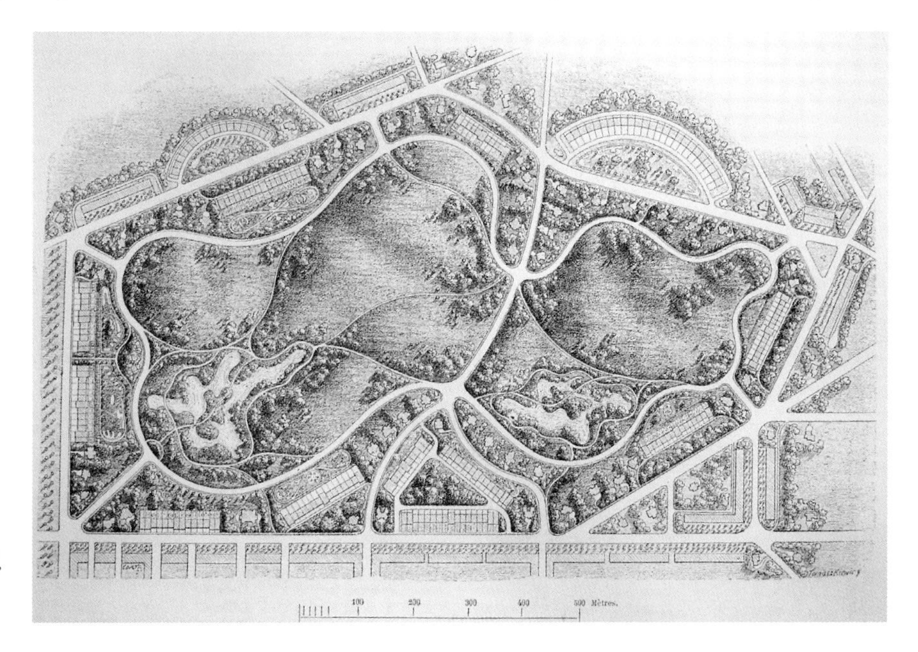

184 Birkenhead Park, from Alphonse Alphand, *Les Promenades de Paris* (1867–73).

185 'An Inviting Dream Valley' in the Concrete Novelty Gardens, Cary, Illinois.

taught through statues and other representations (like the Liberty Tree). These parks – partly memorial, but also educational and even pedagogical – had long characterized public spaces, first with royal and military sculptures, then with a much more varied cast of notables: the small Embankment Garden along the Thames by Charing Cross in London honours the Camel Corps and Gilbert and Sullivan, among several others. While the educational role played out in parks has ceased to be so deliberate and forthright in modern times (memorials being the prime exception[11]), this motive still sustains most modern park designs, as with Lawrence Halprin's general plea in his book *Cities* (1963) for a creative city that has 'great diversity and thus allows for freedom of choice; one which generates the maximum of interaction between people and their urban surroundings'.

Beyond the dedication of parks to public fitness and health, the public park has increasingly in the twenty-first century opted to improve the ecological health of the planet, trying to educate the public to understand the nature and the complex relationships of humans to their environments. They do this both in ways that only imply this perspective and those that make a clear declaration of its public role. Thus Yorkville Park in Toronto displays a cabinet of landscapes and plants from Ontario – rocks, ice walls, marsh lands, botanical plantings (see illus. 237–8) – whereas Fresh Kills Park on Staten Island is designed to reveal some of the layers of this site: once a salt marsh, then farm land, then a major New York landfill (closed, but then reopened to receive the debris from the World Trade Center) and now a public park, where recreational activities, birdwatching and the support of habitats for native plants and birdlife will be the focus for a gradual growth over the next decades (see illus. 139). Greenwich Peninsula, once the site of vast gasworks and then of the wholly forgettable Millennium Dome, is to be regenerated with what its landscape architect, Michel Desvigne, calls an 'intermediate landscape': a forest of native hornbeams and shrubs which will eventually be thinned out, with its clearings dedicated to various, as yet undetermined park activities. It is 'intermediate', too, in that the whole 300-acre site is entirely constructed on new soil to a depth of sometimes six feet, yet it will eventually look as if it is a mature and natural wood.[12]

Among the benefits of new public parks, large or small, highly designed or strongly ecological, has been that they promote and enhance the cities or localities where they were built. Philadelphia urged the gathering of various open spaces along the Schuylkill river to establish Fairmount Park as a riposte to New York's Central Park (its total area is in fact much larger than New York's). The *grands projets* of President Mitterrand in France in the twentieth century produced an astonishing cluster of modern parks – Parc de la Villette (see illus. 163), Parc André Citroën, Parc de Bercy and the Jardin (now Parc) Diderot in the Défense area of Paris (illus. 186) – all of which unmistakably augmented the local culture in different arrondissements and the amenities of the whole city. The scope and elaboration of these were truly astonishing, and the momentum continues with another park, the Jardins d'Éole in the

186 Parc Diderot in Courbevoie, the Défense area of Paris, *c*. 2000.

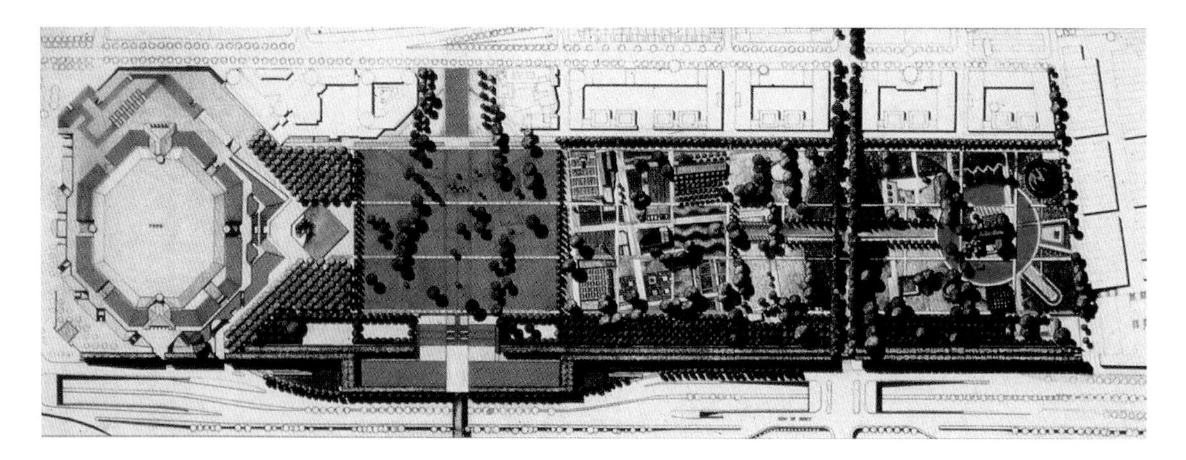

187 Plan of the Parc de Bercy, Paris, early 1990s.

eighteenth arrondissement, on the site of abandoned warehouses and railway tracks. Like the establishment of earlier parks – Birkenhead and Central Park, most obviously – some of these became a magnet for commercial and municipal housing around them. Many in Paris were also created on derelict or abandoned ground: La Villette on the site of the Parisian slaughterhouses (*abattoirs*), Citroën on the destroyed car factory and offices, Bercy on the Depot des Vins (illus. 187). Indeed, the creation of many parklands in the last 30 years has seized upon the sites of derelict or even brownfield sites where after remediation (hopefully) new parks can be laid out, like the early work (1971–88) by Rich Haag at Gas Works Park in Seattle. But Peter Latz + Partner have made the design for Duisburg-Nord Park in the Ruhr district of Germany one of the most successful and influential of these new reclamations of toxic or abused landscapes (illus. 188). It embraced much of the derelict blast furnaces, stoves, ore bunkers and railway lines for new uses rather than leaving them, isolated and unused, like the iconic stacks of the Seattle Gas Works. At Duisburg-Nord these huge and sublime remnants have become climbing walls, sub-aqua tanks, pedestrian walkways, bike paths and children's playgrounds. But along with these are a host of gardens, with flowering cherry trees, pioneer plantings and water retention basins, beds of salvia; burgeoning plants, native and exotic, are growing on the slag heaps, and in enclosed gardens, one of which has a parterre of wavy lines in box (used in Latz's own garden at his home in Ampertshausen).[13]

As Duisburg-Nord design makes very clear with its innovative reuse of garden design and planting, an underlying issue in modern parkland design has been what forms they should take. When Thomas Whately's quintessential book on English gardening was published in 1770, he nonetheless appended to its French translation a year later a note acknowledging that regular, axial and geometrical layouts were better suited to public gardens. Yet this suggestion did not prevent a flurry of sinuous pathways, meandering streams and random planting throughout new parks, like Birkenhead Park, or the re-designed Bois de Boulogne in Paris in the 1860s, where straight avenues of the former hunting park were transformed by

188 Landscape park of Duisburg-Nord, designed by Latz+Partner, 1990s.

Alphand into a mazy whirlwind of paths and carriage routes. Sometimes this format was able to work better in large parks than when employed in small private gardens, where the wiggles and meanders were just silly. But what Alphand did to the Bois was to turn its ancient French avenues into a modish 'English' layout; the stark confrontation of these rival styles has cast its shadow over too much park-making, and the either/or options of nationalist formats still haunt both theory and physical layouts.

It started early on when the American Henry Tuckerman condemned French geometry in parks for perverting Nature and deplored how underestimated was the French sensibility for the 'unadulterated' charms of nature. Similarly, the French author Edmond de Goncourt, hating the axial grid of Haussmann's new Parisian boulevards, thought they were a horrible prophecy of 'some American Babylon of the future'.[14] Conversely, Thomas Hastings thought that a 'rural' parkscape was antithetical to the city and that urban parks and recreational areas should be 'redesigned to conform and harmonize with the rigid lines of the neighbouring streets'.[15] Olmsted cleverly compromised throughout Central Park, which was established within its strict (and now urban) rectangle, but with both pastoral meadows, gentle hills, irregular lakes and a mazy Ramble hidden in the thickets, and a wide tree-lined Mall leading to the Bethesda Fountain and its plaza (he felt the need to defend this straight crib from French design by saying that it did not, in fact, lead to any mansion). Adriaan Geuze thinks that 'greenery' is no longer called for in parks, English or French, and in the spirit of what was, one hopes, merely a provocation he designed the Schouwburgplein (Theatre Square) in Rotterdam without any greenery at all: relatively empty, it is austere and even unpleasing, yet when filled with activity and crowds at night it is something else entirely. His other impressive works, published in *Mosaics West 8* (English edition, Basel, 2008), also make the greenery remark astonishingly bizarre and implausible.

The question of 'nature', though, is not one simply of greenery, plants or perhaps some homage to 'French' or 'English' layouts. As proposed for both Fresh Kills and Downsview Park in Toronto, flora, fauna and geology are meant to have a pedagogical or civic function beyond the mechanics of design. In part, these two drew out whatever the locality has evinced and may once again proclaim on sites that had been taken and ruined by landfill or military installations. And this has frequently marked successful parks: in Vienna, the Türkenschanzpark took its cues from the alpine landscapes to which its local trains took visitors; Buttes Chaumont may have disguised its desolate quarries, garbage dump and gibbets for hanging felons, but it still parades the cliffs and caverns of the original site; the Jardin Atlantique in Paris, constructed over the platforms of the TGV trains that take us to the Atlantic coast, plays with the whole imagined world of boardwalk, cliff caves and pathways, seaside planting and a necessary weather station (see illus. 236); Lawrence Halprin abstracted the scenery of the neighbouring Cascade Mountains for his sequence of fountains in Portland, Oregon (see illus. 250); and Dan Kiley exploited a Texan swampland for his wonderful Fountain Place, hemmed in by offices, in Dallas (see illus. 148).

Though there was a tendency in modernist landscape architecture to eschew locality and plonk down a park that could be anywhere in the world, usually designed by some global-trotting landscape architect who had mastered a homogenous template that can be exported to any international location, this regrettable tendency was never the real dynamics of park-making. Downing and Olmsted looked for precedents to European parks, in England and Germany especially, since there was little else for them to copy; but the success of the latter (Downing died too early to make a similar mark) ensured that he crafted a distinctive American landscape from a selection of ideas borrowed from abroad as well as from distinctive domestic landscapes (like the Hudson Valley). When Olmsted was summoned back in 1865 from his work in the Yosemite to resume his partnership with Calvert Vaux on Central Park, he was called upon to translate 'the republican art idea in its highest form into acres we want to control'. Translation and control were the twin issues, selectively using both the foreign and the wild native landscapes for largely urban forms that provided a theatre of civic responsibility and behaviour. Olmsted knew, but did not borrow, the 'gardenist' ideas from England and the work of J. C. Loudon, because he considered that this style had no place in 'the *park proper*' (his italics). Borrowings from abroad have always inspired and then been translated into American landscape: Japanese garden culture made a considerable impact on the western United States, while a renewed piece of inspiration from the Luxembourg Gardens in Paris made the refurbishment of Bryant Park in New York City between 1986 and 1991 an instant success, with its gravel surfaces and dozens of individual seats (rather than the standard park benches with which Olmsted had lined the Mall in Central Park, and which were still to be imaginatively reused in Battery Park City (by Olin) and in Javits Plaza (by Martha Schwartz).

That is not to say that parks have always been designed with imagination wherever the need for open space and recreation were needed. Too much park-making in the twentieth century has been prosaic and utilitarian, yet is not without its greatly appreciative users. Amenities without imagination – po-faced, earnest, even ardently ecological work – has been challenged by some truly imaginative work, including a suggestive invocation of the theatrical potential of parks and gardens that recalls work in the Renaissance. But I postpone to the penultimate chapter a discussion of what I would call the prose and the poetry of modern landscape architecture.

One development of landscape parks (leaving aside the mania for memorials) that needs recording, since it takes up another Renaissance prototype, is the prevalence of sculpture parks and gardens. Many museums now have sculpture gardens and parks, not least because much modern sculpture is huge, not easily accommodated inside museums, and is anyway designed to be exhibited in the open air. These have elicited a need for a wholly new kind of park, sometimes using agricultural land, botanical gardens or old parks and woodlands: Finlay's 'Sculpture' Park (so-called by the municipality) at Luton; the Botanical Gardens in Wellington, New Zealand (illus. 189); the Canberra Sculpture Gardens in Canberra, Australia; the Kröller-Müller Museum Sculpture Garden at Otterlo in the Netherlands, designed by WEST8 with a chequerboard of different gardens, meadows and woodlands. Sculpture parks like the Sculpture Courtyard at the Museum of Modern Art in New York, the sculpture gardens at the Australian National Gallery in Canberra, Storm King in New York, the Parco di Celle near Pistoia in Italy, the island of Naoshima in Japan, or outdoor exhibition venues like the Fortezza di Belvedere in Florence or Battersea Park in London – all deploy collections of sculpture. Landscape architects have played a big part in designing or modifying these landscapes: Noguchi designed the Lillie and Hugh Roy Cullen Sculpture Garden at the Museum of Fine Arts in Houston; Peter Walker the Nasher Sculpture Center in Dallas, Texas; Tadao Ando's works at the sculpture park on Naoshima. The National Gallery in Washington, DC, has a new sculpture park by the Olin Studio, and the Seattle Art Museum a new Olympic Sculpture Park by Weiss/Manfredi Architects, with a sloping platform cantilevered over roads and a railway line as it descends to the restored shoreline; the Walker Art Center in Minneapolis has new designs by Michael Van Valkenbergh and Michel Desvigne for two different sculpture gardens: one contrasts four garden-like enclosures, allées and a winter conservatory with the naturalistic field dominated by Claes Oldenburg's *Spoonbridge and Cherry* fountain; the second is a new woodland garden by Michel Desvigne. The J. Paul Getty Museum has received the Ray and Fran Stark Collection of twentieth-century sculpture, carefully placed within its grounds

189 Wellington Botanical Garden, New Zealand.

190 Shanghai Botanical Garden, opened in 2010 at Chenshan, southwest of the city, comprises a vast territory (2000 ha), with over 30 different thematic gardens, an anthology of different zones and species, conservatories, and a flooded quarry, seen here, to which a tunnel descends and leads visitors over a floating walkway.

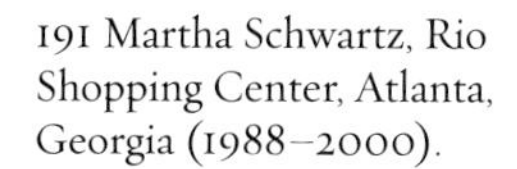

191 Martha Schwartz, Rio Shopping Center, Atlanta, Georgia (1988–2000).

under the auspices of the firm that designed the original campus, the Olin Studio.[16] Indeed, sculpture collections seem to be the current *park du jour*, and this appears to have meant a considerable decrease in random statuary in other public gardens. And there are also some more witty, tongue-in-cheek versions of outdoor sculpture. west8 has also – to the amusement and delight of many a passer-by – installed three huge (8 metres high) inflatable cows in the pastureland along the Dutch autoroute A2. Martha Schwartz, early in her career, decorated a pool in a shopping mall in Atlanta (now derelict) with a quincunx gathering of golden frogs, bought locally at a garden centre (illus. 191). More recently she has taken this motif a little further by peopling a garden in Westphalia with 51 garden ornaments – pigs, windmills, gnomes, miniature well-heads, sheep and flamingos. Not a sculpture garden exactly, but an ironic take on the more serious, even solemn, sculpture gardens that many museums have established; the German garden site was also temporary.

The public park lives historically, metaphorically and symbolically between nature (or natures) and the city; these two conditions were represented extremely early in human history by the Garden of Eden on the one hand, and by the Tower of Babel on the other. Eden was a privileged private landscape for two people, to whom were granted a benign and fruitful plenitude of natural resources and almost unimpeded access (the prohibition of one tree seems a modest injunction compared to the list of prohibitions posted in public parks today). The Tower of Babel was the attempt by scattered tribes, wandering the world since the banishment from Eden, to build a city with a huge tower that would accommodate them all; but the Lord God 'dispersed them . . . all over the earth', without the benefit of a common language. For us, it is the public parklands where we gather in or near cities; places held in common, often without any shared language for either their design or reception, but nonetheless with a shared sense that parks are still somehow sustained by a distant memory of

Eden. How *that* nostalgia is conceived is as various and as democratic as the parks themselves. For some, though, that idea of a pristine, Edenic vision seems less satisfactory in public and above all urban gardens than in the wilder places of national parks, to which we now turn.

16 National Parks and International Exhibition Gardens

National and regional parks for public consumption, along with international exhibitions, were an invention of the nineteenth century and are now spread worldwide. Yet each in its different way involved some transmutation of earlier garden impulses and landscape forms and provided modern experiences of new landscape opportunities. National and regional parks were essentially set aside as identifiable areas, usually places of considerable beauty, though not enclosed terrains as earlier hunting preserves had been; now they were democratic and open to all, and one of their new obligations was supposed to be the preservation of animal and plant habitats. The parks in towns and cities that flourished everywhere by the end of the nineteenth century also had the same democratic agenda, but they were constantly compromised by the social and civic requirements of their clientele; yet even the largest, like New York's Central Park or Fairmount Park in Philadelphia, were far smaller, less diversified and less 'wild' than the territories which came to be earmarked for national and regional parks. Yellowstone and Yosemite, for example, were also differentiated by their remoter locations, often situated far from urban centres. These larger entities were often revered, especially by their early promoters, as places of emotional and spiritual refreshment, and they became for many the sacred places of a secular culture. If a history of the garden begins with the Persian *pairidaeza*, rendered into Greek by Xenophon as *paradeios* (see chapter Two), one of its last manifestations is the large-scale national or regional park.

International exhibitions also manifest a debt to earlier gardens and landscapes, but it took a different form. Their basic structure was a series of pavilions deployed throughout a landscape that was itself part of the exhibit and had in many cases been laid out by professional designers. Their scattering of pavilions, often devised in different nationalistic forms and materials, was a commercial replay of the picturesque vogue for exotic follies and *fabriques* dispersed through an appropriate landscape. The whole point of such exhibitions was their overall promotion of the host nation, as well as the display by individual, participating nations of their own country's culture and industry. Some of these were set in large open parks, like the Crystal Palace in Hyde Park and later at Sydenham, Fairmount Park in Philadelphia for the 1876 Centennial International Exhibition, or Chicago's Jackson Park, the site of the World's Columbian Exposition in 1893. While these exhibitions did not bear much physical resemblance to earlier parks and gardens, not least because of the size of their inserted structures, the motive for their existence recalled earlier designs where some political ruler could show off his prestige and power by displaying a range of various garden buildings or collections, like the Menagerie at Versailles. These ranged from Arabic rulers in Spain, Medici villas in the Tuscan territories, Hampton Court Palace in England and Paleis Het Loo in the Netherlands, to later manifestations by principalities in various German states. Nationality and political power were as much part of the performance of these places as were their fountains, pavilions and collections.

Furthermore, although the parallel cannot be pushed too far, the design thinking behind both national parks and international exhibition sites and how they were used or experienced by the public evolved from earlier expectations and habits in other landscapes. The initial response to Yosemite was that it was like a garden or park (see chapter One); such presumed familiarity not only helped early visitors to cope with its astonishing and demanding sublimity, but also drove the very physical interventions that would give people access to it. Trails and paths, designated places to stop and admire key views, designated routes, the marking and naming of features throughout the valley, and even the gradual establishment of rules and regulations for behaviour were all elements of earlier landscape parks. The control of behaviour had its distant ancestor in the Roman *lex hortorum*, which provided specific exhortations on how to behave, and still has its modern counterpart in notices that instruct visitors how to behave in public parks (no picnicking, picking flowers, skateboarding and so on). What both designers and their visitors in national parks were, of course, not able to count on was the complete control of activities within each designated area: national parks were not museums, and many local inhabitants and much daily life continued within their precincts, while the behaviour of the daily influx of tourists was never the same as that of visitors, however numerous, on privately owned estates. The sheer number of visitors to exhibitions far exceeded that of the tourists who frequented private estates; but their numbers did approximate the crowds who thronged public gathering places and admired the display there of monuments and statuary.

When European nations established national and regional parks, they were critically aware of the head start achieved by the United States a hundred years before in the creation of Yellowstone and Yosemite. Contrariwise, Europe was ahead of North America in its promotion of exhibitions, though this was a development that the States would later seize with alacrity and enthusiasm. Such developments suggest a different narrative logic for this chapter.

ONE

While Yellowstone was the first national park to be created by the Federal Government in March 1872, it was the discovery of Yosemite in the 1850s and the subsequent opening up of its Valley that paved the way for the development of national parklands.[1] Indeed, the handing over of Yosemite and the neighbouring Mariposa Big Tree Grove in 1864 to be managed locally by the State of California rather than by Congress simply delayed its final establishment as a national park until the 1890s. Furthermore, Yosemite has the curious and perhaps not wholly accidental circumstance of being briefly in the charge of the man involved with the establishment of public urban parks throughout North America, namely Frederick Law Olmsted.

In 1863 Olmsted had left New York and his five-year commitment to Central Park to take up the job of managing the Mariposa Estate, a gold-mining operation in the Sierra Nevada of California; it was in the course of inspecting for better water supplies for that enterprise that he came first into the Yosemite Valley. His vision for it was entirely consistent with his experience and thinking as a landscape architect: he saw that the two separate segments (the Valley itself, and Mariposa Grove) would need to be given a context of larger surroundings for their full effect and preservation; he would insist upon ensuring an ensemble of its sceneries ('all around and wherever the visitor goes'). This was not inconsistent with the fashion in which private landscaped gardens were accommodated within larger agricultural estates, or city parks within urban conurbations.

In the Sierra Nevada he studied and expounded the 'peculiar character' of the place (its *genius loci*), and he relied enthusiastically upon artists to shape how its features could be viewed and understood (the photographer Carleton Watkins and painters Thomas Hill, Virgil Williams and Albert Bierstadt). He was fertile in his recommendations for the Yosemite scenery, for he recognized that the art of landscape, whether experienced directly, in poetry or in painting, depended upon the cultivation of associations. He also drew attention to the most interesting views and celebrated the finest trees, as any landscape designer would do in responding to the capabilities of a given site. And in his preliminary report to the California Legislature as the chairman of the first Yosemite Commission in 1865,[2] he laid out a future vision for the national park that, however determined by the innovations required for its size and management, was mediated throughout by his thinking as a landscape architect. In fact, Olmsted was also designing two landscaping projects in Oakland, California, while still serving as a Yosemite Commissioner.

He saw the democratic potential of the national park ('devoted forever to popular resort and recreation'), just as he and Calvert Vaux had envisaged that for Central Park. This continued to be the emphasis in 1890 when the United States Congress was asked to adopt Yosemite 'as a recreation ground for the people as a public park'. And, as with Central Park in New York, Olmsted also saw that it had to be 'presented' and shaped for its visitors, so that they would appreciate its fullest potential. His Commissioners' written report obliged him to use words rather than imagery to describe Yosemite, but this necessity allowed him to make two subtle arguments: first, that no visual image would in fact capture the effects:

> No photograph or series of photographs, no paintings ever prepare a visitor so that he is not taken by surprise, for could the scenes be faithfully represented the visitor is affected not only by that upon which his eye is at any moment fixed, but by all that with which on every side is *associated* (my italics).

Second, given that associations were at the very heart of landscape experience, Olmsted could suggest a variety of them: some of these found Yosemite incomparable and unique, but others required his readers to invoke their best memories of European 'Alps and Apennines' or of Shakespeare's Warwickshire ('pleasing reminiscences to the traveller of the Avon or the Upper

Thames'). And his language, too, is inflected with associative references to designed or ideal landscapes – 'groves' and 'meadows', 'flowering shrubs' – reminding us that for someone like Olmsted, attuned to a culture of the picturesque, a word he also invokes, landscape experiences were by no means confined to the merely visual.

In determining that Yosemite should never run the danger of becoming 'private property', vitiated by an owner's 'false taste [and] caprice' as in the worst excesses of the picturesque (see chapter Fourteen), Olmsted was forced to envisage the commercial opening of the Valley and Grove to popular tourism, and for this he saw European models: Switzerland was lauded for the commercial exploitation of its scenery (above all by its network of access roads and rail services), and Munich was praised for its creation of the Englische Garten. Further, in his long disquisition upon the 'the operation of scenes of beauty upon the mind' and 'the intimate relation of the mind upon the nervous system and the whole physical economy' of the body, he somewhat strangely, given his republican emphases, cites as precedents the health and longevity of British nobility who retire to their country estates, of which there are 'more than one thousand private parks and notable grounds devoted to luxury and recreation', as well as 'the great men of the Babylonians, the Persians and the Hebrews [who] had their rural retreats'. Even as early as 1864, Olmsted realized that his democratic vision ('the pursuit of happiness' for all) should permit only a minimal disturbance of scenery; 'to obscure, distort or detract from the dignity of the scenery' was simply unthinkable, even though such intrusions were already to be seen in New Hampshire (hundreds of hotels) or the Scottish Highlands (railways, steamboats and horses). Olmsted nevertheless argued that since Yosemite had not been purchased for 'a rich man's park', it had therefore to be opened up, with transportation ('the first necessity is a road' into the Valley), moderately priced campsites and then a network of trails 'reaching all the finer points of view', in fact '30 miles more or less of double trail & foot paths'. One such, a 'complete circuit' of the Valley (illus. 192), taking 'a more picturesque course', is similar to circulation in English private and public parkland, as his insistence on reducing 'artificial construction' echoes the elimination of similar effects by European practitioners of natural landscapes on elite estates.

In surveying and proposing how the Yosemite should be opened to public usage, Olmsted clearly had in mind previous versions of private parkland: their formal design, their celebration of natural scenery and presentation of all sorts of plants, their associations with local or indeed national history (including names that referenced specific events and persons), and the role that a landscape designer should play in envisaging the whole as a collaboration of art and nature. In the years that followed Olmsted's return to his landscaping work on the East Coast (never in fact to return to the Yosemite), the Valley found itself in a tug of war between the California Commissioners responsible for its management and various individuals who, even before the 1864 Grant, had exploited the Valley for tourist purposes. This history is not the concern here.[3] But throughout the 42 years before Yosemite Grant was finally assigned to the Federal authorities, both physical and imaginative responses to the Valley reveal lingering assumptions about its park or garden character, though 'Yosemite Forest Reservation' was the designation used by a distant United States Congress. 'Improvement' was the term explicitly used by the California Governor in his 1864 proclamation,

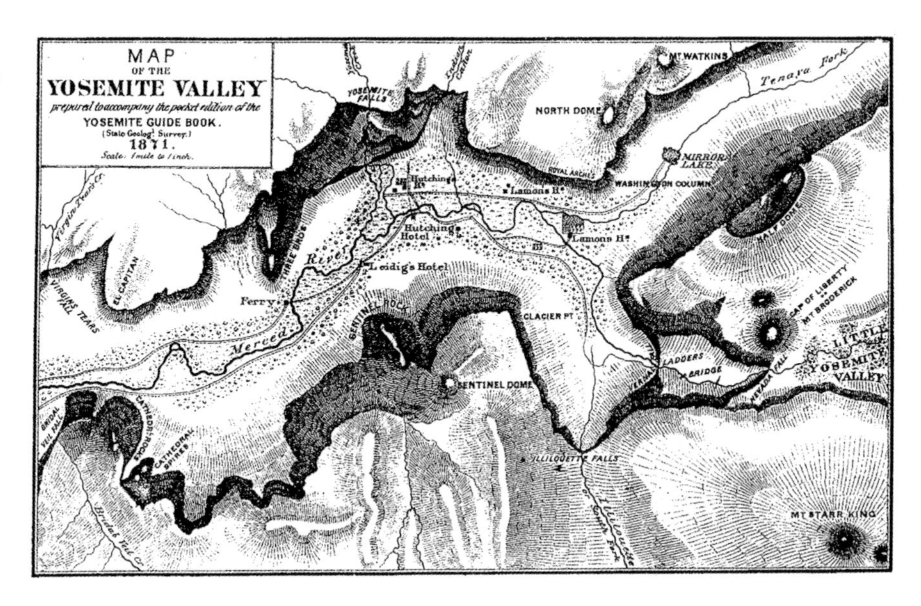

192 Early map of the Yosemite Valley's roads and trails.

a term closely connected with private estate design and its 'improvements'. And the routes by which Yosemite was approached, and more importantly the paths that were contrived within it to bring visitors to view all essential features of the Valley, were similar to the circuits and belt rides of private parklands. Yosemite required a host of different trails for pedestrians, horses, stages and eventually mechanized vehicles; it installed ladders, later stone steps, to reach the Vernal and Nevada Falls, while fixing ropes to help the intrepid up the precipitous Half Dome. There would be a boardwalk, with seats, connecting the Lower to the Upper Village, and bridges across the Merced River. Like many private estates, too, the various features were given names, some of which celebrated individual pioneers, politicians and entrepreneurs in the Valley, while others were devised to guide visitors in their responses, like Inspiration Point, the 'First View of Yosemite' notice at Buena Vista Gap erected by the Modesto Chamber of Commerce, or even the Four Mile Trail; Anglo-Saxon names almost entirely replaced Indian toponyms (Tisayac became Half Dome) in an effort to render the strange and sublime world of Yosemite more hospitable and familiar.

Yosemite also tried, mostly by accident but occasionally by at least unconscious association, to absorb some suitably picturesque elements. Many early buildings were reminiscent of garden *fabriques*, like the Big Tree Room built around a huge cedar trunk, an 'Alpine House', the shack 'perched like an eagle's nest on a very commanding crag' (in the words of John Muir) or the Mirror Lake House (illus. 193). But soon the rustic and picturesque shanties were deemed inappropriate by the Commissioners and many were removed in the late 1880s. More grandiose, bourgeois architecture was substituted, to the displeasure of some, like John Muir, who thought the very Victorian villa-style Stoneman House hotel had a 'silly look amid surroundings so massive and sublime'.

The unequal contest between landscape assumptions and aesthetics on the one hand, and the requirements and demands of a booming

193 The rustic Mirror Lake House, Yosemite, photographed in 1870 while under construction.

tourist business on the other, have taken Yosemite well beyond illuminating comparisons with landscape architecture precedents. Yet the management of the park is still necessarily focused upon how to move visitors around, how to ensure that they appreciate key vistas, and even on promoting suitable emotions and uplifting associations with its geophysical and cultural history, just as owners of a landscape park, including the English National Trust, do on their sites. And in the early years of the twenty-first century it was a distinguished landscape architect, the late Lawrence Halprin, who was called upon by the National Parks Service to review and redesign an approach to the Falls, to choreograph visitor experience through a prime segment of Yosemite scenery.

Other national and regional parks confronted some of the same issues as did Yosemite, even though their profusion and specific needs have lessened the obvious debts to deigned parklands. The European national park probably has its origins in two different kinds of culture. First, there was the territorial identification of scenic areas with a distinct and special character and which were therefore attractive to tourists – in Great Britain it was the Derbyshire Dales, the Lake District, Snowdonia, the Scottish Highlands and later perhaps the Norfolk Broads and the South Downs; in France Chamonix and Mont Blanc, the Pyrenees or the Vosges.[4] Although national parks were established in Britain only in 1949 after surveys by the Ministry of Town and Country Planning to identify appropriate sites, early calls for their creation started much earlier. In 1810 William Wordsworth's *Guide to the Lakes* argued for the Lake District as 'a sort of national property, in which every man has a right and interest who has an eye to perceive and a heart to enjoy'. Much later, on 19 March 1887, John Ruskin, who had been living in the Lake District since he bought his house, Brantwood, at Coniston in 1871, was still making the same demand: a letter to the *Lancaster Observer* argued that 'the whole Lake District should be bought by the nation for itself'. But his was also a preservationist stance: he hated the proliferation of villas ('they are peppering Cumberland out of an architectural caster with white boxes') and he energetically and successfully opposed the extension of the railway line north from Ambleside to Rydal and Grasmere (Wordsworth country) and then over Dunmail Raise to Keswick (through the heart of eastern Lakeland).

The second source of inspiration for national parks was a democratic urge to have spaces for public leisure and recreation that were open to all who could reach them, as huge aristocratic estates and hunting preserves were not. The advent and development of railways played a major role in opening up distant parklands for visitors, but the influx of urban visitors also brought its own problems. One of the claims for British National Parks in the 1949 National Parks and Access to the Countryside Act was their provision of what was called a 'creative geography':[5] namely, the 'creation' of a particular territory with its own character and organization, and a geography that, properly controlled and planned, would 'create' a response. This was as pedagogical and associative in its own way as the repertoire of triggers and prompts contrived for visitors within private gardens and landscape parks. Guidance for visitors unused to parks of any sort had also to be provided; in 1951 the British National Parks Commission drew up *The Country Code* to instruct (especially urban) tourists on how to behave in the countryside, an extension of the public notices that dictate behaviour in most European public parks and still greet visitors at their entrances.

The National Parks had been mooted ever since the height of the Picturesque movement in the United Kingdom in the years around 1800, so it was inevitable that these larger territories were experienced and enjoyed on a similar basis as for large private as well as civic parks. Indeed, as early as Thomas Whately's *Observations on Modern Gardening* (1770), he described both private parklands like Blenheim or Stowe and topographies like the Wye Valley and the Derbyshire Dales as places equally to be visited and enjoyed. Guidebooks for these Picturesque topographies were devised, access and routes laid down, particular viewsheds recommended and sometimes marked,[6] and other facilities provided, so much so that some environmental purists worried that national parks would become too compromised – 'a garden pleasure-ground (small bar attached)', as John Dower put it in the *Journal of the Town Planning Institute* for 1944. When late eighteenth- and early nineteenth-century writers had started to direct attention to these larger landscapes – a key example would be William Gilpin's Picturesque *Observations* on different parts of the British Isles, many of which regions which would turn out to be selected later as national parks: writers were aware that these topographies were different in scale and character from the landscape gardens of private estates (indeed, that was precisely their appeal). Yet in responding to them and in explaining how they could be experienced, familiarity with more compact and indeed more structured sites was often implied or invoked. So the new admiration for the wild and unmediated was nonetheless tempered by a need to assimilate it in the first place to more familiar experiences. It is interesting to realize that the Lake District was itself surrounded by great houses and their estates, which 'encouraged a specifically aesthetic consciousness of the central Lake District'.[7]

Gilpin himself provides an illuminating instance of how a popular appreciation of rural landscapes drew on the basis of experiences honed within private estates. A comparison of Gilpin's early account of his visit in 1747 to the gardens of Stowe with the first of his later observations taken on the river Wye during a journey of 1770 and published twelve years later will suggest something of this mutual experience: in the first, we see clear signs of one of the characters reading the extensively designed landscape of Stowe through eyes that bias him towards what he calls 'rough Nature'; in the later work, we find him responding to the configurations of a non-designed scenery as if he were still, at least occasionally, in a parkland like Stowe.[8]

The character Polypthon in the Stowe *Dialogue* has experience in 'visiting what was curious in the several Counties around him'; while that may include landscape gardens it clearly also takes in larger territories, for he dilates upon his travels in the northern counties. Similarly, Gilpin's narrative of his Wye excursion begins with his stopping to view the designed landscape at Caversham House, where (despite cavils) he notes that 'Capability' Brown's 'great merit lay in pursuing the path which nature had marked out', and then he includes a visit to the landscape of Persfield on the way down the Wye and several other prominent designed sites like Foxley and Hafod. Upon entering the grounds at Stowe Polypthon first responds to an extensive and various 'view', ignoring the immediate foreground of buildings and preferring what he terms 'offskip' or distant backgrounds. While some of his grumpiness is doubtless contrived by Gilpin to activate the dialogue with his companion, the enthusiasm for northern scenery seems genuine rather than provocative. What is more, he seems concerned to negotiate how that taste meshes with the experience of a landscaped garden like Stowe. Not surprisingly, Polypthon argues for bringing the countryside into the garden in a radical fashion, when he opines that were he a nobleman 'I should endeavour to turn my Estate into a Garden, and make my Tenants my Gardiners: Instead of useless Temples, I would built Farmhouses; and instead of cutting out unmeaning Vistas, I would beautify and mend Highways'. And he quotes lines from Milton's 'L'Allegro' in praise of the 'new Pleasures' of an extensive and undesigned countryside. A parallel claim is made by his companion, Callophilus; he expounds the more conventional view that 'the Garden [extends] beyond its Limits, and takes in every thing entertaining that is to be met with in the range of half a County, Villages, Works of Husbandry, Groups of Cattle, Herds of Deer, and a variety of other beautiful Objects are brought into the Garden, and make a part of the Plan'. By the mid-century this taste for such prospects was certainly learnt and enjoyed within designed parkland and gardens; but it soon ensured that this larger nature was explored for its own sake, even if it was still viewed in the light of the experience gained within such places as Stowe.

Switching to Gilpin's *Observations on the River Wye*, we may feel some distinct continuities with his experiences at Stowe. Granted, the landscape is much more extensive, the time taken in its exploration much longer, and granted, too, that his central concern is to instruct a would-be artist in Picturesque techniques and in the choice of subjects, for which unmediated territory is the true source; nevertheless he often seems to respond in a manner that parallels his earlier behaviour at Stowe. Both the landscape gardener and the landscape painter must learn how to 'perfect' nature, so it is not surprising if much advice for one also suits the other. On the Wye trip Gilpin complains of nature's lack of both foregrounds and backgrounds that would frustrate both a gardener and a painter. Throughout his observations he presents Nature as an artist herself, pro-active in 'hanging' the hills, 'correcting' and 'improving' how the topography looks. Furthermore, Gilpin will use landscaping terms – 'a noble terrace' or 'lawn' – about unmediated terrain; he will also write that a view 'unfolds itself', just as incidents presented themselves during a walk through the gardens at Stowe ('what kind of a Building have we yonder that struck our

Sight as we crossed that Alley?'). This merger of experience and response is most acute when Gilpin writes about the landscape at Hafod: its landscape obviously lent itself to his particular enthusiasm for an extended Picturesque that ensured that its beauties were accessible along different circuit paths – 'through this variety of grand scenery the several walks are conducted' (illus. 194). Yet overall, he has conducted his own 'walk' through the Wye scenery as if he were at a site like Hafod, and his route would serve as a model for many tourists treading in his footsteps, *Observations* in hand.

But it is also in his larger, more conceptual concerns that Gilpin reveals his affinities with landscape architectural theory and practice. Responding to a particular territory, he is nevertheless especially concerned to pronounce on the categories that might guide response; one such moment is his sentence 'The *ornaments* of the Wye may be ranged under four heads: *ground, wood, rocks,* and *buildings*'. This echoes Thomas Whately's categories in *Observations on Modern Gardening*, published the same year that Gilpin wrote about the Wye valley: Whately writes, 'Nature, always simple, employs but four materials in the composition of her scenes, *ground, wood, water,* and *rocks*', then adding 'The cultivation of nature [landscape design] has introduced a fifth species, the *buildings* . . .'. Both writers then enlarge upon each of those categories; in Gilpin's case the 'water' theme is adumbrated in his actual mode of travel, by boat down the Wye's 'mazy course'. Both writers address the need for 'character' in a given scene: Gilpin by lamenting that the Wye near Ross provides no 'characteristic objects', that is, structures that would determine the particular 'idea' of a place; Whately by devoting half a dozen pages to distinguishing different characters and how they may be achieved in landscape design. Gilpin does, however, accept that buildings are a better and more useful adjunct for creating character in artificial designs than in natural topographies. One sometimes even wonders whether Gilpin took a

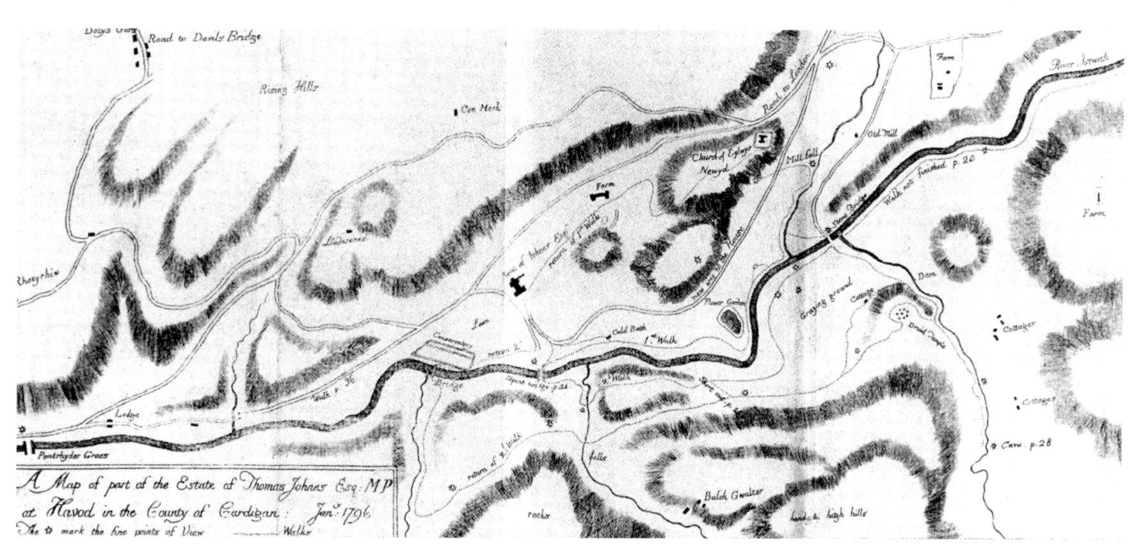

194 Folding plan of the site and walks at Hafod, Ceredigion, Wales, attributed to William Blake, from George Cumberland, *An Attempt to Describe Hafod* (1796).

copy of Whately with him on his excursion, or consulted the book while preparing the piece for later publication, especially the sections on Persfield and Chepstow.

Gilpin's is an early attempt, *avant la lettre*, to guide visitors through a particularly distinctive regional scenery as if it were an extensive parkland. The very term 'park' when applied to national park systems implies something bounded if not enclosed, a special place eliciting some regard, often lyrical, for its intrinsic character, whether topographical or cultural, a place that demands both 'protection' and 'organization'. These were precisely the dual terms of the French decree that promulgated regional parks in 1867. Just as Yellowstone or Yosemite needed to maintain a balance between protecting its natural sublimity and organizing its features for tourism ('as a public park or Pleasuring-ground'), English legislators referred to the English Lake District, among other regions, as facing similarly contradictory objectives; its infinitely smaller size, in comparison with the huge American national parks, coupled with an already peopled and working landscape, even exacerbated the tensions.[9]

One fashion in which a national park like the Lake District differed from the American examples was in the considerable literature that was devoted to this area from the mid-eighteenth-century onwards; this was also the case with French landscapes that eventually became regional centres. This written and graphic response is both paralleled by and indebted to the similar profusion of guides and collections of Picturesque engravings of country seats that flourished during the same period. Indeed, around the lake at Keswick, Gilpin felt it could be improved by being made more accessible, 'cleared of deformities', and 'it might be planted – and it might be decorated'.

195 Title page of J. B. Pyne, *Lake Scenery of England* (1859).

The imagery and written guidebooks (often entitled with titles like *Companion to . . .*) taught visitors how to respond, especially to scenery that at first they found 'horrid', what associations to cultivate, and eventually, with suitable maps, how to explore off the beaten track (illus. 195). Unlike landscaped parks, there were no inserted *fabriques* and follies, but the guides became adept at introducing visitors to the distinctive geology and botany as well as odd features like rocks, waterfalls and ghylls, bridges, 'changes of scene' and other 'objects to be seen'. One of the excitements of different national parks in Europe is the diversity of their culture: building types and materials, local agriculture or crafts and (uniquely, as opposed to private parklands) different inhabitants and local dialects – these give each a distinctive character and appeal. Since nations like France and the United States are larger, the diversity is much greater and more readily appreciated: Yosemite is not New Hampshire nor the Everglades, the Haut-Languedoc is not the Jura nor the Brière, whereas the distinctions within Great Britain are more subtle, in part as a result of the role played by topography (Exmoor is not the same even as its neighbour Dartmoor).

Furthermore, each of their regional areas is acknowledged as a 'joint creation of natural growth and man's cultivation', and so analogous in many ways to landscape gardens. Lord Birkett thought the Lake District an 'earthly paradise' made by nature and man, and its cottages that seemed to spring from the very ground were a device that was widely cited, as they also featured, less obtrusively, on country estates (illus. 196). The Lakes were also a prime destination for at least well-to-do urban residents escaping to the countryside, a long-standing aspect of British aristocratic culture. And the very handling (the 'organization') of this pre-eminent national park observed some of the same criteria that directed landscape architecture: paths were established, significant treescapes preserved, woodland added to mask scars and blemishes, the special character of a mixed and varied topography sustained, associations identified and endorsed, and above all visitors directed (nowadays often through the establishment of visitor centres; but heavily utilized trails in the Lakes have, in my experience, been rerouted, often to avoid the visual blemishes on a prominent hillside). And it was above all a place that had traditionally solicited the attention of artists and writers, who, extending their attention from country seats to countryside, continued to guide others' perception and search for meaning.[10]

196 Robert Hills (1769–1849), *Grange in Borrowdale*, watercolour. Beyond the bridge over the River Derwent can be glimpsed the hamlet of Grange.

TWO

A very early exhibition took place in Paris in 1798, though it clearly modelled itself upon earlier events like Roman triumphs or Renaissance state entries. Its Fête of Liberty paraded artefacts, looted sculptures from Venice and Rome, bananas and other vegetables, lions and a pair of dromedaries through the streets; nine more such events followed in France by 1849. Increasingly in the nineteenth century, national and international exhibitions were held, now on sites earmarked for them, and they have continued to the present day, with the New York World's Fair (1964–5) and those in New Orleans (1984), Seville (1992), Lisbon (1998) and Hanover (2000). The commentary on these events has been almost exclusively on nationalist agendas, the assumptions of sponsors and reactions to them by visitors (notably the role played in both by different minorities), and the anthropological approach to these 'liminal' places where visitors are ushered into a brave new world of modernity in all its forms. Walter Benjamin thought 'world exhibitions are places or pilgrimage to the commodity of fetish . . .'. But while the literature on these topics is immense, there is a dearth of commentary on the landscaping. And this is despite the fact that parks were often commandeered for these events, like the Prater in Vienna for 1873, the Golden Gate Park used for the San Francisco exhibition of 1895, or the Esplanade des Invalides for Paris in 1935, with its display of modern gardens.[11]

However, in materials originally published on the occasion of these exhibitions, landscapes were extensively illustrated.[12] There were bird's-eye views, where the whole, marvellous extent of the site could be truly appreciated. Furthermore, exhibitions were provided with maps and guidebooks, which necessarily paid some attention to different locations, so that visitors could find their way around and get to know the landscape and could take away with them a souvenir of the whole place.[13] Visitors were also eager to view sites from some high viewpoint: there were Brunel's towers erected for Sydenham's Crystal Palace, Ferris wheels (first introduced in Chicago in 1893), the 272-foot Bonet Tower in San Francisco, the Eiffel Tower in Paris, and the so-called Flip Flap in London in 1908 (a dizzy pair of columns that switched to and fro with visitors in their high cages 150 feet above the site). So it is these landscapes that are the subject here, and four in particular are considered: the Crystal Palace Exhibition in Hyde Park in 1851, and its later transfer to the park at Sydenham in south London; the Philadelphia Centennial International of 1876; and Chicago's World Columbian Exposition of 1893. Several themes emerge, notably the eclectic fashion in which a whole world of landscape forms and devices was invoked: lakes, follies and *fabriques*, conservatories with a repertoire of exotic planting, *tableaux*, botanical gardens, zoos, Picturesque landscapes and the plush bedding-out schemes of Victorian gardening.

The choice of the south side of London's Hyde Park for the 1851 exhibition was much protested, not least because the site would have interfered with the fashionable social parade-ground that included Rotten Row with its horses and carriages, and also because the proposed structure was considered far too heavy and cumbersome (13 million bricks would have been needed).[14] Other sites, like the northern segment of Regent's Park, were mooted, but eventually Hyde Park was agreed upon, and Joseph Paxton provided the solution: like two other original submissions, his was for a conservatory of iron and glass, of the sort that he had already devised for the estate of the Duke of

197 Trees inside the Crystal Palace, Hyde Park.

Devonshire at Chatsworth. So a garden and estate connection was clearly established. But Paxton also ensured a strong gardenist interior, incorporating two large elm trees in Hyde Park (illus. 197), an impressive crystal fountain by F. & C. Osler, massed conservatory planting, greenery and innumerable hanging baskets. Cathedral-like it may have seemed, a place of worship for the exhibitors and their visitors; but the sight of a conservatory in the parkland, taken from the Serpentine (illus. 198), and its similarity to the Kew Gardens Palm Stove, also gave the site its distinctly garden appeal.

And like several other subsequent exhibitions elsewhere, an enthusiasm for landscapes and gardens spurred by the 1851 event was directly responsible ten years later for the establishment of the Royal Horticultural Society's new gardens as part of the development of South Kensington. This site, designed by W. A. Nesfield, was one of the largest, most eclectic and elaborate gardens of the period, containing not only a memorial to the original exhibition, but a winter garden, with matching bandstands, encircling arcades, canals, miniature amphitheatres, embroidered parterres emblematically worked in polychromatic gravel, box and coloured plantings – a garden, in short, 'carried on in such high order'.[15]

If the site of Hyde Park spoke of royal estates and posh metropolitan society, Sydenham Park, to which the Crystal Palace was transferred in 1852, offered itself and its much more elaborate site as a more populist venture; the *Art Journal* even thought it smacked too much of Vauxhall Gardens. It had a boating lake, a display of effigies of pre-historical animals ('The Restoration of the Extinct Animals') and ascents in a balloon, and it gradually became more and more popular, with fireworks and bandstands, and eventually in 1902 an amusement park with water rides. During the twentieth century it succumbed to even more unexpected pastimes: cinder-track racing and the cup finals of the Football Association. But from the start there was also a strong pedagogical agenda: its *Guide* summarized garden history from the Middle Ages to the previous century, while the park itself was to be an illustrated encyclopaedia, a 'course of investigation' as important as the interior exhibits, though these too had a distinctly gardenist flavour. Furthermore, beside a historical anthology, the site actually mirrored the graduated pattern of earlier European estates and gardens (see chapter Ten): nearest the Palace were the formal terraces of the 'artistic garden', then came the 'simply gardenesque', then the

198 The Crystal Palace, 1851, colour lithograph.

'dressed picturesque' and finally the rough and wild landscape. It was more than eclectic: it was an assemblage of every European and English garden culture that Paxton could assemble. It lacked any originality in its individual parts, but its repertoire of garden effects was hugely impressive. Created on the former private estate of Penge Place, the new park (illus. 199) combined features of an English landscape garden, featuring an irregular lake and meandering paths, with symmetrical ponds and shapely pools, cascades and water pavilions, an imposing central avenue, hugely colourful planting, a Rosary designed by Owen Jones with an answering 'circle of bowers' on the far side of the central avenue, and pathways that surrounded the formal elements. Its hydraulic system outperformed that of Versailles, with ten miles of iron piping, six million galleons of water in circulation, and two plume water-jets 280 feet high, besides many lesser jets. Many thought Paxton's scheme reckless, many thought it lacked sufficient sculptural decoration in the fountains, and it ultimately failed to achieve the success that had been envisaged.

The exhibition grounds at both Philadelphia and Chicago performed in rather different ways. While the landscapes continued to be the groundwork of the whole, as we shall see, there were now a considerable collection of individual structures. The Fairmount Park site in Philadelphia, laid out by Hermann J. Schwartzmann, boasted 250 buildings of all sorts and sizes; many of these offered themselves in local forms, some in exotic styles (the Horticultural Hall was Moorish; the interior of the Main Building was divided into sections that showed off different states and countries; log cabins at the Hunters' Camp could 'illustrate the various phases of a sportman's life in the backwoods'). In all, the collection of national and international buildings assembled for public consumption seemed to replicate those cultural and imaginative forms that nineteenth-century pattern books purveyed for the design of garden *fabriques* and follies (see chapter Fourteen).

Similarly in Chicago, there were suitably styled buildings from Japan, Alaska, Norway, Brazil and the South Sea Islands.[16] And the extension of its Exposition down the Midway Plaisance that linked Jackson and Washington Parks incorporated several blocks of miscellaneous displays: there was an Irish village with its castle, a German village, streets from 'Old Vienna' and Cairo, Bavarian buildings (suitably Alpine), houses from Westphalia and the Black Forest, an Egyptian Temple, an Eskimo as well as a Lapland village, and a cliff dwellers' habitat.

This miscellaneous and exotic mélange was clearly a much more substantial presentation of cultures and architectural styles and habitats than had been suggested in nineteenth-century publications of *fabriques* and other garden decorations, often fabricated in flimsy materials. The

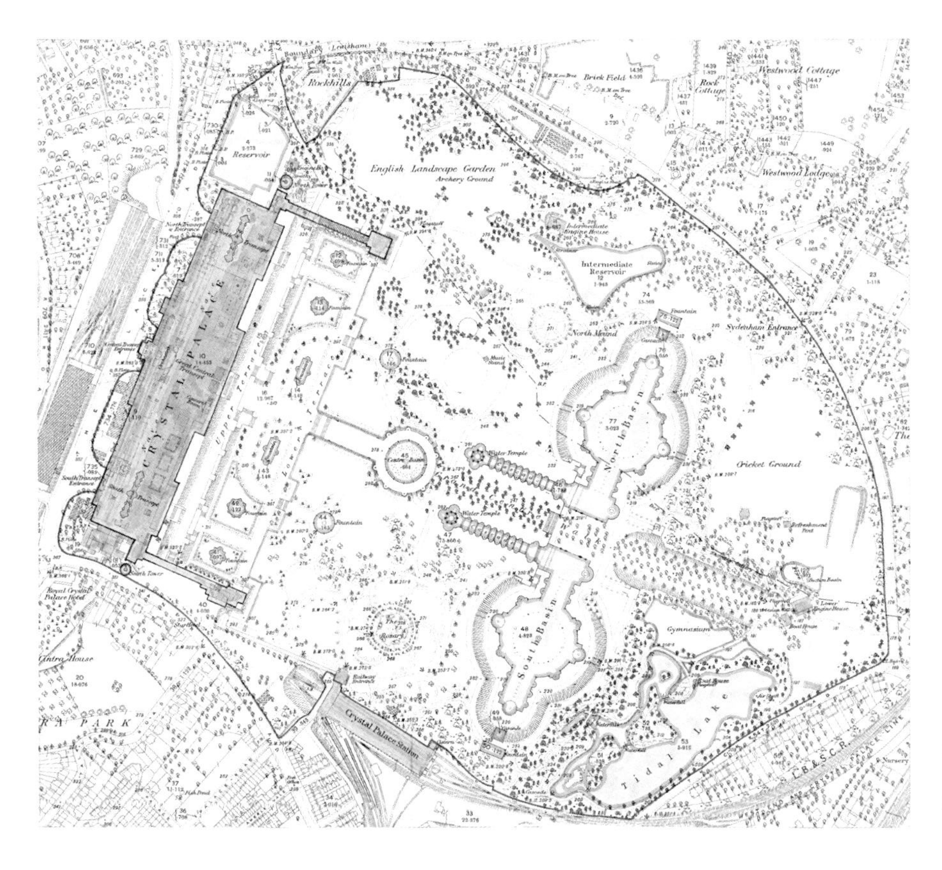

199 Ordnance Survey map of Sydenham Park, 1871, the new site for the Crystal Palace.

combination of pedagogical fervour, curiosity, sheer voyeurism and thrills took these fairs to a much high level of theatrical flair and architectural invention. Sydenham certainly had its tableaux of life-sized extinct animals set in geological replications of early landscapes, but later fairs took matters further: panoramas performed volcanic eruptions in Chicago, famous items of the ancient world were replicated (Nashville in 1897 had a quarter-sized Parthenon, Philadelphia, relief plans of Jerusalem and Paris), whole cultures were presented in life-size but clearly epitomized formats, and even in 'live' performances rather than tableaux (a nudist colony at Zoro's Gardens for the California-Pacific Exposition; Belgium displaying its colonial prestige with a populated native Congolese village in 1894). And all of these displays and expositions, strange and modern encyclopaedias of world cultures, were strangely reminiscent of the projected but never realized idea of Joseph Spence in his *Polymetis* (1747) for a vast landscape dotted with Roman buildings, each of which honoured a specific deity and represented him or her through a collection of statues, coins and busts. Spence, too, had a strong pedagogical agenda, hoping to instruct his compatriots in laws and lore of ancient Rome.

The Centennial Exhibition in Philadelphia relied heavily upon its large and picturesque

200 C. Inger, chromatic view of the Philadelphia Centennial Exhibition, 1875.

Fairmount Park. Guide books and souvenirs lauded the parkland and many included engraved views of areas that had nothing to do with the exhibition site itself (the Wissahickon, for example, and the eastern parkland across the river); Panoramas of the Exhibition Grounds often showed more of the city's spread and river frontage, with its boathouses and waterworks, than the exhibition cluster itself. *The Illustrated History* by James Dabney McCabe particularly praised 'a beautiful landscape extended before the gazer, and afforded a pleasant and grateful contrast' to its 'lines of buildings'. The Fairmount Park Commission transferred 450 acres to the Exhibition's organizers, of which 236 were occupied by buildings. Schwartzmann laid out an impressive and expansive site, adding over 20,000 trees and shrubs, with lakes and fountains, one of which was erected by the Catholic Total Abstinence Union of America (Moses in the centre striking the rock, observed by four distinguished Americans), the usual mixture of formal Victorian planting, parterres with sunken gardens, including what was termed a horticultural plateau, and more meandering byways (illus. 200). Gardeners were always 'busy' through the grounds. As the site sloped sharply to the Schuylkill river, bridges were constructed over the Lansdowne and Belmont ravines. One guidebook account thought that a visit to 'distant parts of the grounds on foot was a slow and tedious undertaking', so a narrow-gauge railway took visitors for four miles through the site; or they could rent 'rolling chairs', with or without attendants, to move them through the landscape.

The landscape for the World Columbian Exposition was also based in one of the early city parks, Jackson Park, created in the aftermath of the city's 1871 fire. The *Chicago Daily Tribune* lamented the loss of this 'piece of virgin ground', its 'oak covered ridges and marshy depressions' succumbing to the 'trim formality of boulevards'. Yet the perceived loss of all vestiges of the picturesque (associations with Native Americans, for example, or native flowers) would be redeemed by 'the splendid pleasure ground'. Its 'chief glories' were to be the lagoons and its islands, features that would be preserved 'intact' after the Exposition. The treatment of the site, it was further affirmed, was 'so artistic that one thinks only of nature'! Some views of the Exposition site published in *The City of Palaces* (1894) did indeed give it a much more natural air (though hardly evincing 'a touch of the country [side]' that its captions claimed). Seen from the roof of the United States Government building, the eastern lagoon was far less monumental (illus. 201) than the more famous view of the Court of Honor lined by the Agricultural Building, the Administration Building and the Manufactures and Liberal Arts Building. On the Wooded Island itself can be glimpsed the elaborate parterre set in surrounding thickets and generous pathways.

Close up, the gardens of the Wooded Island seemed the perfection of modern urban parklands, though the *Tribune* also claimed that the gardens evinced a touch of elite private estates on the East coast. There were benches, winding walks and cosy niches, and 'quaint' Japanese houses, juxtaposed to scenery where the devices of natural selection and planting gave their charms 'alike to the naturist' and to wandering couples (illus. 202). 'The flora', as its caption claimed, 'transplanted from a thousand different and distant places, seemed to thrive here as at home, and nature seemed assisting man to make the whole as nearly a perfect thing as possible'. The 'man' in

201 View of the East Lagoon and the Wooded Island at the World Columbian Exposition, in *The City of Palaces* (1894).

202 On the Wooded Island, in *The City of Palaces* (1894).

203 *American Florist* (January 1896, p. 603), illustrating an article by Frederick Law Olmsted.

question was, of course, the landscape architect Frederick Law Olmsted. His new plantations were described in the proposal, written in 1891, 'Memorandum as to What is to be Aimed in the Planting of the Lagoon District'; this was published, with illustrations, in the January issue of the *American Florist* (1896, illus. 203). His need was to establish plants with rapid growth and to utilize shrubberies and stretches of wild flowers common to northern Illinois. His ambition was to make the lagoon a 'natural Bayou, secluded, shallow and placid', to create the illusion of 'natural scenery' with nothing 'gorgeous, garish or gaudy display of flowers', and to produce a 'mysterious, poetic effect'. This juxtaposed his planting with the grand effect of the Exhibition's buildings seen at distance, thus counteracting 'the effect of the artificial grandeur and the crowds, pomp, splendour and bustle of the rest of exhibition'. Yet, the whole design was to have 'some degree of the character of a theatrical scene, to occupy the exposition stage for a single summer'. Equally the Exposition was also seen as an exhibit for the 'foremost growers of nursery stock' (the *Tribune*) and a lesson of high-end horticulture and garden-making,

Thus both national parks and exhibition grounds were marked by a certain conflict of interest, a paradoxical ambition to be natural and impressive, illusionary yet also 'theatrical'. National parks introduced many to wilder natures, into which they could nevertheless venture in the confidence that these territories would be accessible by foot or by car; thereby they augmented in large measure a garden world. Exhibition grounds, on the other hand, fostered the industry and technology of modernity within parklands that partook of both elite landscapes and newer civic parks. This display of prime garden sceneries encouraged many to devise their own gardens – indeed these exhibition grounds promoted a considerable enthusiasm for both professional and amateur garden-making. The *Century's Illustrated Magazine* in April 1893 advised that the Chicago Exposition 'will stimulate our desire to employ landscape-architects' and be a 'great object lesson in beautiful landscape effects'. Yet it also lamented that professional expertise was very difficult to find and went on to recommend the establishment of a suitable school 'in Boston, or Cambridge, owing to the neighbourhood of the Arnold Arboretum'. And that, in 1901, was precisely what was created at Harvard University's Graduate School of Design.

The display of ideal garden surroundings at international exhibitions soon gathered a momentum of its own, independent of the major attractions: horticultural exhibition and flowers shows flourished, and during the twentieth century gardens increasingly assumed priority, like the 1925 Paris Exposition Internationale des Arts Décoratifs et Industriels Modernes, where Robert Mallet-Stevens and Le Corbusier showed modernist gardens that would have a profound impact on professional practitioners, especially in the United States. Landscape architecture thus came to be the entire focus of garden art – the annual listings of these shows and exhibitions cover many countries, like the German post-war *Bundesgartenschau* concept that aimed to reclaim areas of derelict land in sites after the Second World War. This in its turn inspired the National Garden Festivals in Great Britain, beginning with Liverpool in 1984 and five other places, ending with Ebbw Vale in 1992, leaving in their wake a mix of housing projects and parklands, and a Science Centre in Glasgow, as well as the derelict remains of their original sites.[17] The Chelsea Flower Show in London continues to be a permanent feature, and new events like the Journées des Plantes de Courson in France or the Festival International des Jardins at the Jardins de Métis in Canada provide opportunities for both conventional and experimental planting and design. But with these we are, so to speak, back in the garden proper, and its stories can be taken up in the remaining chapters.

17 Japanese Gardens and their Legacy to the West

I do not know Japanese gardens, but cannot evade their presence in the world of garden-making.[1] I have learnt how much they drew upon Chinese and Korean gardens from the seventh and eighth centuries onwards, yet I recognize how the very look of them in modern photographs distinguishes them from other Asian cultures. I also follow what I can of their beliefs and their aesthetics, yet I also understand why the Dutch specialist and landscape architect Wybe Kuitert insists upon the uselessness of analysing the 'difference' between Western and Japanese gardens. I am also struck by how the Japanese garden, as explained in their treatises, echoes some of what I'd call the essential aspects of all garden-making. And, like Kuitert, I am curious how the traditions of the Japanese garden feed into our own modern designs – their abstract forms, above all. I have a huge admiration for the work of Isamu Noguchi, whose work in California, New York, Connecticut and Paris I know well, and whose dialogue with his Japanese roots is clearly of great importance to Western design. I also register how much the USA has been influenced by its contacts with Japanese culture, notably in the flurry of American writings on its gardens. And of course throughout the United States there are many houses and gardens designed with Japan in mind, like the *three* Japanese gardens at the Brooklyn Botanical Gardens (one a replica of Ryoan-ji), and museums and gardens dedicated to that nation, like the Morikami Museum and Japanese Garden, near Delray Beach, Florida, and the Japanese house and garden re-sited in 1958 in Fairmount Park, Philadelphia, after its exhibition at the Metropolitan Museum in New York. Moreover, Zen Buddhism has enjoyed a considerable influence on modern ideas and living, not least for those with a concern to *design* or replicate a Japanese garden.[2]

I will limit myself to four themes. I will look closely at how photography has represented these gardens, since how we view gardens these days has much to do with the mode or mechanism of our understanding (especially when contrasted with earlier Japanese paintings). I will consult the recent translation and commentary of the unique eleventh-century treatise, *Sakeiki*, partly translated by Marc Keane, an American landscape architect who lives and designs in Japan. I will pay homage, as an unavoidable duty, to the gardens of Kyoto (previously the Heian capital). And I will attend to the roles of abstraction in Japanese gardens, of which Isamu Noguchi said that 'I admire the Japanese garden because it goes beyond geometry into the metaphysics of nature.'

204 Haruzo Ohashi, 'In the gardens of Kyoto Imperial Palace', in *The Japanese Garden* (1986).

The Japanese photographer Haruzo Ohashi's photographs are, as the title of his collection admits, images of serenity;[3] but they draw our attention to much more in the texture and scope of these sites. Nobody else is visible in these gardens, though we see pavilions and benches; for everywhere is serene and empty of human disturbance (except for his own invisible presence, which itself seems barely palpable). Yet the gardener's design and craft are everywhere apparent – in the careful positioning of stones and plants, the raking of the gravel, the lanterns, the trimmed waves of hedging and the carefully graduated distances, where the eye is comforted with layers of ground and foliage. With rare exceptions, the photographs ignore the sky entirely (illus. 204), confronting our vision from top to bottom of the picture with the fullness of the scenery, and (one also senses) filling and fulfilling the mind. They delight in their focus on details, as the eye absorbs its keen appreciation of nuance, but it is also as if we were seeing whatever we touched ('seeing' as in 'I see', that is, understand), our feet crunching on gravel, sensing the bridges underfoot, the hand caressing the surfaces of stones and plants, or the smell of the plants.

One paradox here is that the richness of material details seems at the same time an index of their abstraction. The rocks and the pebbles are, certainly, haptic and palpable, as we would expect them to be; but their juxtaposition above all makes them abstract, as we take in not just each element separately but the relations between them, the ensemble of their carefully contrasting forms; the bamboo fences may be textured, but the patterns seem to escape their mere bamboo-ness and – as the photographs show – make abstract patterns for the photographer. The studied constructiveness of these sites is both richly 'natural' (we read the natural forms, the different shapes and colour of the leaves and, where apt, their colours), but the whole simultaneously allows our realization that this coherence is itself an abstraction. We see nature – rocks, stones, foliage, colours – as a distillation of their own forms. In his book on *Japanese Gardens* Gunter Nitschke says these elements have been 'adopted' (maybe adapted) from nature, yet we also accept how the mind invents their forms; the mind's inventions are in the final resort, if not indistinguishable from the craftsmanship of the gardener, an index of his work.

This concern for abstraction does lend itself readily to modern work – even when it appears in very different situations – hotels, entrances and courtyards of municipal and corporate buildings, or public plazas, like that designed by Isamu Noguchi for his California Scenario in Costa Mesa, California (illus 205 and 206). The forms there – the water falling from the 'cliff' of the parking garage, the 'river' meandering through the 'desert', the 'desert' itself with real cacti, the green 'hillside' walk – all miniaturize, epitomize, the ample topographies and opportunities of California, and the whole is reflected (another abstract displacement) in the glass windows of the surrounding banks that mark two edges of the plaza. Noguchi is indeed a sculptor (or some would call him an 'environmental artist'), and some of his gardens are purely sculpted and abstract, like the courtyard of the Beinecke Library at Yale with its pyramid, the cube, the circle and the incised lines across the marble floor (illus. 207); what they abstract (I would say rather than symbolize) are the geometry of the earth in the pyramid, the circular sun, the moment of chance in the poised cube, the lines of force in the radiating lines. But he also uses what, in contrast, seen 'natural', like the rocks and desert planting at Costa Mesa, or the water gushing from its Energy Fountain. He succumbs to or maybe invites occasionally a more obvious, Westernized symbolism: the Beinecke's geometrical gestures, the piled Lima Beans at Costa Mesa (a homage to the original owner of the site and its previous agricultural activities), and the obvious irony of the gushing 'Energy Fountain' in this dry, southern landscape.

Paintings of Japanese gardens distinguish themselves further from the abstractions already there in actual gardens. In the six-fold screen illustrating the novel *The Tale of the Genji*, dating from the late seventeenth century (illus. 208) , it is easy to recognize the people on the balconies, the maidens wandering in the garden to admire the blossom, the horse in its stable at the bottom left, the paintings inside the houses that themselves reflect (like this painting itself) a world of natural effects, which the novel also illustrates in this image; we can also see how the imagined space in the painting can be navigated – there is the bridge, the intricacies of the stream, and the different moments and trees along its banks. It is an abstract schema for apprehending the garden. The various maidens may even be read as different moments of the same group within the garden stroll, as each group responds to a different moment or view of the whole place.

205, 206 Isama Noguchi, California Scenario, Costa Mesa, California

207 Isamu Noguchi sculpture in the courtyard of the Beinecke Library, Yale University, New Haven, Connecticut.

The eleventh-century *Sakuteiki* is an early treatise on the Japanese garden. While it invokes aspects of both Chinese and Korean culture and gardens, 're-examined and transmuted into a clearly Japanese context',[4] it strives to explain – in the fashion of a do-it-yourself manual – how a contemporary, eleventh-century Japanese garden would be designed and built. A major theme concerns how exactly the world of the garden refers to the larger world of nature. While there is a clear concern for scrutinizing natural elements and utilizing their forms, anything like mimesis or even naturalistic representation in the garden is discounted. Indeed, the third aphorism of the book urges its readers to 'Visualize the famous landscapes of our country and *come to understand* their most interesting points' (my italics). It continues by urging the garden-maker not to re-interpret other landscapes strictly, but rather to 're-create the[ir] essence'. In the segment on stones, the author insists on the need to 'evoke the feeling of a mountain' by setting a stone with others around it; but the need to depict or otherwise 'climb' a hill or mountain is not envisaged. Obviously, the small dimensions of most Japanese gardens or their segments would make the facsimile of any large scale natural scenery impossible, and the instinct of the gardener aspires to invoke the idea of some particular scenery (to be discussed below), so the urge to mimic an actual scene or topography is minimal. But further, gardens would often have been made on the basis of models (p. 153n), just as bonsai

208 Six-fold screen, scene from *The Tale of Genji*, 1650/1700, colour and gold on paper.

'gardens' are models of trees and gardens, and a model of a landscape is already an epitome of the real thing (*epitome* always meaning both a miniaturization as well as a complete conspectus of appropriate materials). *Understanding* – indeed, 'coming to understand' the materials that go to make a garden, the imaginative re-formulating of raw materials – is essential, and it is out of such careful understanding that the garden-maker discovers the *essence* of his own materials. The author insists always on the skill of grasping 'the spirit' of how certain rivers behave as they navigate the land, maybe noticing that 'water bends because there is a stone in the way that the stream cannot destroy'.

Without drawing parallels between Japanese and any Western gardens, it is still worth noting that all gardens essentially select and epitomize their natural materials. The extent of such representation will depend on a variety of local and cultural ideas. Understanding the materials of the natural world – plants, trees, stones, different land formations – will be a true basis for the making of any garden or landscape, but the mimicking of either a particular scene or a particular landform is unlikely, even though certain topographical or even architectural items can be copied (as when the feudal lord Hosokama Tadatoshi so admired the cone of Mount Fuji that he copied it in his own backyard at Suizenji). This epitomizing is what I understand by Ian Hamilton Finlay's previously quoted aphorism that, in England, 'Capability' Brown 'made water appear as Water, and lawn as Lawn': that the essence of lawn-ness or tree-ness is what is in question. Landscape gardeners may well select a certain place as their given, basic site; but, unless it is already a landscape that is uncannily or unusually perfect, already an essence of that particular scenery, they will tweak it, remove things or add more of what is already there; they will thereby affirm what they have glimpsed or seen in that scenery, they will pull out and clarify its essence, even if the keen observer cannot see the changes that have been made.

Intriguingly, the same Japanese characters that denote garden-making in *Sakuteiki* also refer to 'the art of setting stones'. The text reads that 'stones are imperative when making a garden'. This explains Isamu Noguchi's remark that the bones of a garden are its stones. The selection and appreciation of a stone, along with its position in the garden and how it is physically installed, were crucial to the forming of a garden. Stones are, obviously, taken from the larger landscape beyond the garden, and therefore a due appraisal of them as found objects there is essential, but at the same time, it was not necessary to work with them or shape them; indeed, this was deemed inappropriate. So we have here a paradox – find a interesting stone in the 'wild', do not touch or work it, but use this 'found object' as a crucial element of the garden's design. Thus the garden-maker's mind must both appreciate the possibility of recognizing and then using actual stones to create the 'essence' of the garden. He must 'follow the request of the stone', which Keane explains as paying attention to an individual stone as well as following its individuality in making garden layouts. If the garden-maker observes 'waterfalls as they appear in nature', that will dictate, not their mimicry, but an understanding of how suitable dimensions and character for a waterfall might be invoked for the gardenist to employ.

There were no garden-makers as such at this moment of Japanese culture, and it appears that many owners took it upon themselves to design a garden (as the author of *Sakuteiki* notes in one of his Miscellanies); though clearly labourers from the family were employed, especially those with expertise in the management of water in the agricultural lands around Kyoto, where streams abound. The treatise is therefore directed at those who would fashion their own garden, according to the categories of different garden types (see below), observing taboos (*Sakuteiki* contains a whole section on this topic), acknowledging the various strictures of Buddhist doctrines, and following also the direction of local bureaucratic codes and authorities. The various branches of government controlled, among other things, geomancy or ancient forms of divination; they established Bureaux of Gardens, Carpentry (for imperial buildings) and Earthworks, and maintained the trees of the capital.

The treatise takes up and describes different natural formations and topographies as the 'best models' for gardens. The scope and generic classification of these models is detailed. Since 'there are many different ways in which water can be made to fall', its types are then listed as 'Twin Fall, Off-Sided Fall, Sliding Fall, Leaping Fall, Side-Facing Fall, Cloth Fall, Thread Fall, Stepped Fall, Right and Left Fall, Sideways Fall' (171); then each is given a brief description, most of which (like the Stepped Fall with two sets of stone steps) are self-explanatory. The 'Cloth Fall' employed a pool just above the fall, so that the water runs smoothly down the rock 'like a cascade of threads handing off a spool' (174). Similarly, the gardening styles are derived from oceans, broad rivers, mountain torrents, wetlands and reedy beds. One of the earliest, meandering river-style gardens has been excavated and reformulated in Nara, then the Capital of Japan, dating from the eighth century.[5] As there is also a strong instance on securing an island in lakes, however limited, often to use as a platform for musicians, there is a similar taxonomy of islands: their characteristics are now drawn from surrounding adjacent materials (forest, pine, rocky shore, coves, slender stream and so on). Streams are hugely important, not least because they must be organized to flow east from the south and then westward, but also because their formal character requires much attention to both that route and its adjacent banks. Given the detailed classifications of these different garden elements, there is obviously considerable choice for patrons to make, even within the strict categories allowed. Thus the author of *Sakuteiki* occasionally notes how much individual taste or personal preference will dictate the choice, efficacy and success of the chosen garden format and style. (It's rather the way in which formal rhyme or stanzaic schemes in Western poetry or structures in music may impose limitations on a poet or composer, only for the writer or musician to delight in manipulating the sonnet, the *villanelle* or the sonata to find his own best expression.)

The *Sakuteiki* was written during the eleventh century in what is now Kyoto, the ancient capital (before it was moved to Edo/Tokyo in 1867). The city was master-planned (on Chinese models) in the eighth century with a central tree-lined avenue stretching to the Imperial Palace and a grid of streets within which the courtyards were fitted. The city is surrounded on three sides by wooded hills and mountains, and it is largely outside the city that estates and mansions were established and where Zen temples were constructed or took over earlier villas.

Given the length of time during which Kyoto prospered, given too the variety of garden types and uses which came to occupy the hillsides around the city, and not least the modifications that have taken place on various sites, Kyoto remains an extremely rich and complicated world (illus. 209).[6] Some of these garden temples are enormous, others are small or minute. Daitoku-ji is a huge precinct, with over twenty smaller temples and other structures. Within this larger site is the garden at the Daisen-in (Great Hermit Temple), a narrow strip that measures 12 feet by 47 feet, yet it contains a wholly credible landscape of rocks, trees and water, where the scale of every item is scaled to perfection.

Everybody knows, yet does not *know*, the dry garden (*kare sansui*) at Ryoan-ji (*c.* 1500): its sea of raked gravel sand, dotted with 15 rocks, is the ultimate abstract garden, so much so that nobody sees all the rocks at once, or comes to any final interpretation of them. Visitors who sit on its veranda may opt for one 'meaning' or another, but the essence of this experience is to discipline one's mind to the contemplation of the courtyard. Tenryu-ji (illus. 210), from the mid-fourteenth century, also has a display of seven, now largely vertical, rocks; these are here reflected in the lake, their arrangement skilfully composed from whichever direction you look; this is also an early example of borrowed views (*shakkei*), whereby two nearby mountains are called in and merged with the immediate garden forms. Nearby is the garden Saiho-ji, reconstructed at the same time as Tenryu-ji, with over 40 different mosses that carpet the spaces between the rocks, but not on them, and also covers the tree trunks (one might recall Alexander Pope's reflection of human nature – 'as many kinds of mind as moss'). A 'dry waterfall' of horizontal rocks imitates a real fall of water. Saiho-ji is closely related to the style of Pure Land Buddhism, and the gardens of Tenryu-ji, less responsive to that same heritage, serve to exemplify or reflect on earth the heavenly paradise that surrounded Amida, a Buddhist deity. The Imperial Villa at Katsura (seventeenth century) is an early stroll garden and particularly noted for its stepping stones, a Japanese feature that has

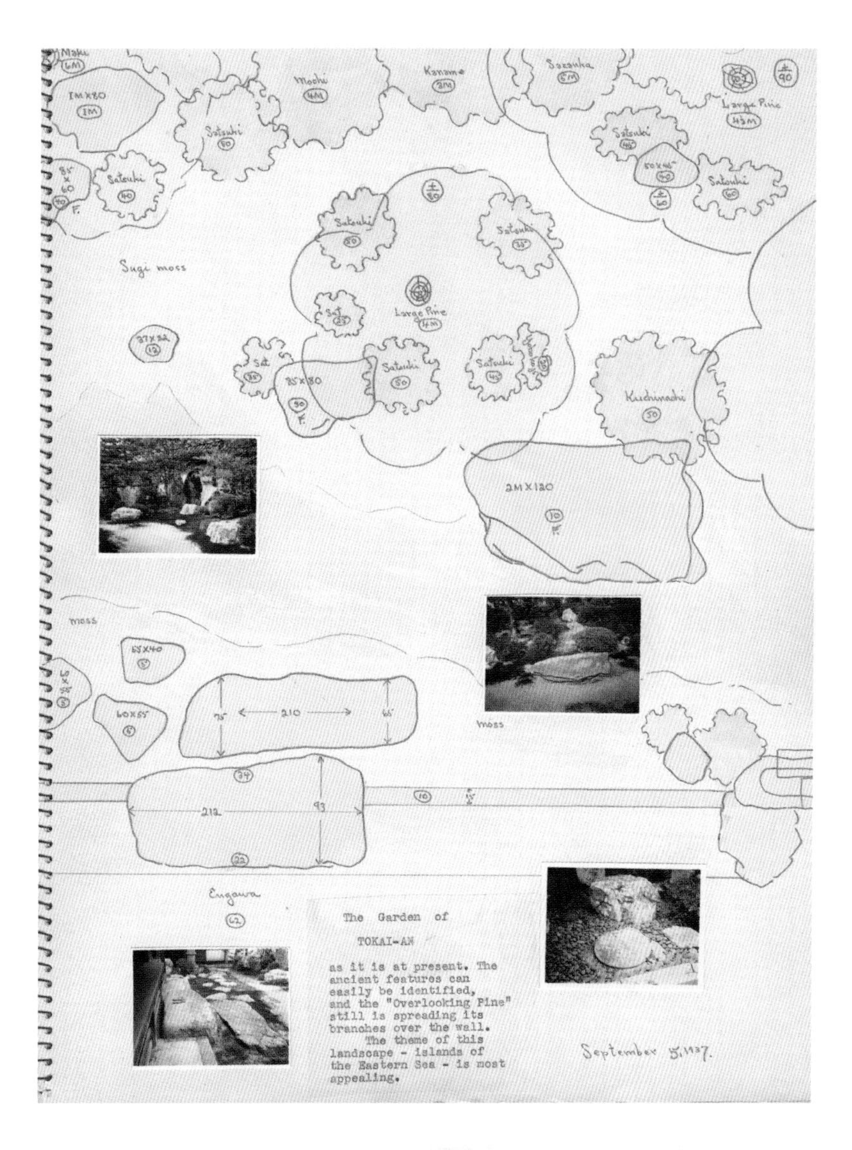

209 Samuel Newson, drawing from his Kyoto and Tokyo Garden Study Books, 1934–8.

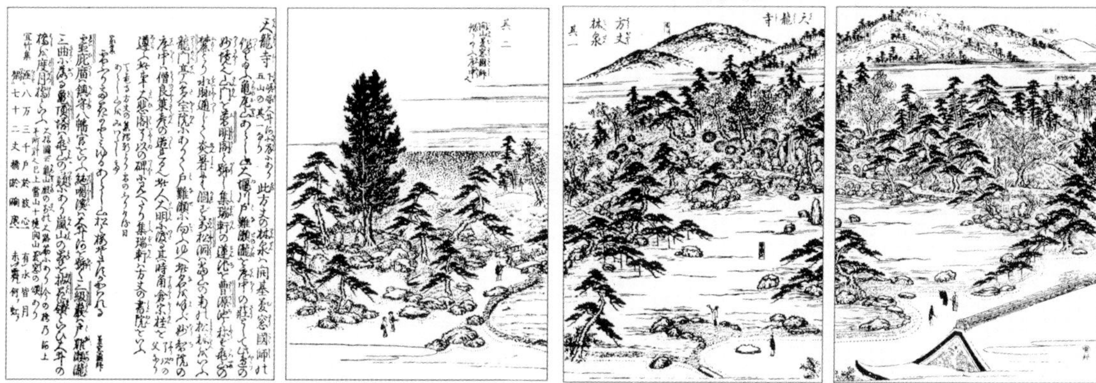

210 Woodcut of the Pond Garden in the grounds of Tenryu-ji, Kyoto, from the *Illustrated Manual of Celebrated Gardens in the Capital* (1799), compiled by Rito Akisato.

inspired designers are different as Lawrence Halprin and Ian Hamilton Finlay (his stepping stones at Little Sparta have inscriptions to alert one to the water that we pass over). Stepping stones, upon which we place out feet carefully, slow up our progress and thus help us to attend more carefully to the immediate vicinity of one's walk; then, as we reach firmer ground, we can look up and take in another view, emphasizing the changes in forms and sceneries as we move. A similar delight in the changes that movement allows comes at Sambo-in (just outside the city, the present site dating from the late sixteenth century, though dating back to the tenth); the opportunity to walk through a series of different scenes, with cherry trees blooming in the spring, is increased by the astonishing display of many hundred stones that make up this garden.

Movement has always been important in Japanese gardens: whether walking, perhaps floating in a boat or by moving the eye while scanning the landscape from a pavilion. The present-day gardens of Saiho-ji (illus. 211), the work of the famed Zen Buddhist priest Muso Kokushi, date from the first half of the fourteenth century, before the time when its famed moss carpeted the ground after various period of flooding. Muso's site requires both movement around the lake and the ascent to a higher garden on Mount Koin. The landscape is designed to be appreciated as it unfolds gradually around the ascending or descending visitor: various structures or buildings may focus attention – rising steps, a cluster of bamboo, a gateway demarcating the lower from the upper garden, foot stones and the increasing demands of the mountain climb. This movement is still an essential experience of Saiho-ji, as we understand where we are from our encounter with its different sections; but for

211 Marc Treib, *Saiho-ji, Kyoto*, capturing the play of moss, light and shadow.

212 A tea garden designed by the tea-master Oda Uraku (1547–1622).

most visitors what has been lost is the Zen Buddhist responsiveness to how movement ensures enlightenment, especially for those who achieved the higher climb ('reserved for special discipline from which the laity was excluded').[7]

It is clear that both geomancy and Buddhism coloured the making of gardens. If the Western mind finds their presence in *Sakuteiki* somewhat underplayed (as opposed to modern commentaries upon Buddhism), this is undoubtedly because these ideas were so much 'in the air' then that allusions and implications sufficed in the eleventh-century discussions. Yet clearly the whole *mentalité* relied upon a regard for nature, the 'innate disposition' of mountain and water, upon their translation into garden form, and upon the mind's response to that. The author of *Sakuteili* will note that 'There are Secret Teachings regarding this' (164) or 'There is a Secret Teaching about this' (188), implying that any aspect of garden-making relied upon the oral training that a pupil would derive from a master and what was called 'skilled ones of old' (31). The Buddhist insight into the true nature of things augmented the earlier, sixth-century Shintoism whereby space was sanctified, a place rather sensed than viewed, and where it was distinguished from the messy world of circumstance and political turbulence. But Buddhism turned away from rituals of shrines and temples, towards simplicity and meditation. The garden world was not only, then, highly valued for its translation of nature, but as a place when the mind would learn peace and quiet. The later sixteenth-century development of the tea ceremony was usually conducted within such a separate space or tea garden (*roji*) in a larger enclosure (illus. 212); here the everyday simplicity of rural life was re-enacted, another remove from the busy world outside and at the same time allowing a moment of greater formality.

We see Japanese gardens as abstract, however much the richness of their materials delights us. Yet this abstraction is neither minimal nor anti-natural. When in 1914 T. E. Hulme, lecturing on 'Modern Art and Philosophy', called for trees to be reduced to cones, or when Wilhelm Wooringer's *Abstraction and Empathy* (1908) argued that early art (Egyptian or Byzantine) created forms as a need to escape nature and that later Romanticism had clouded our concept of abstraction by indulging in an undue confidence in nature, such theories do not at all explain how we respond to Japanese gardens. Worringer, of course, was writing about visual art, where the modes of representation make it easier to say that a mimetic 'rendering of nature' is irrelevant, whereas gardens necessarily utilize natural materials, however they are treated. Yet as Robert Smithson points out, Worringer himself often refers to 'crystalline forms of inanimate nature' and it is precisely this geometry that 'renders' nature; Smithson continues that there is 'no *escaping* nature through abstract representation: abstraction brings one closer to physical structures within nature itself' (my italics).[8]

It is this understanding that helps us to understand a design like Noguchi's for the UNESCO (illus. 213) building in Paris.[9] Here 'nature' is abstracted on two counts. Noguchi relies upon the traditions of Japanese garden-making, discussed above; but his work as a Japanese-American living

213 Overview of the UNESCO garden, Paris, shortly after completion.

in the West also 'abstracts' Japanese gardening for Europe; further, his design for an international location further complicates any translation of Japan garden-making (that it is termed a *jardin japonais* in French compounds that translation). We encounter, then, multiple responses to this garden. It lies behind Peter Walker's general enthusiasm for abstract art that it allows 'different levels of meaning', so that the stark and enigmatic landscape of raked white sand and its boulders at Ryoanji may be understood as islands, clouds, celestial constellations or even animals.[10] And Noguchi, whose minimalist stage settings for Martha Graham in the mid-1930s gave him the opportunity to design for the theatre and who would later invent playgrounds for Riverside Park in New York (with Louis Kahn), was also able in Paris to 'stage' or 'feature' an idea of Japanese gardens for a multicultural audience of unesco delegates and staff. Behind it all is Noguchi's own personality – 'the actual composition' of the various elements 'is my own' – and his difficult collaboration with both international politicians and long-established and traditional gardening masters from Japan.

The garden was begun in 1955 and completed three years later (illus. 213 and 214). It was in the first instance limited to being a delegates' terrace or sculpture patio, but Noguchi pushed to expand the limited space into what is now an extended garden that would connect the main building to a rather uninviting lumpish block to the southeast and at a lower level. Noguchi suggesting linking these with a *hanamichi*, a wooded walkway used variously in both the Japanese Kabuki theatre and Noh drama, with which he would be familiar and which he saw as 'suggestive and essential'. After a variety of 'overcomplicated' models, Noguchi designed a simpler place more in consort with the main modernist building of poured concrete, mounted on pilotes. The garden begins with the stone patio, austere and unchanging, with differently shaped concrete seats, over which a large stele stone presides (carved here with a stylized character signifying 'peace' [*wa*], rather than being an untouched stone). From the stele water descends into a pool and from there it runs down a cascade or trough into a lower garden that is different, promising to be changeful, sensual and green. It is a garden for strolling, though whether it would still be considered a strolling space in the Japanese tradition is doubtful.

That the UNESCO garden promoted strong debate about both its 'Japanese-ness' and its modernity was inevitable in any case, exacerbated by the international scope of UNESCO and the internal squabbles about the garden. However much critics claim its modernity (this claim seems strangely unsubstantiated and the label is affixed without any more explanation than citing its historical moment), or debate how much a 'sculptor' can really make a garden or whether this is 'an artwork rather than a garden', a 'sculptural space' (Noguchi himself called it a 'lesson in the sculpture of place'), or worry about how much it is exactly Japanese by manipulating plans and photographic views that call out those debts or dispute them, the UNESCO garden is miraculously and uniquely itself, despite the various claims about its modernity. It seems beside the point (architectural critics aside) to ask how this garden came into being; I suspect that many famous gardens are experienced less by how and why they were established than by the uses that we continue to make of them.

We need to see the UNESCO garden within the larger and generous world of gardens, beyond its particular genealogy and rather as a site for

214 The UNESCO Garden, Paris.

delegates, staff and visitors who have permission to visit it. The layout is carefully planned so as not to hide the garden's limits (indeed it cannot do this physically), nor are its features concealed for later, surprise discoveries: we get, first, the elevated view from the delegates' patio over the lower garden, then the theatrical gesture of the diagonal walkway down into the lower gardens alongside the cascade trough, and then the left-hand turn over the small bridge onto the irregular platform surrounded by mounds, trees and ponds with their stepping stones. Those who work in the building also obtain a wholly uncharacteristic overview of the site that no Japanese garden would allow. Down below, it is an accessible space, though the trees, grown hugely since its inception, make of it a much more romantic place than the rigour or abstraction for which a Japanese gardener might wish.

The garden, as opposed to the rage for Japan arts generally, had some of its best imitators and commentators in the Western United States. Although it had one British advocate, the architect Josiah Conder (1852–1920), who published an article on 'The Art of Landscape Gardening in Japan', one exceptional student of Japan took his careful study of the art of the garden into careful practice. He was Samuel Newsom. Born in Oakland, California, in 1898, he moved to

Japan during the Depression, learnt Japanese and sketched and studied gardens, with a wide acquaintance among garden designers, before he returned to California in 1938. He produced a series of books, *Japanese Garden Construction* (1939), *A Thousand Years of Japanese Gardens* (1953, with a fourth edition in 1959) and *A Japanese Manual for Westerners* (1965), all of which – as their titles indicate – were dedicated to the careful and detailed study of how Japanese gardens were made, with measured drawings and plans of many sites (see illus. 209). He was convinced, too, that this art could be assimilated into Western garden-making, though he was to face strong anti-Japanese hostility during the Second World War; he was to design, besides his own garden in New Hampshire, the waterfall for the Japanese Garden in San Francisco's Golden Gate Park.[11]

The whole idea of the Japanese garden is as powerful as the modulations it has absorbed. As Bruce Coats reminds us, the Japanese garden has changed in both form and meaning considerably over the centuries, and he instances the garden and temple of Saiho-ji at Kyoto: from a Pure Land garden, a Buddhist paradise, to a Zen temple with a new pond, ruined on many occasions and then repaired, to the well-known Moss Garden with its 120 varieties. The sheer longevity of its careful formulations and yet its generous accommodation of meanings have made the Japanese garden one of the most tenacious modes of garden making. The fourteenth-century Zen master Muso Kokushi urged the reader of his poetry to 'measure how well he sharpens his spirit' against different and various landscapes rather than decide in favour of one or the other. It is thus that tradition and innovation go so happily hand in hand in both the mind and within the garden. Japanese gardens are self-consciously and deliberately complex, not something that can be understood quickly and easily ('The truth never hurries'[12]), and this is perhaps what most distinguishes the Asian garden from Western garden and landscape experience.

18 Arts and Crafts Gardens: The Artist Back in the Garden

After the Italo-French and the 'English' high points in landscape architectural history (sixteenth to seventeenth and eighteenth centuries respectively), one of the next most interesting moments that emerged thereafter was that of the Arts and Crafts garden. It was a style that was widely reviewed and praised around 1900 on both sides of the Atlantic, despite the cultural differences between them. At least in Great Britain, there were aesthetic skirmishes between spokesmen who urged an architectural garden, notably Reginald Blomfield in *The Formal Garden in England* (1892), and those who valued a much more natural if not 'wild' garden, like William Robinson's *The English Flower Garden* (1883) and *The Wild Garden* (1870). We can dwell briefly on these ideas, though it was in some respects a rather futile confrontation: many who did not opt for one side or the other were more productive in both their arguments and, more significantly, in what they designed and built.

The dispute stemmed in part from the legacy of the so-called and misunderstood 'natural' garden (see chapter Twelve) and in part from a dispute as to who should be allowed to design gardens: the horticulturalist, who knows his plants, or the architect, who knows how to design. In the late eighteenth century Jean-Marie Morel in France urged that the responsibility for making gardens be taken away from architects, because they were incapable of understanding the systems of the natural world, and he is claimed to have been the first person to use the term (in French) 'landscape architect'. With the emergence of Humphry Repton in England and later J. C. Loudon, the landscape architect – so named by Loudon on the 1840 title page of his edition of Repton's writings – did become the leading designer of gardens.

The Robinson/Blomfield debate[1] was at best much oversimplified. The latter argued that the architect must be the dominant designer of gardens: 'The question at issue is a very simple one. Is the garden to be considered in relation to the house, and as an integral part of a design which depends for its success on the combined effect of house and garden; or is the house to be ignored in dealing with the garden?' Yet this so 'simple' question was answered somewhat more subtly by Robinson himself, who was never as single-minded as Blomfield, and, more importantly by a group of extraordinary designers, of which the work of Francis Inigo Thomas and the collaboration between Gertrude Jekyll and Edwin Lutyens would be the most impressive.

Another intelligent voice raised in this debate was that of John D. Sedding in his *Garden Craft Old and New* (1891, with a further enlarged and revision edition in 1895). We encountered his work in chapter Ten, where I used him to argue for the gardens that emerged 'between' Le Nôtre and William Kent (in both chronological terms and in the sense of an 'exchange' or mediation between them); he argued, somewhat coyly, for 'a 'natural wildness' touching the hem of artificiality'. Sedding's distinction was to see that the garden was always situated betwixt and between those two poles, not simply being pulled into the magnetic field of either architecture or horticulture. In support of this garden, this 'Betweenity', Sedding shrewdly and surprisingly invoked William Wordsworth, whom he took to be the great poet of English naturalism, celebrating that 'host of dancing daffodils'; one might assume Wordsworth to have been a natural ally of the wild gardening. However, the striking feature of Wordsworth's own home at Rydal Mount in the English Lake District was its 'union of the garden and the wilderness': 'You passed almost imperceptibly', wrote a biographer of the poet whom Sedding cited, 'from the trim parterre to the noble wood, and from the narrow, green vista to that sweep of lake and mountain which made up one of the finest landscapes in England'. Sedding's book seemed to accept that there was no necessary contradictions between various elements or designs of a diverse landscape. On the face of it, this seemed, misleadingly perhaps, to distinguish himself from William Robinson, of whom the architect Edwin Lutyens remarked, after a walk they'd taken together in 1897, that Robinson's 'conversation is wayward and [that he] contradicts himself every two minutes . . . such a foozle headed old bore'. Yet Sedding's book is in one respect at odds with itself: the wood-engravings display what his text termed 'symmetrical planning' and 'distinctly ornamental treatment' rather than the 'undressed parts', the 'open country' and 'wilder effects' that he also celebrated. Perhaps, as his book was posthumously published, its publisher chose wood-engraved images that looked better on the page; the reproduction of photographs of much lesser quality might well have seemed less able to capture how these wilder effects would be allowed to function.

A better sense of the dichotomy in Sedding's arguments is conveyed by designs that other practitioners laid out. An exemplary example is provided by Gertrude Jekyll in the book she co-authored with Lawrence Weaver, *Gardens for Small Country Houses* (1912). They write that the shape of 'Woodgate' in Sussex was determined by the configuration of old highways, and by its moments of regular foci: around the house were lawns, herbaceous borders and a lily pond; and at its northern end, centred around the rose garden, were some geometrical and sheltered fruit gardens and a seating area by the tennis courts. Between these and stretching to the west was a woodland of native oak, holly and silver birch, varied with mountain ash, firs and Spanish chestnut and an understory of bracken and hyacinths according to the season, through which paths wound. Those two regular foci were linked, however, by a broad turf avenue which proceeded straight from the house (yet *not* axially aligned on its centre) and which executed just one bend in the midst of the woodland. From either end of the property, then – lily pond in the south, tennis court seating at the north – garden visitors would see both the continuation of the regularity in which they were at that moment, but also a regularity that surrendered in the distance or at its edges to less rigid planting and forms.

215 Woodgate, a photograph from Jekyll's and Lawrence Weaver's *Gardens for Small Country Houses* (1912).

As Jekyll and Weaver explain, 'Woodgate' was also extensively planted, creating a profusion of effects which their three photographs also confirm (illus. 215). However, what needs to be stressed is their own emphasis upon the calculated *design* that underlies and sustains that apparent wild fecundity – from notes on how the sand and gravel soils were augmented with leaf-mould and lime (natural to the woodland, but also by importations into the site), to the range of plant material (Canadian lilies, South African ixias and sparaxis, hardy calochorti from North America or Californian poppywort, all of which were inserted into native flora). The creation of 'Woodgate' owes as much to its horticultural technology as to natural happenstance; in praising the vegetal and floral casualness, Jekyll and Weaver also insist upon the *effects* of design. And it is in fact not Robinson, but Reginald Blomfield whom they approvingly cite, when they express their own contempt of 'impressionism', which he had considered the equivalent of ill-thought out and incomplete ideas, urging instead the need to think through a design to its 'utmost' ('the incomplete phrase in our case, is no phrase at all'). When nature has been, as Jekyll and Weaver put it, 'less prodigal of growth', it is assisted; and the full repertoire of both natural and architectural effects is presented by carefully determined sightlines, pathways and vistas. There was nothing new in this aesthetic, as the 'improvement' of nature had been, variously in different cultures, true of all design; but it properly reinforced the essential requirement of artistry in a country where too much emphasis had been placed upon uncritical naturalism.

Three earlier gardens are also worth more discussion: Barrow Court (1892–6) by F. Inigo Thomas (who incidentally drew most of the images for Blomfield's book), Buscot (1904) by Harold Peto, and, perhaps the Arts and Crafts masterpiece, Hestercombe, by Lutyens and with Jekyll's planting (1904–9).[2] There are certain shared elements that recur and mark these designs: the careful mix of architecture and horticulture, self-conscious and self-evident craftsmanship, a wonderful variety in both the different elements of a garden's segments and in their relationship to the surrounding topography, some strong sense of geometrical control yet at the same time no instinct to parade these 'formalities', and , above all, a keen play between old ideas (historical forms and reminiscences) and the strong sense that these gardens are clearly, distinctly modern. But beyond these shared characteristics, all three sites are in fact very different.

Thomas had already designed Athelhampton, Dorset, when he came to Barrow Court in Somerset (just south of Bristol), and for both he provided new gardens for existing houses. Barrow Court combines a terrace with a sunken pool nearest the house, a grove near the church, and a large lawn with stone balustrades and a wonderful hemicycle, looking out over the sloping countryside (illus. 216). There are hints of Italy here, stone

216 Hemicycle with herms at Barrow Court, North Somerset, 1894–6.

217 The end-of-terrace archway with steps that lead down to the great lawn at Barrow Court.

urns and obelisks, a stone table for outdoor meals, the hemicycle and its herms comprising twelve busts, but the rest gives a feel of the English seventeenth century: the generous expanse of Somerset countryside that falls away beyond the great lawn, pavilions (somewhat modelled on those at Montacute), the herms which represent the twelve months of the year and are carved to represent twelve generations of womanhood, with a dual acknowledgement of both seasons and the passing of human time. Over all preside the Jacobean house and church and an old barn (barn and house have been remodelled and are divided into apartments). The grove (with a handsome alcove and stone table) and its yew allée offer secluded spaces (illus. 218[3]), and this contrasts boldly with the two other elements – open lawn and the terrace with its pool and flower beds.

Buscot Park (now in Oxfordshire) is totally different, with a water garden, passing through woodland, that links the Georgian house to the large lake below (illus. 219). It is eclectic, but it draws all the various elements – a Mughal-like water course, a Roman boy playing with a dolphin

218 The 'double staircase' in the grove at Barrow Court.

in a fountain, different geometrical-shaped pools, stone edges bordering the channel, a mix of sculptures, a tiny bridge across the channel before it runs into the lake and a garden temple on the far side of the lake – into an accomplished and wholly new form. It is, in more ways than one, single-minded – both as a garden ribbon threaded through woods and by its firmly modern gesture.

Hestercombe had, in the eighteenth century, to the north of the house enjoyed a landscape (now restored) around a lake, with its usual collection of Picturesque effects seen up the combe ('Cascades Water and Root House . . . Chinese seat . . . Gothic seat', in the memorandum of a visitor of

219 The water channel designed by Harold Peto for Buscot Park, 1904.

220 The Great Plat at Hestercombe, Somerset.

1761), all created by Coplestone Warre Bampfylde, a watercolourist. Lutyens in the early twentieth century was called to the same Queen Anne house, poorly remodelled by the first Viscount Portman, and, for the second Viscount, was able to turn attention away from the building and its meagre terrace and create a garden that would, first, be a joyful and richly worked parterre, with an Orangery and 'Dutch' garden off to the north-east, and then allow the sight to be swept over the countryside to the Blackdown Hills.

The careful geometry of the Great Plat, as Robert Williams has shown, is studied ('a mathematics of design'), but at the same time refuses to impose itself; it repays what any great garden needs, careful attention. Lutyens exploits the six levels on the site; uses local stone for the paving and rubble for the walls; allows plants to grow in their joints and down the stairs, and by making entrances into the parterre or Great Plat at each corner he divides it with diagonal grass walks, breaking up its otherwise immense 125 square feet area. A similar handling of the stone edges of the rills that run down each side of the Plat is also broken by three pairs of circular tanks filled with plants (Jekyll saw the 'tanklets' as being 'after the manner of the gathered ribbon strap-work of ancient needlework', illus. 221). It is never a busy garden, yet the eye constantly delights in its details. Lutyens thought that flowers were

221 The 'tanklets' (Gertrude Jekyll's word) at Hestercombe.

'untidy', 'whimsy' laid on top of the built work or backbones of gardens; but Hestercombe is not whimsical or untidy, for the details that catch our eye are always absorbed into its larger gesture. The rotunda to the east of the highest terrace is a shallow niche, ribbed with different layers of stonework, with a pool in front and convex steps that expand as they descend to another, smaller pool: it is not quite the double-staircase Lutyens would have known from Serlio and used elsewhere, but it echoes its platform-like theatre.[3] The area before the niche allows us a moment of quiet intimacy and private performance before the stairs release one into another circular pool space, and then into the larger garden, with the landscape beyond (illus. 222).

Such places combine the work of professional architect and amateur plants-person, of vernacular materials and formal structures, of overall relationships and lively planting details, and allusions to old garden elements within a distinctly new garden. Beyond these dialogues, these Arts and Crafts designs posed a larger and more exciting question: namely, how were you to use the resources of a given site, and whatever extra resources that could be brought to it, as a means of expressing locality? That 'locality' could be the particular site itself, a new confidence in these designers to make good Alexander Pope's early plea for the 'genius of the place'. But it was also the site of England and Englishness. Gertrude Jekyll had a particular love of the county of Surrey, on which she wrote a book (*Old West Surrey*, 1904), and she was none too happy when, for a while, her family moved up to Berkshire. This preference, combined with the fact that Robinson, too, settled in Surrey, having refused a property in Hertford, had enormous consequences for this kind of garden, socially, environmentally and horticulturally; though as we have seen, the Surrey/Sussex border did not have an exclusive hold upon the new gardening. Generally speaking, these gardens catered to a new social class, and to a new ownership of country property; even work for gardens of aristocratic clients seemed modest by comparison – Robinson's

222 The 'double staircase' at Hestercombe.

thousand acres at Gravetye was small in comparison with old established estates.

These gardenist debates were part of a larger concern with national identity at this period.[4] This was reflected in novels like Ford Madox Ford's *Parade's End* (1924–8), with its constant references back and forth between society, culture, landscape and architecture, or the earlier novel by E. M. Forster, *Howards End* (1910), where the eponymous house seems to be contemporary with much of the development of Arts and Crafts gardening. *Howards End* is preoccupied with the theme of England: who owns, who 'runs' England, and exactly which *England* is in question; is it the world of Howards End itself with its ancient wych-elm and ambient gardens, or the slow-approaching suburbs, encroaching like a creeping rust on its horizon, let alone the crowded housing of distant and unseen metropolitan London, a world of 'telegrams and anger'? Forster's evocative invention of Howards End makes clear how this matter of England could be focused, without being unduly simplified, in terms of landscape. And as late as 1936 the American poet T. S. Eliot, also meditates implicitly upon the condition of England in the garden at Burnt Norton, Gloucestershire, a garden – though his literary critics miss the association – that brings together an ancient house and a modern garden ('Time past and time future') that point to 'one end, which is always present'.[5]

The English question in the 1890s, as Sedding, realized, was a question of 'betweenness' – mediating plants and architecture, near and far within the larger landscape, Englishness *and* foreignness, 'old' houses and new gardens for a modern society and its culture, professionalism and amateurism. While many professionals were engaged in these garden designs, the importance of amateurs, who resented professional input, like at Snowshill Manor (illus. 223), and even of professionals' debt to local and therefore amateur craftmanship, made a distinctive contribution. And the great distinction of this kind of garden at its best, why it appeals strongly to gardenists today, is that it was able to make fresh connections *between* these apparently rival traditions and ideas, that it was 'wise enough to freely employ old experiences and modern opportunities' (Sedding, p. 127). 'Englishness' was not, then, a simple notion. It was not the discovery of what was endemic in the nation either by objective historical enquiry or by horticultural searches of what was local; nor was it a question simply of atavistic 'style'. It was rather something that had to be invented or re-invented, an 'Englishness' to be contrived and fabricated, part of the largely nineteenth-century 'invention of tradition'.[6] Along with the Victorian Christmas and the Scottish tartan, the 'English garden' of around 1900 was an attempt to imagine and then to create gardens that answered to assumed traditions of English nature and English gardening. That is interesting enough, I'd argue; but furthermore this invention of an English gardening tradition was, by the very fact that it was an invention, also a very modern move (it was modern aesthetically, socially and technologically). Perhaps it took an Irishman like Robinson to articulate this dream, this re-invention of an 'English' garden that was also up to date and contemporary.

Robinson, as Lutyens assumed, was somewhat paradoxical if not contradictory. He took his stand, as Walpole had done 150 years earlier, against the invidious invasion of English ground by foreign effects. His dislike of carpet bedding, the dotting of lawns with fancifully shaped flowerbeds – a revulsion he shared with Sedding – as well as his anger against hothouses and elaborate

223 Snowshill Manor, Gloucestershire. The garden was created by Charles Paget Wade in the early 1920s, and its succession of enclosures nestle in a larger countryside, like those at Barrow Court.

stonework were all an attack upon design that was not premised upon a proper understanding of Englishness. In one way he was absolutely right: the Victorian age had seen a plethora of pseudo-historical garden styles – especially Italianate or so-called Dutch – with their elaborate architectural interventions and repertoire of new, commercially produced plant species. Against these 'foreign' effects Robinson urged the model of the English cottage garden with its display of flowers from the surrounding English countryside, which Jekyll had also explored and admired in the Surrey countryside and published in her 1925 book, *Old English Household Life: Some Account of Cottage Objects and Country Folk*, and which earlier artists like Helen Allingham or Kate Greenaway had celebrated.

But then in *The Wild Garden*, even when Robinson opposed hothouse floriculture, he invoked the 'beautiful hardy plants from *other* countries [sic] which might be naturalized' in England; so 'wild' begins immediately to be relative and is certainly not simply a plea for indigenous or 'native species'. Similarly, his 1883 book, *The English Flower Garden*, is full of material on foreign importations. And his *Parks, Promenades and Gardens of Paris*, published just one year before *The Wild Garden*, had praised the capital's horticultural technology, in particular the large-leaved, half-hardy plants that could make bold statements in large spaces. These examples from Robinson suggest that we must see part of the distinctive 'Englishness' that he appeals to in his planting advices as that of a long-established imperial and global power, capable of drawing into its horticultural orbit the wealth of the entire world in much of which its colonial presence was still a reality. Sedding described Robinson's design principle as an 'epitome of the flower-garden of the *world*' (my italics). In those days much of the world was coloured red in British atlases.

We can see this more fully if we look at the sites that Robinson and Jekyll created for themselves in the county of Surrey: Robinson at Gravetye Manor, an original Elizabethan mansion that he purchased in 1885 and to which estate Robinson eventually added hundreds of extra acres; and Jekyll, in fact, on two adjacent sites – Munstead House, an undistinguished lump of a building, around which she made her first garden, and then on some newly purchased land across the road where Edwin Lutyens created her more famous home of Munstead Wood, where she made her second garden.

Jekyll's two properties at Munstead concentrate their regular and more architectural effects near the building or buildings (illus. 224, 225). Axial avenues, paths or rides stick out like spokes from these central areas, connecting them with surrounding woodland through which further, smaller pathways meander (what Stephen Switzer had nicely called 'private and natural turns'). At Munstead House, which Robinson may have had a hand in designing, those woods are however scattered with particularized areas – some formally marked off, like the amphitheatre of lawn flanked by a parterre on its northwest and a pergola along its northeast edge, the traditional oblong *potager*, or the Baroque-shaped vegetable ground; others simply distinguished by concentrations of planting (azaleas, rhododendra, auriculae, primroses, rock plants or alpines).

At Munstead Wood, by contrast, Jekyll played down the gradations of form as the house was left behind; instead, the different woodland spaces were marked, as in parts of the first garden, by concentrations of plant materials. But Jekyll's distinctive move at Munstead Wood was less to

224 Munstead Wood, Surrey, designed for Jekyll by Edwin Lutyens, 1895–7.

225 Gertrude Jekyll's garden at Munstead Wood, 1900s.

rework the spatial gradations of effect but to do so over time, through their seasonal variety and through colour effects. She would bring out seasonal changes through her plantings – narcissus and lily-of-the-valley in the woods on either side of the Green Wood Walk giving way to a natural growth of bracken, ornamental brambles and groups of *Lilium auratum* as the year wore on. This now became a very well-remarked feature of her planting schemes, which she discussed in her book, *Colour in the Flower Garden* (1908); but this attention to seasonal variation spreading through the smaller areas of these properties also serves to capture the illusion of larger spaces. Time and space work again in conjunction, another more subtle instance of collaborative betweenity. In her famous hardy flower borders, colour schemes also manipulated the spaces: one at Munstead Wood was 200 feet long, 14 feet deep and backed by an 11-foot wall; its complex and intricate colour scheme – we know her concern with these from her detailed planting plans with their characteristic cartoon bubbles – was predicated on the principle that 'blue conveys a sense of distance, while warm colours bring objects closer'.

For all his rhetoric against unnecessary architecture in the garden, Robinson's first task at Gravetye (illus. 226) was to terrace the slope around the house, thus connecting the building with the distant fields, as Sedding had urged. (And one of Robinson's first reactions to his future home had been that the view of fields was blocked by muddled planting.) Both Robinson and Jekyll relished the old road systems, remnants of an almost vanished England, within which their properties were established: in Robinson's case, a so-called 'Smugglers Lane' ran across the property at Gravetye and he maintained it with careful planting, as he did with his determined preservation of a sixteenth-century building. Jekyll appealed to the flora and fauna of native woodlands in her *Wood and Garden: Notes and Thoughts, Practical and Critical, of a Working Amateur* (1899), although her garden at Munstead House was marked by a pine tree imported from Ravenna, and when she came three years later to write about the more enclosed and regular areas near buildings in *Wall and Water Gardens* (1901) she did urge attention to wind-blown seeds that bring 'the most unlikely kinds' of 'some common native plant' from woody hollows into the cracks of drystone walling. Sedding also urged whatever native plantings that would additionally attract native songbirds (p. 162).

Yet from the very beginning Robinson received gifts from abroad: cyclamen (that did not flower) from Edouard André in France, Moccasin flowers (*Cypripedium spectabile*) from a 'Mr Eliot of Philadelphia' that also failed to thrive. Just as Jekyll had travelled widely throughout Europe, Robinson had of course visited the United States as well as France, and a love of American native trees certainly guided his hugely extensive planting at Gravetye (hundreds of thousands by the end of his life, yet always in the 'forest way', not as specimens). However, he also took down several Wellingtonias as being unsuitable to his sense of English locality.

Robinson's work at Gravetye, spread as it was over a huge area of varied terrain, was not always in the spirit of his more rhetorical battles on behalf of the 'wild garden'. But his work is better understood, I would suggest, in the light of Sedding's shrewd sense of seventeenth-century garden-craft as a 'taming' of 'wild things', the insistence upon 'graduated formality': 'Everything in a garden, has its first original in primal Nature', so that a lawn 'is only a grassy, sun-chequered,

226 Gravetye Manor, Sussex.

woodland glade in, or between woods, in a wild country idealized'.

To track Arts and Crafts garden-making into other countries risks losing the distinct quality of this 'betweenity'. What survives are the deliberate, sometimes *sotto voce* tones of the artist-designer, and a determined interest in atavistic effects. Jekyll and Robinson may have disliked the newfangled Impressionism (Jekyll herself, before her sight failed, was a competent watercolourist), but their emphasis on careful design and finish by the artist-gardener was always evident, and it is a similar awareness of the artistic garden-maker that pervades other countries. Furthermore, whatever allusions to earlier garden forms and structures were made, the results were still 'new'.

Some places, like the Tuscan countryside in Italy, managed better in drawing out the traditional potential of old villas and farmhouses and the distinctive quality of an ambient nature; they relished their skill in melding older items (sculpture and certain formal structures) with modern techniques. Yet nobody can visit the three designs by Cecil Pinsent at I Tatti (from 1909), Le Balze (1912) and La Foce (after 1924) without realizing that they were emphatically *modern* gardens, and that they were distinguished precisely by his attention to form and detailing.[7] Pinsent, like Lutyens, was much taken by geometry, which some will claim as inherently traditional, and they used circles, lines and squares to draw out the materials of the natural materials, stone and plants; it is the balance between geometry and detailing that marks both designers, At La Foce, the repertoire of pergolas, terracotta vases and garden compartments are 'old', some of the allusions to Peruzzi and Piranesi telling, but the precise shaping of the hedges and compartments and the detailing and cutting of the stonework are modern (illus. 227). So too is the studied inwardness of these gardens in their dialogue with the landscape beyond. It is undoubtedly the precision of their maintenance that colours these gardens – but that capacity is itself a distinctly modern emphasis. It recalls the words of Geoffrey Scott, with whom Pinsent worked closely, on the dialogue of what he called the 'materials' and 'poetry', or the 'eye' and the 'fancy', whereby it is what you see, notice and feel that releases the imagination. Happily and symbolically, 'La Foce' in Italian means such a meeting-place.

227 La Foce in southern Tuscany.

These encounters played less surely in Arts and Crafts gardens in the United States, no doubt because the movement lasted longer, responded with more attention to wider European models, and was spread over a much more extensive geography. If 'Arts and Crafts' gardens is still a useful designation, then it is to its great variety that we need to attend. Arts and Crafts gardens in the United States lacked old and traditional architecture on the one hand, and on the other, focused its own revival on colonial styles without the compelling model of English country gardens, and sometimes without much knowledge of what colonial gardens actually looked like. Furthermore, the work of garden-making was increasingly professional, and frequently designed and written about by women (Beatrice Farrand and Ellen Biddle Shipman, in particular, and authors like Rose Standish Nichols). Many of these designers, and many of their clients, travelled abroad to study European models, in Italy above all, but also in France and Great Britain, and to bring back both garden structures (fountains, garden furniture and notably terra cotta) and well as design ideas. Edith Wharton, Charles Platt and Fletcher Steele were particularly vocal about these travels and their foreign resources, but the period-

228 Wilson Eyre, 'Little Orchard Farm', Camp Hill, Pennsylvania, 1903, watercolour and gouache.

icals and magazines of this period are also full of articles proclaiming the European debts.[8] Exposure to Europe during the First World War also stimulated the architectural imaginations of many who fought there. And given that the War played a far less dramatic role in cultural life back home, the life of the Arts and Crafts gardens was much extended. The variety of climates and topographies (geographical and social) in the United States was larger, as were in very many cases the sheer acreage of American gardens, in part a question of the wealth of clients but also of the sheer amount of land available. The gardens at Dumbarton Oaks, Wharton's The Mount and Faulkner Farm, for instance, were bigger than most comparable English ones, though there were some significant exceptions: as in Cornish, New Hampshire, where Charles Platt was able to match house with garden (in form and extent) and both to the views over the Connecticut River and towards Mount Ascutney.[9] Another garden at Cornish was created by Stephen Parrish at Northcote, designed by a Philadelphia architect, Wilson Eyre; Eyre's country houses around his native city were small in scope, but evinced considerable visual grace (illus. 228). Platt had an affinity with Italian gardens, but others now paid more attention to the Cotswolds than to Surrey/Sussex. In this tradition of understated American gardens, Mrs Mellon's Oak Spring Garden at Upperville, Virginia, is a late example.[10]

Distinctive American work began obviously in the eastern United States, and its European inspiration were interestingly documented in *The Practical Book of Garden Structure and Design* by Harold Donaldson Eberlein and Cartwright Van Dyke Hubbard (Philadelphia, 1937). Its bibliography is extensive and wide-ranging, and the photographs are eclectic (images of Italian gardens are much more grandiose than its English examples). Yet

229 Frontispiece to Harold Donaldson Eberlein and Cartwright Van Dyke Hubbard, *The Practical Book of Garden Structure and Design* (1937).

its frontispiece (illus. 229) captures the required tone exactly: an octagonal stone gazebo with steep slate roof, beside a Lutyens-like pond, with an extended farmland behind. Photographs throughout match mainly English sites with local work, notably those in Chestnut Hill, Philadelphia, though the flower beds, interesting in plan, are often fussy in execution and the gardens themselves larger and more impressive than what is shown in English ones. Drawings of garden details – pillars, wall copings, pergolas, paving, terra cotta pots – make more sense because they omit any overall view or context. Gardens offered in plans, in this book and elsewhere, are careful and simple; but when they acquired their planting the geometry is often overwhelmed. The Colonial garden, invented in the 1910s by the Colonial Dames of Philadelphia for Stenton, an eighteenth-century house to the north of Philadelphia (now within the city limits), is a case in point: the original ground plans were rectangular and observed a simple ground plan, similar to those in English seventeenth-century books by Thomas Hill and Gervase Markham; but the eventual planting, under the direction of people with enthusiasm and a bountiful supply of plants, but no real understanding of what the original gardens might have looked like, is overblown; exactly what Lutyens might have called whimsical and untidy.[11] The ratio between formality and plantings is miscalculated, and such work loses sight of how Jekyll's work in courtyards or strongly geometrical gardens like Hestercombe was quite different from her famous proposals for hardy flower borders.

What was interesting – and perplexing – was the desire to find 'a type particularly American [which] embodies the poetic and artistic sense of our country', as Louis Shelton wrote in 1915.[12] But 'our country' was big, and regional preferences, dictated by climate, populations and social clusters, began to dominate and complicate any 'American' typology. The East Coast gardens were varied in size and amenities, though not as much by social class.[13] Different 'types', while clearly in the same traditions of well-crafted and artistic gardens, emerged in the Midwest (Frank Lloyd Wright and Jens Jensen) and California. The Greene brothers, architects but also craftsmen in wood, were much impressed by the work of Mackintosh on a visit to Great Britain in 1901 and were in their turn praised by C. R. Ashbee for being in the spirit of 'our best English craftsmanship'; they created California houses and gardens with nods to Italian, English, Spanish and Japanese culture.[14] Yet their planting and their celebration of local views are distinctive, regional and very modern, which enlarges once again Robinson's plea for a 'Beautiful *Art* of Gardening' (my italics).

230 Paoli Bürgi, the Geological Observatory, Cardada, Ticino, Switzerland.

19 The Prose and Poetry of Modern Landscape Architecture

Landscape architects exist and must function in societies divided between practical demands and imaginative possibilities, between the prose and the poetry of our lives. Some professionals dedicate themselves to – or maybe find their readiest work and even satisfaction in – the world of pragmatic concerns, contributing to the basic infrastructural necessities of a complex community; to do this, at their best, they must learn the grammar, vocabulary and syntax that sustain all good prose; much recent landscape architecture is required to confront places stripped of soil or vegetation, often derelict, or even toxic, for which engineering and knowledge of planting are essential. Others, however, may fret at such apparent restraints upon their creative powers and grasp at whatever opportunities their professional existence allows them through which to push beyond the merely functional and prosaic. These competing dedications are not, of course, mutually exclusive: poetry should itself be as dependent upon good grammar and rhetorical skills as prose, and much prose may itself stimulate a response beyond acknowledging its pragmatic virtues.

Yet it seems useful to make something of this distinction between poetry and prose. Their combination is often difficult to achieve amidst the insistent demands, on the one hand, of clients, budgets, legal and social rules and regulations and, on the other, of designers' own creative aspirations, inspired in great part by the considerable tradition of famous predecessors over many centuries. Furthermore, the desire to impress can elicit mere pretension or absurdity, when designers strain for too much effect. Yet it is precisely in this difficult collaboration between the prose and the poetry of their work that the best landscape architects have thrived and continue to do so. We may well look back and see only the 'poetry', say, of the Villa d'Este, Versailles, Castle Howard or Wörlitz, rather than the practical issues of hydraulics, engineering, earth-moving, political iconography and client agendas that each of those sites entailed. Yet it is by that same symbiosis of pragmatism and poetry that contemporary professionals distinguish themselves.

In the past even 'prosaic' work was necessarily done for elite personages in making parks and gardens, fulfilling all the necessary infrastructure to make lakes, hills, terraces, groves and grottoes; in all of this, the experts were required to know their stuff, to function efficiently as hydraulic engineers, agricultural workers, foresters, plants men (they were usually men), or to be used to working with the given topography (maybe as military engineers). Functional elements were necessary, and often remained simply functional, even if set in a garden like Versailles (as Kafka remarks in *The Trial*, the 'steps explained nothing, no matter how long one stared at them'). Even those who 'decorated' a garden with staircases, sculptures and other devices – the poets and antiquarians who provided inscriptions – were required to know how to do the necessary work and to provide it well and efficiently. Out of such combined and talented expertise were created places like Villa Lante, Rousham or Birkenhead Park. That these places were lauded by their own contemporaries for being rich in significance – 'poetic', if you like – was as much as anything else because people responded to the *ensemble* of the different talents involved. When in later years such sites acquired a further reputation for an augmented richness of response, this was largely due to how well they had sustained themselves over the years. History lends its own patina – its own mode of 'poetry' – to sites, even when they undergo changes; the gardens of the Villa Borghese or of Courances, for example, have suffered or survived with skill and dignity whatever was done by human restorers, social changes or even natural disasters. Central Park or the Parc de Buttes-Chaumont, both from the second half of the nineteenth century, though their original formations were hugely dependent upon technologies and techniques, have both since acquired respect and acclaim for their highly successful creation, from the patina that time has given them, and from their continued use as public parks.

But with very contemporary landscape design we cannot invoke any sustained life and growth of a designed place – such analysis is not possible and, often, less necessary for many people (though people are still nostalgic for 'old' places). If this or that park is useful, good to use for a variety of purposes (jogging, taking the children to play, sunbathing, walking or other more active sports), we probably don't want more. Yet in this instance I am not concerned with these fundamental and pragmatic issues: we know, collectively or individually, a good design when we see and use it – paths that make sense, steps that are easy to climb, seats in the right places (shade, or with views), interesting and delightful plants, safe, pleasurable and stimulating places, sites that have learnt to present themselves engagingly and are able to demonstrate sound ecological principles. (Nor am I eager to point out failures: again, these are usually places that people dislike and eventually shun). Fletcher Steele, writing on 'Public splendour and private satisfaction' in 1941, emphasized this need for all civic undertakings: 'Street layouts must first be practical. Light, water, and all modern conveniences must be for use'.[1]

What is interesting, then, are those places that, for whatever reason, deliver something more: I have used the term 'poetry', though other words have been invoked, like 'beauty', *genius loci* or sense of place, the 'magic' of space, 'visionary' design or aesthetic satisfaction. (However, it not a question simply of 'beautifying things', as some contemporary landscape architects dismissively argue, even where they are responding to urgent and infrastructural needs). Others see the distinction as one between '"the Soul and the Body" of Landscape Architecture', with the role of the social, rather than private, designer as the key to such distinction. The geographer Denis Cosgrove urged 'the idea that under certain conditions humans can touch the deepest rhythms of creation and achieve a unity between their own spirit and that of the living universe'.[2] That idea is difficult to seize, even by those who are touched by it, and its terms are infinitely vague. But it deserves to be explored, so I will use a cluster of sites and their designers to do so (again, remembering that there will be other candidates and even other criteria for such an account and that this is, in the end, a very personal account).

A key issue today for landscape architects is how to respond effectively to ecological demands. Yet this often does little more than ensure or imply compliance with environmental obligations: the French landscape architect Bernard Lassus notes the

> confusion that reigns between the concepts of environment and landscape . . . The fact that we are solving the various problems of pollution that face us does not mean we are bringing landscapes into being. It is all too easy to imagine having perfectly clean environments that are not attractive at all . . . a healthy environment is a 'degree zero' – obviously a child who is playing in a meadow must be able to drink the water of the stream . . . and building on this substratum alone will we be able to invent landscapes, that is say, our 'culture'.[3]

Thus after the early advocates of ecology – notably Rachel Carson and Ian McHarg – came a fresh focus upon how designers, focusing 'on the known world', could instigate 'aesthetic experiences that reduced barriers between humans and the natural world' (Meyer) and who (now in Cosgrove's phrase) were able to appeal to 'the deepest rhythms of creation'. Many examples come to mind and many sites seem to have responded to these new demands.

In a series of projects by the Swiss designer Paolo Bürgi, landscape is allowed to reveal itself as well as to work upon the visitor: in an interview he stated that 'the task of every landscape architect should be to address the well-being of humankind through the shaping of our environment'.[4] On the summit of Cimetta above Lake Maggiore there is a Geological Observatory (at 1,670 metres), a platform through which local rocks protrude and where there are two lines of shaped geological blocks (illus. 230). Visitors are invited to gaze into the far distance, where the eye may readily notice a cleft or long fissure in the hillside; this Insubric line marks a time, millions of years ago, when the African and European tectonic plates shifted and the fracture was formed. The blocks on the Observatory floor are examples of the different geologies from the two continents and are arranged so that the younger European rocks are closer to the Insubric line. The hand rail to the Observatory contains information about the geology.

Visitors arrive at this platform from a lower funicular station, designed by Mario Botta, where Bürgi was also concerned to make them do more than take photographs of the usual alpine view and move on to an adjacent cafe (what Bürgi calls a 'landscape for consumption'). A promenade (illus. 231) juts out above the forest and into thin air, and (at least for me) this vertiginous thrusting of oneself into the void, far from the mountainside, ensures a wholly different sense of the place and how one responds to it. On the floor of this promenade are also etched a variety of signs – DNA shapes, maple leaves, a buzzard's footprint and signs of the Fibonacci sequence (where each number is a sum of the two previous numbers – a kind of expanding ecosystem). Nothing explains their meanings; the insistence is upon how we might be intrigued, how the whole experience – looking out, looking down – lends itself to the mystery. Other clues alert one, stimulating curiosity and enhancing how one views the site: as we emerge from the funicular station there are granite chevrons leading one across the plaza into the wooded mountainside, and the gap between them grows wider and is filled with grass, a signal that one has arrived on the mountain from the city below; the widening chevrons also point the visitor in different directions. There is also a simple fountain, where children play in the trough, hewn from a single piece of a hundred-year oak, and dogs drink from the water emerging from one end of it.

And if we walk from the plaza upwards towards the chairlift that takes one up to the Geological Observatory (from the lift we can look down on the grazing cattle in the meadows below), Bürgi has instituted what he calls a 'game [or ludic] path'. About one hundred metres apart are a series of games that explore the laws of physics – parabolic mirrors used to communicate at a distance, a swing to experience centrifugal force, a seesaw that plays sounds as you try and balance on it. The path is gentle, 'possible for all ages to walk on it', and there are seats along the route. It rises through a series of zones: silver fir trees, beech trees, larch, Norway spruce and hazelnut trees, each in their different environments from cold to mild. Unusual benches provide places from which to enjoy the forest.

Ludic games are perhaps the contemporary equivalent of eighteenth-century follies or *fabriques* in older picturesque gardens that once tempted visitors to contemplate times past or unfamiliar (see chapter Fourteen), though these days they strike us merely as exotic curiosities rather than poetic stimulants and with which we are less concerned to get involved. Bernard Lassus has also enjoyed the stimulation of games.[5] He organized a game where he asked passers-by to respond to a large wall covered with red dots inside rectangles and convert the given red dot (with black felt tip pens) into whatever image they wanted on sheets of paper – the red dot stares out of faces (as an eye), a belly button, a traffic light, sunshine through prison bars, a pompon on a French sailor's beret, a dot on an 'I' or a letter 'O'. This was clearly not a landscape, but it allowed Lassus to reveal that the large wall with regular dots – a visual effect – is changed when a visitor comes closer and sees an individual drawing – a tactile effect. Participation was at once a visual game for the passers-by, and a lesson on how we might respond to the sight of a large landscape or to its close-up stimuli. Almost thirty years later, in 1996, Lassus played another, a 'Garden Game' now, at the château of Barbirey-sur-Ouche in Burgundy. On the glass windows that overlook the untouched site Lassus engraved sixteen images of different garden items that alert us to the duration and future of the site (Stephen Bann compares these to Humphry Repton's overlays, where a flap is raised to show the watercolour of what Repton was proposing). But at Barbirey we are not asked to transform the existing site; as we look through the glass and then at the images on its surface, fresh ideas emerge from the landscape beyond.

Lassus also envisaged a similar game by juxtaposing different landscapes, not now palimpsest-like glimpses on window panes, but juxtaposed as layers across a real space between the rue de Rivoli and the river Seine. He proposed a narrative to make clear the sequence of historical gardens that had occupied the Tuileries from the sixteenth century to the present day (illus. 232).[6] His idea was that, as the site could neither be 'restored' to any one particular period nor plausibly redone in a wholly contemporary way, he would recreate

231 The Promentory, designed by Mario Botta, Cardada, Ticino, Switzerland.

layers of the previous gardens: the lowest stratum was the earliest Medici garden (80 cm beyond present grade), then the Claude Mollet garden (20 cm below), then André Le Nôtre's garden at the current grade, a nineteenth-century garden now raised by 50 cm, and finally a contemporary garden, with water basins, at 170 cm above grade. And further, the new/old garden now corrected the original distortion of Le Nôtre's garden by aligning itself between I. M. Pei's Pyramid in the Louvre courtyard with the arches of both the Carrousel and the Arc de Triomphe down the Champs Elysées and, on the horizon, with the new arch at La Défense. In short, what the garden proposed was to rethink both near and far, present and past; our imagination was called upon to connect time and space, just as we read the red dots and their fresh reformulations.

In many contemporary landscapes we see, not perhaps 'games' as such, but the play of wit that imaginatively enlarges the found landscape. Wit can be, certainly, an awkward or even uncomfortable tool in public places; but in the still older sense of wit as intelligence or perception, it can be useful. There was a wit (in both the newest and the oldest sense) in Lassus's project for the Tuileries.

232 Bernard Lassus, project for the Tuileries. The different historical moments begin on the right and finish with a modern area alongside the river Seine.

Similarly, through with a radically different sensibility as well as very different materials, Bürgi took the grounds of a rather derelict neuropsychiatric clinic in Mendrisio, Italy, to make a park, what he called another 'bosco ludico': a path wanders through the site, already well stocked with trees and bushes, where the visitor encounters six 'follies'. These are all made from trees, to entertain and awaken the reveries of the walkers. Bürgi compares them with the carvings in the wood at Bomarzo, or the Parco Durazzo Pallavicini in Genoa, with its built follies. The tree follies are named according to the effects that each attains: the linden tree house is a familiar device, though with a window looking up to the sky; a row of fastigiated hornbeams left to grow and twist around their stems; a cathedral of evergreen oaks; a grove of oaks that will eventually, after many decades, grow so close that we cannot enter the circle; a place of pleasure and play (hornbeams again) within a labyrinth of paths; a perspective view in which cypresses on either side of an imaginary line depend on the direction in which we look (a three-dimensional *trompe-l'œil*). These green follies are simple, and Bürgi says ethical, resisting any fashionable hedonism or the lure of consumerism: they perform the park with economical wit, and will do so for decades. We view them in the light of the surrounding woodland, and in turn see the parkland through the 'window' of the new follies.

Bürgi has also contributed to what are called 'field studies'.[7] Vestiges of former agricultural land have survived in the Ruhr Metropolitan Area in Germany, hemmed in by roads and housing estates; farmlands have become urban playgrounds and farmers, threatened by the encroaching developments, survive at an uncomfortably low subsistence level,. Yet, as Bürgi argues, people's desire for 'beauty in a landscape' led him to propose a fieldscape, an 'aesthetics of agriculture', where it was at once an agricultural and a designed field, a game that gradually delights as much as it perplexes. The fields change with the seasons (over a two-year cycle), stripes of ornamental flowers within the differently coloured crops draw out the beauty (and usefulness) of these, contrasting and changing endlessly with the colours and times of year; the very mechanism of the ploughing and seeding marks the contours of the land, as they in turn produce new visual experiences. These interventions are entitled 'Venutas et Utilitas' (Beauty and Usefulness'), 'because beauty should not be at the expense of the harvest [for the farmers], an attitude that cannot be supported from an ethical point of view'.

Another large landscape intervention began in 1986 with a plan for the 700th Anniversary of the Swiss Federation. The Swiss Way, a 35-kilometre path around Lake Uri, was divided among the 26 cantons, each of which, independently, would have taken on a separate sector. Georges Descombes, with other collaborators, assumed the two-kilometre segment from Brunnen to Morschach on behalf of the Canton of Geneva.[8] Given the cultural richness of the whole environment, a minimal intervention was called for, eliciting the 'historical thickness' of this segment by focusing upon the path itself and an attentive reading of the territory and its rocks, trees and plants. The erratic rocks were sometimes cleaned of lichen and moss to reveal the stark white granite below. Stones were set along one side of the forest path; others mark its centre from time to time; various new elements – metal strips and galvanized tubing – were added to railings and steps, the contemporary materials clearly signalling modern interventions that guaranteed routes, views and safety. Steps were constructed to connect the path to a former railway line; horizontal wooden blocks terrace the hillside, with grass growing between them, and these can also be used as seating. A wall is restored below the former grand hotel Axenstein, which with other hotels traditionally dominated the sharp elbow bend of Lake Uri. One last insertion was made – a circular, metallic belvedere overlooking the lake far below. The belvedere is a familiar feature in Swiss mountains, where views can be taken and picture postcards verified against the surrounding landscape. But just as Bürgi needed to frustrate this habitual 'landscape for consumption' by intensifying the experience at the Cardada mountain, Descombes moved to enlarge the customary 'picture postcard' view by cutting a gap in the curved, double scrim of the *Chanzeli* belvedere, and transforming the conventional postcard into a 'diorama'. The Swiss Way is a very functional, prosaic itinerary that also draws out the rich elements of the topography as we negotiate its route. The poetry is in the place.

The range of poetry to be discovered in landscapes is more various than we imagine and not all of it is conventionally 'beautiful' or 'lyrical'.[9] So we might think of 'lyrical', less as some quality of a place, but as a way of using the languages of landscape materials to enhance and even startle us. Bürgi's fieldscapes do just that. Another Swiss landscape architect, Dieter Kienast, thought that 'poetry' (his word) could be communicated when the ground of urban squares at night was illuminated, and the night lighting by WEST8 on Rotterdam's Schouwburgplein transforms the rather bleak platform. The night-time lighting at the park of Duisburg-Nord also transforms the already astonishing structures of the former steel mill into a fantastical medley of coloured surfaces.

Another way of finding poetry in the very basic materials of landscape was seized by a variety of land artists; their main claim was that they used the very materials of the earth and its land forms to make art, and they often did this in remote areas where visitors would go only with difficulty (see chapter Twenty). A similar, but nicely ironic version, smaller and readily visitable, can be discovered in the Ratcliffe Institute quadrangle at Harvard University: here Chris Reed stacked different materials for his 'Stock-Pile' (illus. 233), a stock-pile being 'something vital or indispensable' (in the words of *Webster's Dictionary*). These are heaps of various materials used in the making of landscapes, such as stones, sand and rubble, two of which heaps are also planted with ancient ferns. As the piles have been made impossibly steep, they will change over time, subsiding in keeping with the inherent physical and structural characteristics of each material; no 'tidying up' is allowed.

Our attitudes to 'poetry' change and – in a startling renewal of the Picturesque – landscapes that would earlier have been ignored and distasteful have caught our attention. What have been termed 'drosscapes',[10] or 'brownfield sites', the wasted and derelict, even toxic, industrial landscapes can startle us with their unexpected forms and associations; Alan Berger, who introduced the term drosscapes, quotes the Elizabethan poet Edmund Spenser, 'All world's glory is but dross unclean'. His aerial views in particular suggest their sublimity, their 'glory'. Equally, even close-up photography renews our ways of looking at items that we usually ignore. Robert Smithson, on 'A Tour of the Monuments of Passaic, New Jersey' in 1967, took his Instamatic ('what the rationalists call a camera') to record the sandboxes, industrial pipes like gushing fountains and car lots that dot this worn and clapped-out landscape. For the 'rationalist' these were presumably the detritus of what Berger now calls 'vast, wasted or wasteful land surfaces'; but for Smithson and his alert camera they were 'monuments' – 'ruins

233 'Stockpile', Harvard University, designed by Chris Reed, STOSS.

234 Gardens in the Landscape Park of Duisburg-Nord, designed by Latz+Partner.

in reverse'. Thus he responded to the associationism of the Picturesque by seeing these industrial leftovers as 'monuments', at the same time quoting a passage of the Picturesque theorist Uvedale Price on the destructions wrought on a piece of ordinary land. Yet at the same time Smithson refuses the usually Picturesque moves, mocking any pretentious discovery of extravagant sexual meanings in the place, refusing 'crass anthropomorphism'. It was enough for him to conclude 'I will merely say "it was there"'.

Smithson, of course, was not there to remedy or to 're-design' the topography of New Jersey (he does that elsewhere with his 'earth art'). But other landscape architects have confronted these waste lands, sometimes indirectly, sometimes face-on. Berger argues for new initiatives that combine the role of an inventive landscape architect with those who can grasp the potential of all the empty spaces of urbanism ('an archipelago of lacunae') and can coordinate work to 'decontaminate, re-regulate or otherwise transform' land for re-use over time. The work of Michael Van Valkenburgh uses a variety of plant materials to redeem toxic sites, and ponds to collect and process run-of from buildings or hard surfaces, but does so with elegance and wit (see next chapter). Others have responded to wasted landscapes, relics of industry, by confronting aspects of sites and renewing them. Many of these prosaic responses can be instinct with a new poetry (illus. 234).

In the eighteenth arrondissement in north-east of Paris, the municipal Direction of Parcs, Jardins et Espaces Vertes took over 4.2 hectares from the redundant marshalling yards of the French railways to create a much needed park in a somewhat disadvantaged area of the city. The resulting Jardins d'Éole (dedicated to the gods of the winds) takes over the long, narrow area of the former railway tracks, protected by a raised walkway to the west, from which we overlook the still remaining lines coming out of the Gare du Nord, and by open fencing and wooden walls along the street to the east with housing beyond (now receiving considerable refurbishment, since the new park will attract new residents). The design by Georges Descombes takes its cue deliberately from the former rail lines (illus. 235), stretching narrow lawns, a slim canal, an elevated promenade, tree-lined allées and gravel sections, a mist fountain along the central podium, a line of tennis courts one after the other down its length. The planting is new and will eventually provide shade, but much of the rest is fairly low-key, ecologically responsible and caters to folk who can now enjoy, in this immigrant neighbourhood, amenities like sporting facilities, children's play areas, spots for quiet seclusion and plots for growing vegetables. Ramps and entries also take their direction from the long latitudinal space. It is obviously a place carefully designed for a variety of daily activities, but it allows not only a sense of the garden's origins, but also the pleasures of a happy place without any gesture to some more conventional 'pastoral' enclave.

One of the most resourceful and at the same time witty responses to railways in another part of Paris, the Jardin Atlantique (illus. 236), also offers little park space. This garden is suspended over the platforms of the TGV (high speed trains) at Gare Montparnasse, from which travellers leave for the Atlantic coast. While there is a dull and stuffy waiting room near the tracks below, this garden provides a different place to wait for the departing trains. Though somewhat difficult to find, stairs from the station emerge onto the garden level where we discover a whole agenda of seaside events: a weather station, giving details of wind and rainfall;

235 Jardins d'Éole, Paris.

a boardwalk, slightly tilting towards the 'beach' across which 'waves' advance. There are two cliff-like caves, a raised walkway behind them, as if we are exploring a cliff top, seaside planting and a host of children's games. Surrounded by tall apartment buildings, it also functions as a public park for the residents. The Jardin Atlantique plays with its visitors, not expecting them to be in any way lured by some mimetic scenery, but allowing them to enjoy hints of what would await them after the train journey to the Atlantic coast, or maybe consoling them for staying in Paris. It is suitably theatrical (the cliff-like caverns are simply theatrical flats supported by battens behind), the waves are wonderfully abstract, and it recalls Charles Baudelaire's complaint that scenery painters for the theatre were liars ('menteurs'), because they were not inventive enough. But at the Jardin Atlantique the creators of this theatrical decor are properly inventive and the 'lie' is much appreciated.[11]

Two American designers in particular make strong and convincing claims for going beyond the necessary prosaic moves for their site programmes; they can also make much with little or diminished resources, and they also bring a sprightliness, even witty, feel to what they propose. In Toronto Ken Smith and Martha Schwartz, along with David Meyer, produced the Yorkville Park, where a row of Victorian houses was demolished for a subway beneath and a parking lot (illus. 237). The long oblong block mirrors the footprints of the houses and, though far less apparent, the designers have

236 Jardin Atlantique, Paris.

used the lots' spaces to form a cabinet to display a collection of plants, rocks and weather effects from the surrounding Ontario countryside. Among other items, there is a huge rock, a pergola of vertical wires down which water drips and freezes in winter, a botanical cabinet with the plants neatly labelled, different groups of trees (pines, birch, alder), and a criss-cross boardwalk raised over a wetland garden.

Smith himself has come to move with ease between public parks, installation art and private gardens, and his materials range from plastics to artificial stone to cones of glowing light. Sometimes the proposal is temporary, sometimes destined for long (if not permanent) use; sometimes on private penthouses or on the 'public' roof top of the Museum of Modern Art, the work is highly (pun intended) elitist. The overall mixture of this work is both, or at first, conventional or ordinary, then alarmingly eccentric, depending upon closer perusal of the materials or the spaces to be used. He responds warmly to community gardens, with a particular passion for vertical planting in tight spaces, but panels along the sides can be torqued and respond to the patterns of the skewed planting beds. For the much larger Santa Fe Railyard Park (illus. 239), where he collaborated with Mary Miss and the architect Frederic Schwartz, he deferred with enthusiasm to local tastes and materials: it is educational and playful, celebrating water and irrigation practices, with the ancient Acequia Madre irrigation ditch running

237 Yorkville Park, Toronto.

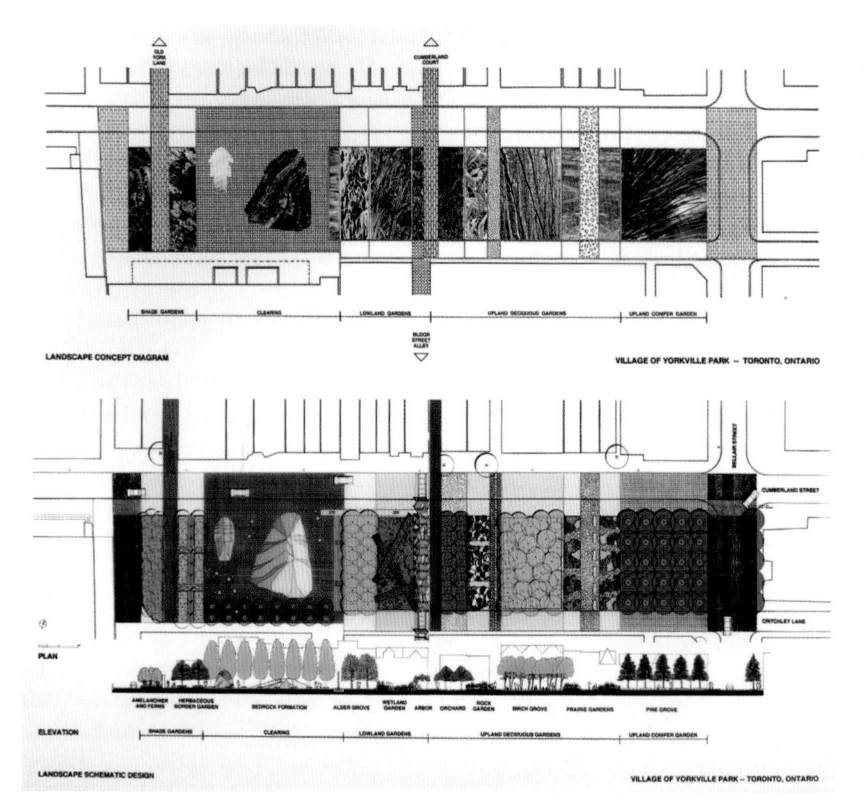

238 A design for the layout of Yorkville Park, Toronto. The criss-cross paths seen at the site today do not appear on these plans.

through the park, rust-coloured adobe walls and seating, scattered with rocks, large wooden benches with metal backs and modernist pergolas which the vertical vines will gradually climb. Yet it is in its turn a far cry from the Wall Flower installations he has worked on for years, where grids hold an apparently random array of flowers: flat, cut-out, large-scale fabric flowers are mounted with 3-D silk petals, as was done on the facade of the National Design Museum. This tapestry of flowers, decorating the building, turned the building temporaily into a vertical garden.[12]

Martha Schwartz, too, takes particular delight in playing with modern forms, accommodating both the oddities and challenges of modern circumstance, including limited funds or spaces over garages where no deep planting is possible, and the demands of many who are no longer particularly happy with, or do not even need, the habitual 'pastoral' lawns of nineteenth-century public parks. She has been innovative in the display of 'sculptural' additions in parks and gardens, whether white tyres, wafer candles, globes, gilded frogs, plastic topiary, bagels, boxes of all sizes, plastic palm trees or pink plastic rocks, or hay bales wrapped in red plastic, along with glacial mounds or huge earthforms. She works with aplomb, ingenuity and considerable wit, playing games with the programmes of their sites and the choice of materials. Schwartz, of course, made an early name for herself with lines of purple bagels regularly spaced in the garden of her mother's house, then with a quincunx of cheap garden frogs, painted gold, in a pool of an Atlanta shopping mall (see illus. 191). When funds ran out for the installation of royal palm trees in Florida, she invented canopies of coloured vinyl mounted in steel poles to gesture to the trees formerly gracing the Everglades. Schwartz risks offending many who cleave to 'conventional' gardens, with classical sculpture and the usual palette of grass and flowers. Yet her range of work has grown since the early days of being just provocative and funny: she responds well to both private grounds, where clearly her imagination and reputation allow her to invent with flair for sympathetic patrons, and in a cluster of varied public work that challenges even as it bemuses visitors. There have been Japanese car parks, with what Schwartz calls 'whales' swimming across the surface of the parking garage below, a community space (also in Japan) with a host of coloured outdoor rooms and fountains in the form of boats, and now in the city plaza of Minneapolis, glacial mounds and sawn logs that honour the founders of Minnesota.

But perhaps her most successful plaza is that for Exchange Square in Manchester, from 2000 (illus. 240). Where the IRA had torn apart the city fabric with a bomb in 1996, Schwartz drew together the old and the new city, the precincts of the cathedral and the new Marks & Spencer store. At the top, in front of the modern storefront, is a striped, austere pavement, with seats in the form of railway trolleys on fake tracks (originally designed to be moveable); then follows an arc of curving stone benches set in a warm yellow puddingstone

239 Ken Smith, Santa Fe Railyard Park, Santa Fe, New Mexico.

240 Exchange Square, Manchester, by Martha Schwartz.

that overlooks the water feature that marks the boundary of the old medieval Hanging Ditch. This is fed by three gushing pipes that seem to bring the water from below and send it down through the curving riverbed, filled with large and irregular stepping stones on which children can play. This ample and various amphitheatre allows people to be spectators and 'performers', to seek their own bit of space where they can 'hang out'. As Schwartz says, it is designed to 'encourage traditional activities and inspire new ones'. Designers like Smith and Schwartz have worked deliberately to extend our understanding and use of garden-making, even while invoking older forms. This, then, is the theme which the final chapter will address.

20 The Once and Future Garden

es mucho haber amado,
haber sido feliz, haber tocado
el viviente Jardin, siquiera un día

it means much to have loved,
to have been happy, to have laid my hand on
the living Garden, even for one day
JORGE LUIS BORGES, *trans. Alastair Reid*

T. H. White wrote his *The Once and Future King* about King Arthur. It is unlikely that any future garden will be a Camelot, or could take us back there. Camelot, like the idea of the garden, is essentially a mythic place, a metaphor of perfection. Actual gardens have existed in many places and some have been flawed, like the 'original' Camelot itself, but none can ever be that once perfect place. What this final chapter confronts is the omnipresence or language of gardens in many times and places, along with whether we can think of them as having some collective significance over and above the culturally generic. May we generalize about the world of gardens beyond bland and empty gestures that make us feel good about them? And can we envisage the world of gardens in the future? Bernard Lassus writes that 'the garden, the hypothesis on what, from yesterday and today, will be perpetuated in the sensory approach of tomorrow and its new ways to touch and be moved – is not such a garden before all else philosophical?' That is the theme of this last chapter.[1]

I like to think of myself as a cultural historian. That is, I attend to how various cultures have made their particular marks on the world and how those material manifestations – gardens, in this instance – shape culture in their turn and are used by their inhabitants and visitors. But I also believe to a large extent in the permanence of certain human dreams and desires, their ascendancy over the quotidian and merely circumstantial world, even if that also shapes and changes them. So I need to deal with both the permanence and the alterations in garden-making, with the once and future of them.

The history of the garden continues even as I write and as these chapters are being read. While it would be foolish to propose a narrative of the future, it is nonetheless tempting to peer into the state of the art as it emerges from the first decade of the twenty-first century. Forecasts and prognostications are not lacking, since the world is full of books and magazines on garden-making and so much landscape architecture also surfaces in professional journals and on websites – the plans, projects and proposals of designers that have not yet been built and perhaps never will be, hence the wonderful world of their unconstrained imaginations. Indeed, the coloured and inviting computer screen with its dazzling arsenal of digital imagery seems to have usurped the site as it is customarily considered (the actual location for building). Unconstrained, so it seems, by the immediate pragmatics of financing, planning and building, the virtual landscape and its

241 A pair of 'Before and After' views in Humphry Repton's Red Book for Stoneleigh Abbey, Warwickshire (1809).

gardens flourish more than ever; with a click of the mouse and the convenience of Photoshop come visions of fresh landscapes, already populated with convincing human occupants (see illus. 137). We may think of the Red Books by Humphry Repton, where his patron lifted a flap to see the brave new world of his landscape and imagine the capabilities of his property, even if he couldn't afford them (illus. 241, 242). Ever since (and earlier, too, before there were landscape architects), professionals like Repton have tried out ideas with no client or site in mind, like the anonymous author and illustrator of the *Hypnerotomachia Polifili,* or the compilers of pattern books in the late eighteenth century, or the brothers André and Paul Vera in the early twentieth century.[2]

That there will always be gardens and designed landscapes is not in doubt. The private garden in particular maintains a considerable hold upon individuals whose imaginations are drawn, even nostalgically, to the enclaves of others. Garden magazines and the garden pages of 'posh' newspapers promote these places, opening up private places to the gaze and perhaps envy of others, who generally cannot enter them except through photographs and descriptions. These publications love to reveal people busy in their garden, potting shed or greenhouse, in suitable gardening attire – sporting floppy hats against the sun's glare – or relaxed, in a deck chair with friends on the terrace. Creators of these gardens have a wonderful confidence in their own effects, often with a good eye for other's creations

242 Here, with the moveable flap lifted, Repton presents his design project for the gardens and the house.

and a ready skill at pillaging older ideas, even relying happily on mail-order catalogues, the ubiquitous garden centres or the temporary garden shows at Chaumont sur Loire or the Métis Gardens in Canada. There are always hints to be gathered, moments to be imitated for our own places, if we have them; given the readership of these magazines, there is often a desire to look out for some particular element that could be copied – how best to envisage a pergola, enhance a yard or back garden, or improve an unprepossessing sidewalk. For the enclosed world of the medieval garden is still with us. And an owner's self-promotion and self-presentation, albeit much diminished, are undoubtedly flourishing, even for the public park and garden on behalf of its local community.

Our fascination with these glimpses into other gardens, which if not picturesque are always photo-esque, are also increasingly gathered for publication in books, and the garden shelves of bookstores boast a litany of titles: accounts of a 'year in my country garden', or the 'magic spell of gardens', publications like the Royal Horticultural Society's *New Classic Gardens* by Jill Billington (a title that nicely appeals to the brand-new along with the security of 'classical' design), Udo Weilacher's thoughtful *In Gardens*, Katherine S. White's writings from *The New Yorker* in *Onward and Upward in the Garden*, or *The Passionate Gardener* (a rare treasure by the German Rudolf Borchradt, now translated by Henry Martin). Other books take one into usually private places: guides to '101 Best Gardens' in the western United States,

Christopher Lloyd's *Other People's Gardens*, or the omnibus *Garden Visitor's Companion* by Louisa Jones. They mix the pleasures of the voyeur, the knowledgeable commentator and the instant survey of a site's best features.

There is, though, a widening gap between the professional and the amateur that can be disconcerting ('amateur' being used properly, as somebody passionate for his/her work, but not trained or licensed to practise in a professional capacity; there are, to be sure, many garden designers). Publications for professional landscape architects rarely address the work of the private gardener or garden-maker. And my impression is that professional designers, much (and rightly) preoccupied with large public work, do work that does not lend itself to being pillaged by amateurs, who look to their own publications for inspiration. Perhaps this varies from country to country. Yet while many of us need to have information on how best to be gardeners, our activity itself seems straightforward; landscape architecture, on the other hand, is not easy to explain (unlike 'architecture'); the late Lawrence Halprin confesses that it was a term difficult to explain and always 'ambiguous'. The owners of small and modest gardens since at least the sixteenth century have been impressed with larger and more prestigious designs and may have aped bits and pieces of their work, sometimes to the merriment of more sophisticated folk (see illus. 185); I suspect that this exchange happens much less these days.

One continuing and sometimes perplexing question is what, these days, is the connection between the garden per se and landscape architecture, when this is largely thought of as the making of 'big parks' or of contributing to infrastructure? As an environmentalist, Ian McHarg argued fiercely against gardens, and others have thought it both anti-ecological and, usually, elitist. Yet this prejudice has not, in the end, been very effective: many landscape architects seem willing, if not eager, to see that the garden impacts their work. Garrett Eckbo was quoted in chapter Fifteen as saying that 'garden design is the grassroots of landscape design . . . Private garden work is really the only way to find out about relations between people and environment'.[3] Others have seen similar connections. Lassus argued for the garden as 'in all else philosophical', as well as providing the basis of all future work in the laying out of ground and the choice of its forms. Critics, like Marc Treib, seem happy to involve the garden in their discussion of such a place as Duisburg-Nord, which he describes as 'the garden, the park, and the plaza'.[4] That certainly implies a distinction between the enclosed garden there and the larger parkland, along with its place for performance, but as other contributors to the book on Duisburg-Nord make clear, there are other exchanges between park and garden, and many other examples – by Bürgi, Smith, Schwartz – discussed in the last chapter also suggest that the garden has acquired a new vitality. The 'new' garden may not be enclosed, but it still acknowledges a liminality, a sense of a special place to which we gain access. It may not be small-scale, but it still serves to focus itself around specific moments (the invocation of 'follies' at Duisburg-Nord is an example), and to provide other triggers and prompts for human interaction. Indeed, the garden has become an interesting parallel to what Rosalind Krauss termed the 'expanded field of sculpture': the long traditions and assumptions of garden-making have been absorbed by creators of parks and even national parks, or themselves have been colonized by being taken over (the countryside leaping the fence, or the recognition that huge tracts of industrial 'lacunae' that call for remediation have become a new site of making).

Sometimes, it seems that gardenists and landscape architects deliberately confront each other, and these battle lines often are drawn along the official boundary of licenced practitioners. Yet the conflict is not really between garden designers and licenced architects, but far more a resumption of the old Renaissance confrontation and challenges of the different arts: the *paragone*. Instead of disputes between painting and sculpture, or images, words and music, we now have conflicts of scale, materials, audience reception, between ornament and master plan, between installations and place-making. A classic example of this current situation can be seen on the grounds of the Getty Museum in Los Angeles.

The Philadelphia firm Olin Partnership designed the campus of the new museum by Richard Meier and subsequently helped to lay out modern sculpture from the Ray and Fran Stark Collection. Sculpture gardens have indeed become a major site for landscape management (see chapter Fifteen), and they continue in a modern mode the Renaissance presentation of classical antiquities in villas and courtyards (see chapter Six). In fact, Meier designed the Getty Center as 'a reinterpretation of a classic Italian Renaissance villa and garden'. But what is interesting about the new Getty Museum is this confrontation between how professionals like Olin and an 'amateur' designer, albeit famous as an installation artist, Robert Irwin, go about making a place.

The Getty campus is designed to provide ample space around Meier's hillside cluster of museum buildings. It welcomes museum visitors and scholars, yet gives them equally a chance to relax and explore the interstices of the buildings and contemplate the sculptures, often poised against the empty sky and horizon.

It features entry plazas, stairs, Italian stone pines and a bosque of London plane trees, pools and fountains, terraces, openings onto the surrounding landscape, and a wonderful cactus and succulent garden, where its planting re-lives the desert world of southern California, and projects itself into the void at the very edge of the hillside. It was into this campus that another, very different, 'subversive' garden by Robert Irwin was inserted down a canyon leading from the high point of the Museum plaza and ending in a 'sort of bowl', all bursting with plants everywhere (illus. 243).

Irwin had no garden experience: he tells how he bought a thousand dollars worth of horticultural books, cut them up and made collages of them. The stones that line the descending watercourse were also dumped on site and then put together as if to 'compose a painting, a little more complicated maybe' (Irwin); the water emerges from a sort of *chadar*, common in Mughal gardens. The effect is indeed dizzy, though 'they just add up to make an experience', Irwin again. Yet his garden appeals to and even feeds the appetites of many gardeners, for whom a focus upon careful, logical site plans and the subtle palette espoused by professional gardeners (rather than 'all sorts of weird plants' and 'clashing colours') holds little interest. The overall busyness of this 'Central Garden', that leads down the 'Stream Garden' to the 'Bowl Garden', seems a direct affront to the austerities of Meier's campus. The *paragone* is calculated, the challenge deliberate. The entry down the zig-zag path (required to enable wheelchairs to descend) presents different sensations at each of its five turns, and it leads to an azalea labyrinth in the pool (from which entry is excluded, except for maintenance) and from which at this level the views outside are largely hidden.

243 Robert Irwin's Central Garden at the Getty Center, Los Angeles.

Irwin did however exclude the sculptures that were installed elsewhere on the campus and that would have made it even busier and unsuitable as a setting for sculpture; indeed, Irwin's garden (as he himself proposes) may best be seen as a painting, to compete with those hanging within the museum itself, or (as himself reported) a 'sculpture in the form of a garden aspiring to be art'. The garden has caused much dispute.[5] If, as one commentator (David Marshall) says, 'contemporary amateur gardening culture' is concerned only with 'mere plants', then Irwin's garden should delight with its intricate layout and eclectic use of colours and plants. But it is also somewhat clumsy, as one feels that professional landscape designers might have been more skilled at accommodating even this particular confection within the overall grounds of the museum. It is indeed an extravagant piece of installation art; but it is not a question of money and the engineering and horticultural skills that are unlikely to be available to non-professionals (these the Getty could pay for), nor is it a question of aesthetics, trying to oppose Irwin's flamboyant place to the dour and austere regime of the campus design. Beyond the debates over this particular garden, it offers some perspectives on the future of the twin fields, the intersections between garden-making and landscape architecture.

The public garden and park of the future are inevitably, or inevitably *need* to be, created for public use and consumption, which means in the hands of the professionals, in collaboration with the institutions and municipalities who commission that work. But that need not exclude what the long traditions of garden-making have taught humans about the pleasures and excitements of place-making. There have been several attempts to inject landscape architecture with new forms and ideas, and some of these repay attention.

Some have come from the work of land artists, which Udo Wilacher chronicled in his interviews with landscape architects in *Between Landscape Architecture and Land Art*. While land artists have been much cried up as fresh ways to vitalize our experience of land and its materials, most landscape architects themselves repel their advances.[6] The land artist augments or modifies wild, even largely inaccessible and undesigned territory, but more often than not does not actually provide places as sites that we can enter (many are presented anyway, and better, in photographs); we respond to the artist's insertions rather than what has been done to reformulate the landscape. There are exceptions, often ones that allow visitors to respond and, in some form, inhabit the place, and they are also sites that seem to respond to the opportunities of garden-making. The very moving homage to Walter Benjamin in Dani Karavan's *Passages*, or place of remembrance, at Portbou on the Costa Brava, where Benjamin committed suicide in September 1940, is one example (illus. 244).[7] On a site next to the local Catalan cemetery is a platform of corten steel in the midst of which sits a simple cube and beyond it, a wire fence that seems to block any hope of escape or freedom. Nearby are a solitary olive tree and a rectangular, hooded opening through which descends an intimidating, steel stairway, cut through the rocky

244 Dani Karavan, *In Memory of Walter Benjamin*, Portbou, Spain: the hooded entry to the descending stairwell and the solitary tree.

coast; this passageway leads to the brink of the cliff and through a glass wall, engraved with Benjamin's words, we see and hear the restless waves breaking below. His words speak of honouring the memories of the nameless rather than those of the renowned. Karavan wished this to be a monument to all the nameless ones who crossed the Pyrenees into Spain to flee the Third Reich. Those who know Benjamin's writings and the conflict that engulfed him will undoubtedly derive more from this landscaped event, more than the conventional epitaphs for Benjamin in the cemetery itself; for others, the landscape and the insertion into it are eloquent enough with stillness and thoughtfulness.

But there are other losses to confront in the modern world, where tragedies from the Holocaust to terrorists attacks at Mumbai, London and New York need to be memorialized. Not all of them are as eloquent or as *sotto voce* as Karavan's homage to Benjamin, yet architects as well as landscape architects have been eager to offer designs, and something of a memorial mania has set in, rivalling the sculpture park as a genre for the present. None the less, these designs pose particular problems in navigating both the conflicting demands of victims' families and other interested parties or communities, and the rival calls for realistic or abstract imagery. In fact, how well these features, that draw in great part upon the repertoire of gardens, will survive in the future when memories become less precise and immediate is not obvious.

But there are other losses to confront, not immediately personal or indeed of immediate human comfort: the derelict spaces of distraught and disfunctional industry, where loss of jobs and empty sites leave often vast lacunae in the fabric of cities, suburbs and their populations. Yet these have become one of the most exciting challenges for landscape architects: not, if at all, to erase an industrial site and install a new park and gardens – as with the Parc Citroën in Paris – but to reconfigure existing remains for a variety of purposes that do not always want or receive the pastoral treatment of nineteenth-century parklands. Some spectacular successes have been achieved here, with Duisburg-Nord in Germany setting a magnificent example of absorbing its derelict steelworks into a contemporary world of activity, performances and revitalized civic life. Brownfield sites and other toxic territories or 'drosscapes' have also been identified as places where the remedial work of good landscaping can heal both social and built disruptions (see chapter Nineteen). The much trumpeted 'landscape urbanism' (at least in the United States) seeks to take up that particular challenge of re-envisioning how landscapes may take charge of urban infrastructures and the large-scale spread of cities throughout the world. This means refusing to let the architect set the agenda, letting landscape and buildings honour each other's presence and usefulness, or even empowering the expanded field of landscape itself to set the agenda.[8]

For my own part, I think there is a cluster of further ideas and objectives that may direct new landscaping, in addition to sculpture parks, memorials and renovations of industrial wastelands. If we draw upon the past world of garden making for inspiration for the future, which is all we *can* do, an agenda for landscape architects might need to deal with the following: (1) reviewing the 'non-places' of modern life – railways and freeways; (2) confronting issues of locality and globalism, and the possible ways in which different cultures and even different times may reassert their own gardens today; (3) acknowledging that much public life needs more drama, more interaction between individuals, for landscape architects can be particularly good at exploiting what are long-standing debts of garden-making to theatre; (4) being well positioned to invoke history and historical memory, especially in sites where renewal is needed, yet the past need not be lost.

Beyond the fascination for renewing derelict sites, there is also the need to make more interesting the basic infrastructures of modern life, like railways and motorways (autobahns, autoroutes or freeways), malls and airports, where we spend so much of our time, what Marc Augé calls 'non-places'. Airports have a hard time of it, with security and the enormous discomforts of air travel; malls have certainly tried and their exterior spaces fare better than do airports.[9] Railways have managed far better: not just the destination dreamworld of a Jardin Atlantique (see chapter Nineteen), but the new stations that have responded throughout France to the advent of the fast train (TGV). The loss of huge unused marshalling yards have also released space for new gardens (see chapter Nineteen). Yet what tends to have been lost are the local gardens that graced suburban stations, welcoming the *pater familias* returning from the office on his way to his own surburban garden; in many European countries railway cuttings and express trains now bypass local halts, where passengers might have waited for a local train and taken pleasure in flower and vegetable plots, which vied for prizes for the most attractive railway gardens, but not any more.[10]

However, if railways have suffered a decline in horticultural showmanship and gardenscapes, motorways in France have not, and the French Government has worked hard from the 1980s to obviate criticisms of how the system of autoroutes was destroying the French countryside. Autoroutes now provide an immense and highly efficient network of roads. Their owners (since autoroutes are owned and maintained by private companies), as well as their users, are beginning to appreciate how these interminable journeys need both stimulation and respite. Some real success has been achieved with the creation of rest areas that provide more than just the necessary toilets and eating places as well as with re-thinking adjacent landscapes along their routes. That last move of course challenges the designer to invite lateral glimpses without distracting from fast driving; yet careful screening on the one hand, and contriving perspectives towards distant mansions or other notable buildings on the other, make long European car journeys much less boring. And the creation of rest areas, where the traveller may spend some time out of the vehicle by exploring local natural features or learning something of the locality through which he passes, also enlivens long journeys. Invitations to turn off and explore the rock formations and caverns at Crazannes on the Saintes-Rochefort autoroute, with its strange hollowed-out stone quarries and its rare ferns, or the chance to stop in the Provençal landscape near Nîmes are richly repaid. At Nîmes-Caissargues, taking a break from the interminable drive from Italy to Spain, visitors can walk on a huge *tapis vert* that stretches on either side of the autoroute, making one wonder which came first, the sunken roadway or the huge lawn; there is also a garden-landscape that features a folly in the form of a facade from the former nineteenth-century opera house in the city and two belvederes that resemble the famous Roman Tour Magne in Nîmes itself, placed at each side of the roadway (illus. 245). These allow a distant view of the city

245 Autoroute rest area outside Nîmes, by Bernard Lassus.

and include miniature versions of the Tour; a model of the ancient city was also planned for display there. In each place local inhabitants can now visit and enjoy these rest areas for motorists: exploring the quarries (to which an access road brings them without the need to enter the autoroute), or taking a distant view of their own city of Nîmes.[11]

Historians need to respect the cultural contexts of local gardens and parks, but there are some fascinating occasions when diachronic gestures to other forms and ideas across centuries and cultures enlarge our attention. We can all think of examples, whereby our response to a specific time and place is augmented or intriguingly sidetracked by other impressions. The Chinese, Turkish or other exotic folly, once so ubiquitous and yet increasingly empty as a trigger in gardens and parks of the late eighteenth century, has returned as a 'strongly identifiable landscape of leisure' in derelict landscapes: the Picturesque has not lost its hold upon our associations, though what prompts us is new. Christopher Girot sees the 'dilapidated steel plant' of Duisburg Nord as 'a highly emblematic folly'.[12] A modern Islamic park in Bradford may be designed to appeal to a local immigrant population (see chapter Four), but it also intrigues us with memories of Mughal gardens in the midst of the foreign fields and factories of Yorkshire. In France, a Garden of the Noria relies upon this ancient mechanism for drawing water from wells – the noria was established very early under Arabic agriculture (illus. 246 and 247); but here in Provence, the device is augmented with drainage culverts, a miniature pool in the manner of the Generalife in Spain, and modern seats and huge flower pots that echo Islamic garden art but in modernist forms. Noguchi's garden at the UNESCO Headquarters in Paris may be Japanese in inspiration and in its formal devices (see chapter Seventeen), but it too reads as a modernist space and one that also evades a merely Japanese identity. Conversely, Lawrence Halprin's Heritage Park on the banks of the Trinity River draws upon Japanese forms and ideas to mark the site of the first Fort Worth that gave its name to this Texas city. Heritage Place is a meditative spot, with a series of rather abstract concrete walls that outline the first fort's footprint, and it was originally (though the water has been turned off, at least for the moment) filled with the sound of water, falling over the initial wall, then flowing in runnels from a concrete pavilion: both the water channels that flow from a stone basin reminiscent of the Japanese *chozubachi* (where one purified oneself before entering a shrine) and the stepping stones (or rather concrete slabs) are clear gestures towards Zen gardening, what a critic has described as a 'sequence of carefully evoked sensual experiences'.[13] The precise cultural reference to Japan may be missed by many visitors, especially in the absence of the water effects; but Halprin's design invites meditations beyond the immediate locality, even while it celebrates another time and place.

Borrowings from other garden cultures raise the interesting problem, in a world that is much connected and global in its reach, of how and to what extent we confront the local in landscape architecture. A provoking appeal to connect 'place, memory, and identity' is outlined in a book by Philip Sheldrake (see chapter One). His appeal to the work of Gerard Manley Hopkins recalls another writer, the Jesuit-trained James Joyce, who also appeals to the *integritas* of something as simple as a butcher's basket and the analysis that will apprehend it as 'complex, multiple, separable, made up of parts . . . and their sum, harmonious'; we need only attend to Joyce's own understanding of the specificity of

246, 247 Two views of the Jardin de la Noria, near Uzès, Provence.

Dublin to realize the powerfulness of such a thick description of place.

One of the sustaining interests in a volume like Weilacher's *In Gardens* is that all the sites discussed and illustrated reveal the power of a specific place. This is in part simply a refusal to consider anything that does not have this particular quality; from Scotland to Switzerland, Berlin to Chandigarh, Amsterdam to Graz, designs reveal their distinctive response to materials, form and context. They are often wildly different: Ian Hamilton Finlay's 'close walks' (a term I draw from eighteenth-century garden vocabulary); Charles Jencks's expansive landscape with its apotheosis of fractals, wave transmission and chaos theory in the Garden of Cosmic Speculation; the Tarot garden at Garavicchio, Tuscany – a intricate and colourful garden whose images outdo the carved figures of Bomarzo; a similar confection of indigenous cement sculptures and rubble work in the ever-expanding Rock Garden in the Punjab (ironically in the vicinity of Le Corbusier's modernist city, where its creator, Nek Chand, worked as a street inspector). The range of sites, spaces and occasions makes this anthology of garden-making essays hugely exciting: each performs its own role, with sculpture, water forms, sharply shaped grass plots, occasional in-

scriptions (but not too many), historical gestures (Christophe Girot's Invalidenpark in Berlin), and sharp confrontations of busy, rococo planting beds with austere canals. Some places do not appeal – Karavan's Garden of Memories in Duisburg fails to do this for me, with its fussy shapes and whitened remains of the old harbourside; but other quasi-insertions of land art in the Park of Magic Waters at Bad Oeynhausen work well. In the end, it is something like Kienast Vogt Partnership's street garden and courtyard in Zürich that most commends itself – simple, playful (children dipping their fingers in the water troughs outside the building) and an acknowledgement of the process by which algae, lichen and moss will colour the small courtyard (illus. 248).

That ecological gesture, the fondness for relatively small interior courtyards and the renewed appeal to garden forms and motifs are also a leitmotif in much recent work. The work of Michael Van Valkenburgh maintains a strong ethical and ecological approach[14] – retention ponds, marsh lands fed by runoff from buildings, car parks marked by panels of constructed wetlands – yet melds it with an attention to the history of a place and an understanding of what gardens contribute. Mill Race Park in Columbus, Indiana, can tolerate widespread flooding, and elsewhere he encourages

248 Courtyard for the Basler + Partner Building, Zürich, by Kienast Vogt Partnership.

ice to form down rocks or garden screens. He is not afraid of utilizing landforms to enrich our discovery of small valleys, or of invented, even mimic, geology, like the rocky slabs in Teardrop Park in lower Manhattan, an abstracted wilderness that recall sections cut through the New York Catskills Mountains when roads were made there (illus. 249). He also loves entrances or thresholds – portals in Teardrop Park, the entry to the Children's Museum plaza at Boston, or other layouts that play with the excitement of arrival.

Amateur gardenists by necessity tend to pay attention to how a site derives its character from a combination of something inherent in the topography and its accumulation of human histories. Landscape architects have been less committed: they are, of course, invested in their own imaginations, but the need to push past the usual forms and manoeuvres of public landscaping to apprehend a certain identity of place is increasingly needed today, not least because, as Sheldrake admits, 'Place is no longer simply local'. One way in which locality has a chance to inflect public places is by engaging in a proper respect for the theatrical possibilities of their designs; not, if at all, merely scenographic inventions that 'stage' the place (like some theme park), but ensuring that the places themselves perform well and that people are also able to see themselves in public spaces and, in their turn, able to enjoy others doing the same. The old links between theatre and gardening that go back to Greek theatres, to Renaissance gardens, to the links of designers like William Kent to work for the theatre (see chapter Nine), to the rebuke that Charles Baudelaire delivered to boring scene-painters in the theatre that they were not inventive enough, all these need to find their own contemporary forms and uses. J. B. Jackson as early as 1979 argued for the landscape to be thought of as a theatre in both physical and metaphorical terms, but we still do not respond fully to those possibilities.[15] Modern graphics always present us with people Photoshopped into the scenery (see illus. 137), but rarely do we think of these same folk as *engaged* in the exchange of the promenade, as participants in the theatre of park or garden.

In many instances, the theatrical event is there already, assumed into the design. The park at Duisburg-Nord actually has a stage for performances, but as Girot saw it, the whole park itself was 'staged'.[16] By this, I take him to say that the place not only allows many activities to be watched as well as undertaken, but also that it presents its best and interesting face to the crowds who come there. Similarly, Martha Schwartz's Exchange Square in Manchester is an invitation to observe other people doing a variety of things and to be observed oneself, and to do so in a setting (scenery) or place of some significance (see chapter Nineteen). Any stage or hint of amphitheatre invites that dual activity of people watching people watching people. But there are many other, more subtle ways in which we instinctively respond to theatrical opportunities.

Entry into public plazas or squares from surrounding streets and buildings may be heralded by grandiose gateways, inviting exedras or simply the act of emerging into another space. There is nothing, as John Ruskin realized, quite like bursting into the Piazza di San Marco in Venice from the surrounding *calle*, as recounted in his second volume of *The Stones of Venice*; but he also had carefully contrasted that apotheosis with a different excursion through a quiet English city and, under 'a low grey gateway', emerging into a cathedral close. The comparison with Venice makes very clear the wholly different sense of each place, for which his detailed and alert descriptions prepare one.

Such thresholds or portals, whether palpable, subtle, prepared for by entrances or suddenly confronting us as we emerge into an unexpected square, are a dramatic form of grabbing our attention and easing us over the entry. And the architect has always to assess and plan movement through parks and cities; even small gardens can respond to a thoughtful sequence of moves. Lawrence Halprin may have learnt much from his talented

249 Teardrop Park, Manhattan, New York, by Michael Van Valkenburgh.

dancer wife, Anna, how to choreograph places and plot movement through them; but he also learnt from the stepping stones of Japanese gardens how to step and when to stop and look. It shows in his detailed notation of how to negotiate civil space (illus. 250), how to pause and observe; but it also worked in his much smaller private gardens.[17] Another very simple device was pioneered by Serlio (see chapter Nine), the opening and ascending pair of stairs that lead the visitor into a new space. It functions, too, in reverse, and is a form that we find in many places: it reappears several times in the Arts and Crafts garden (see illus. 218 and 222), and has been imitated many times since then, as well at the entrance to the Turner wing of the Tate at Millbank, London.

The modern hostility to Picturesque, notwithstanding, has also a key role to play in a theatrical response to places: not by the trite journalistic praise or design of some pretty spot, but in how the designer encourages us to respond in both rural and urban sceneries, triggering our memories and associations through the activity of movement. The Picturesque was never, in practice, a stationary event, even if it is captured in engravings, paintings and photographs. The Picturesque, on the contrary, struggles against the restriction of any place to the confines of a picture frame or a lens viewfinder, refusing to see 'all terrains [reduced] to the flatness of a sheet of paper' (in the words of the eighteenth-century French gardener René-Louis de Girardin). It practices what the

250 Skyline Park, Denver, Colorado, by Lawrence Halprin, before its partial destruction.

251 Pieter Andreas Rysbrack, *The Orange Tree Garden, Chiswick*, late 1720s, oil on canvas.

eighteenth century termed 'parallaxis', from the Greek, 'a displacement of the apparent position of a body, due to a change of position of the observer'.[18] As a pedestrian walking through a garden or park, we encounter how the place presents itself to us and how we in turn respond to it as if in a *moving* picture. Hence the simple but effective role of the Serlian double staircase, a miniature moment of how we negotiate a site.

In landscape we live by and in the process of this tripartite experience: we find or chance upon a site, we see it in perambulation, and we absorb its significance, an inner movement of the mind like that of the feet. A sequence of site, sight and insight. This process is difficult to capture except in practice and on real ground, but just occasionally we can intuit this experience even in a fixed image, and the eighteenth century is precisely the time when we find good examples, given that it was the high point of a properly understood Picturesque. Pieter Andreas Rysbrack painted the Orange Tree Garden at Chiswick in the late 1720s, and he shows a party emerging through the far hedges, one of them gesturing to what is just then to be encountered in the gardens (illus. 251, 252). This is an experience we all have of responding to landscape, often without realizing its dramatic possibilities.

All these possibilities rely upon the resources of the past to implement the present and shape the future. Georges Descombes calls design the 'most evident prism of sedimented culture';[19] from a beam of light, the designer separates and declares the spectrum of forms and meanings. Being alert to what lies hidden in the sediment of different cultures can draw upon the rich language of gardens and landscapes, so that the future may be as striking as its past has been.

252 Detail from Rysbrack's *The Orange Tree Garden.*

The eighteenth-century naturalist and garden-maker Gilbert White, author of *The Natural History and Antiquities of Selborne,* is invoked in chapter 17 of the medieval world of T. H. White's *The Once and Future King* (1958) as a spokesman for the recurrence of important things, future and past. Though the context concerns ornithology rather than gardens, the 'language' of gardens is also curious and stimulating, 'large', and 'ancient'. It is an apt note upon which to end.

> 'Do you know', asked the Wart, thinking of the thrush, 'why birds sing, or how? Is it a language?'
>
> 'Of course it is a language. It is not a big language like human speech, but it is large.'
>
> 'Gilbert White,' said Merlyn, 'remarks, *or will remark, however you like to put it,* that "the language of birds is very ancient". . .'.

REFERENCES

1 Sacred Landscapes from Delphi to Yosemite

1 For Finlay see my discussion in *Nature Over Again: The Garden Art of Ian Hamilton Finlay* (London, 2008), in particular pp. 82–7. For Bomarzo and Retz, see below in chapters 7 and 14.
2 Leon Battista Alberti, *On the Art of Building in Ten Books*, trans. Joseph Rykwert, Neil Leach and Robert Taverner (Cambridge, MA, 1988), Book Six in particular.
3 I am indebted here and in what follows to Lothar Ledderose, 'The Earthly Paradise: Religious Elements in Chinese Landscape Art', in *Theories of the Arts in China*, ed. Susan Bush and Christian Murck (Princeton, NJ, 1983).
4 Yungang Grotto in Central China has a network of grottoes, with over 1,000 niches and 51,000 statues, some of which are the size of a thumb, others more than 50 feet tall. I owe this reference to Senta Burton.
5 See the translation of Pausanias' guide by Peter Levi (1971), and the article by Patrick Bowe, 'The Sacred Groves of Ancient Greece', *Studies in the History of Gardens and Designed Landscapes*, XXIX (2009). The literature on this theme, though not much focused upon landscape architecture per se, is immense: see particularly Susan E. Alcock and Robin Osborne, *Placing the Gods: Sanctuaries and Sacred Space in Ancient Greece* (Oxford, 1992), also Hugh Bowden, *Mystery Cults in the Ancient World* (Princeton, NJ, 2010).
6 For a detailed and fascinating reading of the landscape of Delphi, Apollo's shrine, see Vincent Scully, *The Earth, the Temple, and the Gods: Greek Sacred Architecture* (New Haven, CT, 1962), chapter 7, from whom some phrases are quoted here.
7 See Alfred Runte, *Yosemite, The Embattled Wilderness* (Lincoln, NB, 1990), *Such a Landscape! William Henry Brewer*, his 1864 survey, ed. with modern photographs by William Alsup (Yosemite Association, 1998), and Ethan Carr, *Wilderness by Design: Landscape Architecture and the National Park Service* (Lincoln, NB, 1998).

2 Hunting Parks to Amusement Parks

1 I am again indebted here to Lothar Ledderose, 'The Earthly Paradise: Religious Elements in Chinese Landscape Art', in *Theories of the Arts in China*, ed. Susan Bush and Christian Murck (Princeton, NJ, 1983). See also Edward H. Schafer, 'Hunting Parks and Animal Enclosures in Ancient China', *Journal of the Economic and Social History of the Orient*, XI (1968), pp. 318–43.
2 I am grateful to my colleagues in the Chinese section of the University of Pennsylvania Library for help in deciphering the materials here. See also Che Bing Chiu, *Yuanming Yuan. Le jardin de la clarté parfaite* (Paris, 2000), p. 59 for notes on Chengde and p. 60 for a Yuanning Yuan map of 1860 which shows Jingyi Yuan at the far left. Philippe Forêt, *Mapping Chengde: The Qing Landscape Enterprise* (Honolulu, HI, 2000), discusses in passing the hunting parklands, but without any descriptions or imagery.
3 See J. K. Anderson, *Hunting in the Ancient World* (Berkeley, CA, 1985), from whom I draw material on Ashurbanipal, Xenophon and royal hunting parks. See also D. B. Hull, *Hounds and Hunting in Ancient Greece* (Chicago, IL, 1964), and Robin Osborne, *Classical Landscape with Figures: The Ancient Greek City and its Countryside* (London, 1987).
4 See Jacques Aymard, *Essai sur les chasses romaines, des origins à la fin du siècle des Antonions* (Paris, 1951), but much here is devoted to the techniques rather than the specific sites for hunts.
5 The key article here, re-interpreting much earlier erroneous material, is by Anna Hagopian van Buren, 'Reality and Literary Romance in the Park of Hesdin', in *Mediaeval Gardens*, ed. Elizabeth B. MacDougall (Washington, DC, 1986), pp. 115–34, which is quoted here. The author relies also on a little-known survey from 1906.
6 A somewhat similar Burgundian painting in the Dijon

Musée des Beaux-Arts shows a similar pavilion raised over marshes.

7 *Filarete's Treatise on Architecture,* translated with an Introduction and notes by John R. Spencer, 2 vols (New Haven, CT, 1965). The following reference to Giorgio Martini is to *L'Arte dei giardini*, ed. Margherita Azzi Visentini, 2 vols (Milan, 1999), I, pp. 86–9.

8 See David R. Coffin, *The Villa in the Life of Renaissance Rome* (Princeton, NJ, 1979), chapter 4, 'The Hunting Lodge and Park'.

9 See D. R. Edward Wright, 'Some Medici Gardens of the Florentine Renaissance: An Essay in Post-aesthetic Interpretation', in *The Italian Garden: Art, Design and Culture*, ed. John Dixon Hunt (Cambridge, 1996), pp. 34–59, who draws upon both archival descriptions and books of hunting lore.

10 The map is on p. 13. See also issues of *The London Gardener or the Gardner's Intelligencer*, with its very useful coverage of similar projects: an index for the years 1995–2005 was published in 2005. Duck Island was illustrated in an article in the third volume. I have been somewhat less restrictive in using the term 'pleasure gardens' than Jonathan Conlin, ed., *Grounds for Pleasure: The Pleasure Garden in Britain and North America 1660–1914* , forthcoming (Philadelphia, 2012), since I wish to enlarge my topic to include other amusement parks.

11 Wroth devotes a considerable section of his book to both Vauxhall and Ranelagh; see also the forthcoming volume *Grounds for Pleasure*, ed. Conlin, with wide-ranging essays, including American sites that recalled the name or the entertainments of the original London site, and bibliography on Vauxhall; and my own publications, *Vauxhall and London's Garden Theatres* (Cambridge, 1985), and 'Theaters, Gardens and Garden Theaters', in *Gardens and the Picturesque: Studies in the History of Landscape Architecture* (Cambridge, MA, 1992), chapter 2.

12 See Terence Young and Robert Riley, eds, *Theme Park Landscapes: Antecedents and Variations* (Washington, DC, 2002).

13 The discussion of the bourgeois effect of public parks, and its debts to theatre, are set out in Heath Schenker, *Melodramatic Landscapes* (Charlottesville, VA, 2009).

14 I am much indebted to the lively account by John F. Kasson, *Amusing the Million: Coney Island at the Turn of the Century* (New York, 1978), with extensive references and images; and to 'That Was Coney As We Loved It . . . ', chapter 7 in Jon Sterngass, *First Resorts: Pursuing Pleasure at Saratoga Springs, Newport, and Coney Island* (Baltimore, MD, 2001). The days of Coney Island, long since under threat, have lost further icons of its past: the *New York Times* (11 December 2010) listed a hotel and a dance hall for demolition.

3 Ancient Roman Gardens and their Types

1 Materials are extensive and I give here just a few further references, to which I am indebted: Elizabeth B. MacDougall and Wilhelmina F. Jashemski, eds, *Ancient Roman Gardens* (Washington, DC, 1981); Elizabeth B. MacDougall, ed., *Ancient Roman Villa Gardens* (Washington, DC, 1987), especially essays by A. R. Littlewood and Nicholas Purcell; John Henderson, *Hortus: The Roman Book of Gardening* (London, 2004); Linda Farrar, *Ancient Roman Gardens* (Stroud, 1998), here at p. 163; Maureen Carroll, *Earthly Paradises: Ancient Gardens in History and Archaelogy* (Los Angeles, CA, 2005). An early magisterial survey, untranslated, is Pierre Grimal's *Les Jardins Romains* (Paris, 1969). On Greek gardens in particular, not discussed here, see Robin Osborn, 'Classical Greek Gardens: Between Farm and Paradise', in *Garden History Issues, Approaches, Methods*, ed. John Dixon Hunt (Washington, DC, 1992), pp. 373–91. A discussion of modern Greek vernacular gardens is to be found in Martine Landriault, *L'Espace et le temps au jardin grec* (Paris, 2010); it is always an interesting perspective to see how conservative have been local gardens and their gardeners.

2 H. H. Tanzer, *The Villas of Pliny the Younger* (New York, 1924), surveys later reconstructions; also taken up by Pierre de la Ruffinière du Prey, *The Villas of Pliny from Antiquity to Posterity* (Chicago, IL, 1994).

3 The essential volumes here are W. Jashemski, *The Gardens of Pompeii and Herculaneum and the Villas Destroyed by Vesuvius*, 2 vols (New York, 1979 and 1993); but see also Bettina Bergman, 'Visualizing Pliny's Villas', *Journal of Roman Archaeology*, 8 (1995), pp. 406–20, with extensive notes, and Michel Conan's essay 'Nature into Art: Gardens and Landscapes in the Everyday World of Ancient Rome', *Journal of Garden History*, 6 (1996), pp. 348–56.

4 Farrar's book *Ancient Roman Gardens*, as does A. R. Littlewood's 'Ancient Literary Evidence for the Pleasure Gardens in Roman Country Villas', in E. B. MacDougall, *Ancient Roman Villa Gardens* (Washington, DC, 1987), usefully notes all the literary references to gardens discussed; for an extensive discussion of one very famous garden of a Roman historian and how over the years it came to be discovered and interpreted, see Kim J. Hartswick, *The Gardens of Sallust: A Changing Landscape* (Austin, TX, 2004).

5 The essential reading here is William L. MacDonald and John Pinto, *Hadrian's Villa and Its Legacy* (New Haven, CT, 1991), to which I am much indebted.

4 Islamic and Mughal Gardens

1 I am indebted to William Tronzo for commentary on these two Palermo buildings. On early gardens in Palermo, see Henri Bresc, 'Les Jardins royaux de Palerme (1200–1460)', *Mélanges de l'école française de Rome, Moyen-Age*, 84 (Paris, 1972), pp. 55–127.
2 While I have myself visited Granada, Palermo and Istanbul, for the rest of my work I rely upon the two excellent books by D. Fairchild Ruggles: *Gardens, Landscape, and Vision in the Palaces of Islamic Spain* (University Park, PA, 2000) and *Islamic Gardens and Landscapes* (Philadelphia, PA, 2008), with extensive lists of further reading. Also Luigi Zangheri, Brunella Lorenzi and Nausikaa M. Rahmati, *Il giardino islamico* (Florence, 2006), with excellent plans and photographs. I have also been helped by the entries in *The Dictionary of Art*, edited by Jane Turner. Other debts are cited in later notes.
3 Taken from the translation in Oleg Grabar, *The Alhambra* (London, 1978), p. 123.
4 *Clavijo: Embassy to Tamerlane 1403–1406*, trans. Guy de Strange (London, 1928). Other useful accounts are in Norah Titley and Frances Wood, *Oriental Gardens* (London, 1991), especially sections on Iran and India, to which I am much indebted, and Donald Newton Wilber, *Persian Gardens and Garden Pavilions* (Washington, DC, 1979).
5 These later descriptions can be found in Martin J. Price, *The Seven Wonders of the Ancient World*, ed. Peter Clayton and Martin Price (New York, 1988), pp. 42–6, and for subsequent attempts to image the Gardens see Jan Pieper, 'Die Natur der Hangenden Garten / The Nature of Hanging Gardens', *Daidalos*, 23 (1987), pp. 94–105.
6 See James L. Wescoat Jr, 'Picturing an Early Mughal Garden', *Asian Art*, II/4 (1989), pp. 59–79, for a fundamental essay in the reading of garden paintings and their relationship to other materials. For Babur's memoir, see *The Babur-nama in English*, trans. A. S. Beveridge (London, 1969). See also James L. Wescoat and Joachim Wolschke-Bulmahn, eds, *Mughal Gardens: Sources, Places, Representations, and Prospects* (Washington, DC, 1996).
7 Prabhakar B. Bhagwat, 'The Gardens of India', at www.international.icomos.org/publications/93garden5.pdf.
8 For this see Eugenia W. Herbert, *Flora's Empire: Garden Imperialism in British India* (Philadelphia, PA, 2011).

5 Western Medieval Gardens: From Cloister to Suburban Backyard

1 On traditions of old and new enclosed gardens see Rob Aben and Saskia de Wit, *The Enclosed Garden: History and Development of the Hortus Conclusus and its Reintroduction into the Present-day Urban Landscape* (Rotterdam, 1999) and Francesco Nuvolari, *Hortus Conclusus* (Milan, 1986).
2 A most useful, further collection of colour images of medieval gardens can be found in the small book by Marie-Thérèse Gousset, *Eden* (Paris, 2001).
3 I am indebted to the translation of Book 8 and the discussion of it by Johanna Bauman, 'Tradition and Transformation: The Pleasure Garden in Piero de' Crescenzi's *Liber ruralium commodorum*', *Studies in the History of Gardens and Designed Landscapes*, XXII (2002), pp. 99–141.
4 Such directions are still at the centre of modern works, like Garrett Eckbo in *Landscape for Living* (New York, 1950), where he lists 'surface, enclosure and enrichment' as the main elements of any landscaped area (p. 6).
5 See the discussion of the *making* of late medieval and early Renaissance gardens in Raffaella Fabiani Giannetto, *Medici Gardens: From Making to Design* (Philadelphia, PA, 2008).
6 The planting discussions of medieval gardens are superbly explained by John Harvey, *Mediaeval Gardens* (London, 1981); but see also Teresa McLean, *Mediaeval English Gardens* (London, 1981).
7 As early as the *Epic of Gilamesh* (2700 BC) the Babylonians described their paradise garden as 'beside a sacred fount the Tree is placed'. There are similar references in the Qur'an to both water and shade.
8 There is an engraving of a Carthusian monastery in Venice that shows all these features: see my book *The Venetian City Garden: Place, Typology, and Perception* (Berlin and Boston, 2009), figure V.23, and where in my fifth chapter I discuss the varieties of ecclesiastical and monastic gardens in that special city.

6 The Renaissance Recovery of Antique Garden Forms and Usages

1 I have elaborated on these English visitors to Renaissance and Baroque Rome in the second chapter of my *Garden and Grove: The Italian Renaissance Garden in the English Imagination, 1600–1750* (London, 1986; new paperback edition, Philadelphia, 1996), 'Classical ground and classical gardens'; here I am more concerned to see how the Italians revisited

Roman culture and made it new.

2 For modern discussions of the gardens of Sallust, see Kim J. Hartswick, *The Gardens of Sallust: A Changing Landscape* (Austin, TX, 2004). For the Villa Medici, G. M. Andres, *The Villa Medici in Rome*, 2 vols (New York, 1976)
3 The Belvedere Courtyard and its sculpture court have been much studied. The most recent publication, with excellent colour images and an extensive bibliography, is by Alberta Campitelli, *The Vatican Gardens* (Vatican and New York, 2009). Other useful English essays are by J. Ackerman, 'The Belvedere as a Classical Villa', *Journal of the Warburg and Courtauld Institutes*, 14 (1951), and Hans Henrik Brummer, *The Statue Court in the Vatican Belvedere* (Stockholm, 1970).
4 This is clearly shown in the anonymous plan of the courtyard in the Soane Museum (H. H. Brummer, *The Statue Court in the Vatican Belvedere*, Stockholm, 1970, fig. 5) at the very top of the courtyard; Campitelli seems to suggest, somewhat misleadingly, that it was situated at the mid-point between the divided ramps.
5 See Elizabeth Blair MacDougall, *Fountains, Statues, and Flowers: Studies in Italian Gardens of the Sixteenth and Seventeenth Centuries* (Washington, DC, 1994), and David R. Coffin, *Gardens and Gardening in Papal Rome* (Princeton, NJ, 1991).
6 See my *Garden and Grove*, pp. 195ff. It is ironic that on two occasions when I have tried to illustrate this engraving of the Villa of Servilius Vatia, publishers have deleted the caption that identified the landscape as being that of a villa!
7 I draw here upon the extended discussion of water in these gardens by Claudia Lazzaro, *The Italian Renaissance Garden* (New Haven, CT, 1990).
8 I have explained this extension of Cicero's ideas and terms into modern gardening in my chapter 'The Idea of a Garden and the Three Natures', *Greater Perfections: The Practice of Garden Theory* (London, 2000), pp. 32–75.
9 The subject is much studied: Elizabeth Blair MacDougall, ed., *Fons Sapientiae: Renaissance Garden Fountains* (Washington, DC, 1978). A modern edition of Switzer was published by Garland in 1982.
10 The claim is in J. Mordaunt Crook, *The Greek Revival* (London, 1972), p. 62.

7 The *Paragone* of Art and Nature in the Renaissance and Later

1 *L'Arte dei giardini*, ed. Margherita Azzi Visentini, 2 vols (Milan, 1999), I, p. 68.
2 I am indebted here to the detailed discussion by Anatole Tchikine, 'Giochi d'acqua: water effects in Renaissance and Baroque Italy', *Studies in the History of Gardens and Designed Landscapes*, XXX (2010), p. 57.
3 For a fuller discussion of the water effects at the Villa Lante (and the Villa d'Este) see Claudia Lazzaro, *The Italian Renaissance Garden* (New Haven, CT, 1990); but for an earlier and rather different narrative of the sequence of waters, see her 'The Villa Lante at Bagnaia: An Allegory of Art and Nature', *Art Bulletin*, 59 (1977), pp. 553–60. On Pegasus and the Muses here and in many other gardens of the period, see my 'Pegaso in villa. Variazioni sul tema', in *Villa Lante a Bagnaia*, ed. Sabine Frommel (Milan, 2005).
4 Quoted in Lionello Puppi, 'Nature and Artifice in the Sixteenth-century Italian Garden', in *The Architecture of Western Gardens*, ed. M. Mosser and G. Teyssot (Cambridge, MA, 1991), p. 49.
5 See the discussion and editorial commentary in Taegio's modern edition of *La Villa*, ed. Thomas E. Beck (Philadelphia, PA, 2010)
6 The fascination of English visitors for Italian gardens is explored in my *Garden and Grove: The Italian Renaissance Garden in the English Imagination, 1600–1750* (London, 1986; new paperback edition, Philadelphia, 1996).
7 Puppi, 'Nature and Artifice in the Sixteenth-century Italian Garden', p. 56, but many others have been excited by the fantasy world of Bomarzo.
8 Luke Morgan, *Nature as Model: Salomon de Caus and Early Seventeenth-century Landscape Design*, Penn Studies in Landscape Architecture (Philadelphia, PA, 2007), p. 188, also for a delayed analysis of the Hortus Palatinus, on which I have briefly drawn.

8 The Botanical Garden, the Arboretum and the Cabinet of Curiosities

1 I owe the Chinese account to Loraine Kuch, *The World of the Japanese Garden* (New York and Tokyo, 1968), p. 19; for Stowe, George Clarke, 'Where Did All the Trees Come From? An Analysis of Bridgeman's Planting at Stowe', *Journal of Garden History*, 5 (1985), pp. 72–83.
2 The St Gall map, as well as the woodcuts *c.* 1500, are illustrated in Anthony Huxley, *An Illustrated History of Gardening* (New York and London, 1978).
3 See my *The Venetian City Garden: Place, Typology, and Perception* (Berlin and Boston, 2009), pp. 105–10.
4 See M. Azzi Visentini, *L'Orto Botanico di Padova* (Milan, 1984), and Fabio Garbari, Lucia Tongiorgi Tomasi and Alessandro Tosi, eds, *Giardino dei Semplici / Garden of*

Simples (Pisa, 2002), with large bibliography.

5 John Prest, *The Garden of Eden: The Botanical Garden and the Re-creation of Paradise* (New Haven, CT, 1981) is still one of the best introductions to the topic of origins, myths and assumptions about Eden and Paradise and their possible whereabouts. For a large bumper survey, see Edward Hyams, *Great Botanical Gardens of the World*, with photographs by William MacQuitty (London, 1969), with a gazetteer of world botanical gardens on pp. 284–8.

6 See Garberi, Tongiorgi Tomasi and Tosi, *Giardino dei Semplici*, pp. 1–34, and her 'Geometric Schemes for Plant Beds and Gardens: A Contribution to the History of the Garden in the 16th and 17th Centuries', in *World Art*, ed. Irving Lavin (University Park, PA, and London, 1989), vol. I, pp. 211–18. See also similar discussions in books listed in notes 3 and 9.

7 See Oliver Impey and Arthur MacGregor, eds, *The Origins of Museums: The Cabinet of Curiosities in 16th- and 17th-century Europe* (Oxford, 1985), from which I have drawn here on my own contribution, 'Cabinets to Adorn Cabinets and Gardens', pp. 193–203.

8 See John Evelyn, *Elysium Britannicum: Or the Royal Gardens*, ed. John Ingram (Philadelphia, 2000).

9 See my *Garden and Grove: The Italian Renaissance Garden in the English Imagination, 1600–1750* (London, 1986; new paperback edition, Philadelphia, 1996), chapter 6.

10 My sources are: Joel T. Fry, 'An International Catalogue of North American Trees and Shrubs: The Bartram Broadside, 1783', *Journal of Garden History*, 16 (1996), pp. 3–66; Nancy E. Hoffmann and John C. Van Horne, eds, *America's Curious Botanist: A Tercentennial Reappraisal of John Bartram* (Philadelphia, PA, 2004); Thomas P. Slaughter, *The Natures of John and William Bartram* (New York, 1966). The writings of William Bartram are collected in the Library of America edition (New York, 1996). For discussions of the Bartram legacy to English gardening, see Douglas D. C. Chambers, *The Planters of the English Landscape Garden* (New Haven and London, 1993), and Mark Laird, *The Flowering of the Landscape Garden: English Pleasure Grounds, 1720–1899* (Philadelphia, PA, 1999), where the exchanges between Bartram and Collinson are set out (pp. 69–78).

11 The bibliography would be large: but in particular see the work in Lucia Tongiorgi Tomasi, *Livorno e Pisa* (Pisa, 1980), or the fine colour illustrations of the Pisan collections in my *The Venetian City Garden* and *Garden and Grove*; see also Lucia Tongiorgi Tomasi, ed., *An Oak Spring Flora: Flower Illustration from the Fifteenth Century to the Present Time* (Upperville, VA, 1997) .

12 David Sturdy, 'The Tradescants at Lambeth', *Journal of Garden History*, 2 (1982), pp. 1–16.

13 Frances Harris and Michael Hunter, ed., *John Evelyn and his Milieu* (London, 2003); Therese O'Malley and Joachim Wolschke-Buhlmahn, eds, *John Evelyn, the 'Elysium Britannicum' and European Gardening* (Washington, DC, 1998), and Evelyn, *Elysium Britannicum*.

14 I am indebted here and in the next paragraph to the discussions in Chambers's book *The Planters of the English Landscape Garden*, chapters 3, 6 and 7.

15 *Westonbirt Arboretum. Catalogue of the Trees and Shrubs*, compiled by A. Bruce Jackson (Oxford, 1927) provides a definitive record of the Arboretum in its best years. This chapter is not, however, concerned with development of trial plantings for national forests, such as those maintained by the Forestry Commission and by Key at the National Pinetum in Kent, though these endeavours clearly had effects upon the world of garden design and elaboration.

16 Melanie Louise Simo, *Loudon and the Landscape* (New Haven, CT, 1988), especially chapters 10 and 11, to which I am indebted.

17 Barbara Rotundo, 'Mount Auburn: Fortunate Coincidences and an Ideal Solution', *Journal of Garden History*, IV (1994), pp. 255–67. Also Blanche Linden-Ward, 'Strange but Genteel Pleasure Grounds: Tourist and Leisure Uses of 19th-century Rural Cemeteries', in *Cemeteries and Gravemarkers: Voices of American Culture*, ed. R. E. Meyer (Ann Arbor, MI, 1989), pp. 293–328.

18 Latz's plan is illustrated in Udo Weilacher, *Between Landscape Architecture and Land Art* (Berlin and Boston, MA, 1999), pp. 134–5, and Mosbach's in Peter Reed, ed., *Groundswell: Constructing the Contemporary Landscape* (New York, 2005).

9 Garden as Theatre

1 See Jean Jacquot, Elie Königson and Marcel Oddon, eds, *Le Lieu Théatral à la Renaissance* (Paris, 1964), Jean Jacquot, ed., *Les Fêtes de la Renaissance* (Paris, 1956 and 1960), and the catalogue of the exhibition *Il Luogo Teatrale a Firenze* (Florence, 1975), all of which have useful imagery. There have been many other publications in French and Italian; for English examples see Stephen Orgel and Roy Strong, *Inigo Jones: The Theatre of the Stuart Court*, 2 vols (Berkeley and London, 1973), and John Harris, Stephen Orgel and Roy Strong, *The King's Arcadia* (London, 1973) and A. M. Nagler, *Theatre Festivals of the Medici* (New Haven, CT, 1964). On the Vatican Belvedere courtyard, R. Leone, 'Torneo nel cortile del Belvedere', in *La Feste a Roma dal*

Rinascimento al 1870, ed. M. Fagiolo, exh. cat. (Turin, 1997).

2 I have myself focused upon the garden aspect of this development in various writings: I draw here on both chapter 5 of my *Garden and Grove: The Italian Renaissance Garden in the English Imagination, 1600–1750* (London, 1986, republished Philadelphia, 1996), and *Vauxhall and London's Garden Theatres* (Cambridge, 1985). See also the essay by J. B. Jackson, 'Landscape as Theater', in *The Necessity for Ruins* (Amherst, MA, 1980), pp. 67–76.

3 See my *Garden and Grove*.

4 John Nichols, *The Progresses of Queen Elizabeth I*, 5 vols (London, 1823), and *The Progresses . . . of James I*, 4 vols (London, 1828), and Ian Dunlop, *Palaces and Progresses of Elizabeth I* (London, 1962).

5 The most useful texts are Orgel and Strong, *Inigo Jones*, and Harris, Orgel and Strong, *The King's Arcadia*.

6 These are illustrated in Per Bjurström, *Giacomo Torelli and Baroque Stage Design* (Stockholm, 1961)

7 Many plans and engravings are available in Peter Willis, *Charles Bridgeman and the English Landscape Garden* (London, 1977; new and augmented edition Newcastle upon Tyne, 2002). For instances of Kent's theatrical spaces shown in his drawings, and for a discussion of his theatrical work, see my *William Kent: Landscape Garden Designer: An Assessment and Catalogue of his Designs* (London, 1987).

10 The Garden of 'Betweenity': Between André Le Nôtre and William Kent

1 This occurs in his chapter entitled 'The "Landscape-Garden"' (p. 122). I had not encountered Sedding's work when I was working on the similar period for my *Garden and Grove: The Italian Renaissance Garden in the English Imagination, 1600–1750* (London, 1986; new paperback edition, Philadelphia, 1996), a book that did not entirely pass muster with those who clung to a Walpolean narrative! But I did briefly salute a similar interest by Alicia Amherst, *A History of Gardening in England* (1895, 2nd edn 1896) who, at the same time as Sedding, devoted an unusual amount of attention to the pre-landscape garden era (in fact 234 pages out of the remaining 314).

2 I am indebted here to what David Leatherbarrow taught me in 1984 in the *Journal of Garden History* and subsequently republished in his *Topographical Stories: Studies in Landscape and Architecture* (Philadelphia, PA, 2004), chapter 5, where he also reprints the memorandum by Shaftesbury, pp. 266–9.

3 I have discussed and illustrated these in 'Castle Howard Revisited', in *Gardens and the Picturesque* (Cambridge, MA, 1992).

11 Leaping the Ha-ha; or, How the Larger Landscape Invaded the Garden

1 There are several republications of Walpole's essay: quotations in the text here are taken from the edition with my introduction (New York, 1995), p. 43. See also the critical edition of his essay by Isabel Wakelin Urban Chase, *Horace Walpole: Gardenist* (Princeton, NJ, 1943).

2 William Gilpin, *Dialogue Upon the Gardens ... at Stowe* (Augustan Reprint Society, Los Angeles, CA, 1976); page references are included in the main text.

3 These sightlines are plotted in diagrams by Hal Moggridge, 'Notes on Kent's garden at Rousham', *Journal of Garden History*, 6 (1986), pp. 187–226;one is also reproduced in my essay 'Verbal versus Visual Meanings in Garden History: The Case of Rousham', in *Garden History: Issues, Approaches, Methods*, ed. John Dixon Hunt (Washington, DC, 1992), pp. 151–81.

4 Quotations in the following paragraphs can be found in John Dixon Hunt and Peter Willis, eds, *The Genius of the Place: The English Landscape Garden, 1620–1820* [1975] (Cambridge, MA, 1988).

5 See here the splendid account of British self-construction in Linda Colley, *Britons: Forging the Nation, 1707–1837* (New Haven, CT, and London, 2005).

6 See my essay on Kent's landscape architecture in the forthcoming catalogue to the exhibition of Kent's varied career at the Bard Graduate Center for Design (New York, 2013).

7 See Kent's sketches in my *William Kent: Landscape Garden Designer: An Assessment and Catalogue of his Designs* (London, 1987).

8 *Observations on Modern Gardening* (London, 1770), p. 50.

12 The Role of the 'Natural' Garden from 'Capability' Brown to Dan Kiley

1 I refer to the imagery in David Dillon and Gary R. Hilderbrand, *The Miller Garden: Icon of Modernism* (Washington, DC, and Cambridge, MA, 1999); but also Dan Kiley and Jane Amidon, *Dan Kiley: The Complete Works of America's Master Landscape Architect* (Boston, MA, 1999).

2 *Horace Walpole's Correspondence with George Montagu*, II (New Haven, CT, 1948), p. 44.

3 Quoted in a Brown exhibition review by Robert Williams, 'Making Places: Garden-mastery and English Brown', *Journal of Garden History*, 3 (1983), pp. 382–5, to whose essay I am indebted. See also Dorothy Stroud, *Capability Brown* (new edition, London 1975), Roger

Turner, *Capability Brown and the 18th-century English Landscape* (2nd edn, Chichester, West Sussex, 1999), and *Capability Brown and the Northern Landscape*, exh. cat., Tyne and Wear County Council Museums (1983).

4 The major work on Morel is by Joseph Disponzio: see his 'Jean-Marie Morel: A Catalogue of his Landscape Designs', *Studies in the History of Gardens and Designed Landscapes*, XXI/3–4 (2001), from which I quote here, and 'Jean-Marie Morel and the Invention of Landscape Architecture', in *Tradition and Innovation in French Garden Art*, ed. John Dixon Hunt and Michel Conan, Penn Studies in Landscape Architecture (Philadelphia, 2002), pp. 137–59. Both essays were derived from his unpublished PhD dissertation on Morel for Columbia University (2000).

5 See Udo Weilacher's elaboration of these materials in his introduction to *Between Landscape Architecture and Land Art* (Basel, 1996).

6 *Transforming the Common/Place, Selections from Laurie Olin's Sketchbooks* (New York, 1996), and *Olin Placemaking* (New York, 2008). He refers briefly to and illustrates one private design in 'Regionalism and Practice of Hanna/Olin', in *Regional Garden Design in the United States*, ed. Therese O'Malley and Marc Treib (Washington, DC, 1995).

13 The Chinese Garden and the Collaboration of the Arts

1 It is, nonetheless, puzzling to a non-specialist that garden discussions jump across centuries willy-nilly. For my debts to Chinese garden writings I am obliged to Maggie Keswick, *The Chinese Garden: History, Art and Architecture* (new edition, London, 2003), with a new introduction by Alison Hardie, and to Osvald Siren, *Gardens of China* (New York, 1949), not least for its compelling photographs. I have also used the translation by Alison Hardie of Ji Cheng's *The Craft of Gardens*, with an introduction by Maggie Keswick (New Haven, CT, 1988), with its own good selection of photographs. See also Joanna F. Handlin Smith, 'Gardens in Ch'i Piao-chia's Social World: Wealth and Values in Late-Ming Kiangnan', *Journal of Asian Studies*, 51 (1992), pp. 55–81. There is an exhaustive bibliography of secondary sources on Chinese gardens by Stanislaus Fung, *Studies in the History of Gardens and Designed Landscapes*, XVIII (1998), pp. 269–86, contained in an issue of that journal devoted to Chinese garden art. Fung has edited two issues of this journal to give Western readers a chance to appreciate the complexity of Chinese garden arts, *Studies in the History of Gardens and Designed Landscapes*, XVIII/3 and XIX/3 and 4, to which I am also much indebted.

2 I owe this point and others in this chapter to Alison Hardie; see also Craig Clunas, *Pictures and Visuality in Early Modern China* (Princeton, NJ, 1997), chapter 5.

3 That is true even of a writer like Chen Congzhou: though mindful of different historical moments, his analyses frequently seek to draw out some essentialist position. See note 18 below.

4 Stanislaus Fung, 'The Imaginary Garden of Liu Shilong', *Terra Nova*, II/4 (1997), pp. 15–21. See, further, his 'Notes on the Make-do Garden', *Utopian Studies*, IX/1 (1998).

5 A discussion of the tensions within the novel and its characters is available in Xiao Chi, *The Chinese Garden as Lyric Enclave: A Generic Study of 'The Story of the Stone'* (Ann Arbor, MI, 2001), especially chapter 8.

6 David L. Hall and Roger T. Ames, 'The Cosmological Setting of Chinese Gardens', *Studies in the History of Gardens and Designed Landscapes*, XVIII/3 (1998), p. 185. See also John Makeham, 'The Confucian Role of Names in Traditional Chinese Gardens', *Studies in the History of Gardens and Designed Landscapes*, XVIII/3 (1998), pp. 187ff.

7 Derek Bodde, introduction to Bradley Smith and Wang-go Weng, *China: A History in Art* (London, 1973); the second remark is quoted by Keswick in her introduction to *The Craft of Gardens*, p. 23.

8 These paintings are illustrated and explained in an album by Che Bing Chiu, *Yuanming Yuan. Le jardin de la Clarté parfaite* (Paris, 2000). For the scroll see Philip K. Hu, 'The Shao Garden of Mi Wanzhong (1570–1628): Revisiting a Late Ming Landscape through Visual and Literary Sources', *Studies in the History of Gardens and Designed Landscapes*, XIX (1999), p. 314 with attached scroll.

9 I refer to the work of Professor Chen Congzhou, *On Chinese Gardens*; see note 18 below.

10 Andong Lu, 'Deciphering the Reclusive Landscape: A Study of Wen Zhen-Ming's 1533 Album of the Garden of the Unsuccessful Politician', *Studies in the History of Gardens and Designed Landscapes*, XXXI/1 (2011), pp. 40–59.

11 I cite Michael Sullivan, *The Three Perfections* [1974] (New York, 1980), pp. 7 and 17.

12 G. W. Robinson, *Poems of Wang Wei* (Harmondsworth, 1973).

13 I am indebted here and in what follows to Robert E. Harrist Jr, *Painting and Private Life in Eleventh-Century China* (Princeton, NJ, 1998).

14 Translated and with an introduction by Philip Watson, 'Famous Gardens of Luoyang, by Li Gefei', *Studies in the History of Gardens and Designed Landscapes*, XXIV/1 (2004), pp. 38–54. Zhu Changwen's text is not translated.

15 Kenneth J. Hammond, 'Wang Shizhen's Yan Shan Garden Essays: Narrating a Literari Landscape', *Studies in the History of Gardens and Designed Landscapes*, XIX (1999),

pp. 276–80, giving a detailed account of the possession of one particular garden; Joanna F. Handlin Smith, 'Gardens in Ch'i Piao-chia's Social World', *Journal of Asian Studies*, 51 (1992); Craig Clunas, *Fruitful Sites: Garden Culture in Ming Dynasty China* (London, 1996).

16 Wanggo Weng, *Gardens in Chinese Art* (New York, 1968), p. 5.

17 Harrist, *Painting and Private Life in Eleventh-Century China*, described three gardens along these lines, pp. 50–60.

18 Chen Congzhou, *On Chinese Gardens* (Shanghai, 1985), with texts in both Chinese and English and with a cluster of illustrations, not alas captioned.

19 I own this interesting observation to Carl Steinitz, who proposed the discussion of failure in 1947, though it met with a certain studied indifference among designers at Harvard's GSD (personal communication)

20 John Minford, 'The Chinese Garden: Death of a Symbol', *Studies in the History of Gardens and Designed Landscapes*, XVIII (1998), p. 258.

14 Follies, *Fabriques* and Picturesque Play

1 I have in this instance relied upon two of my earlier accounts: the relevant pages of *The Picturesque Garden in Europe* (London, 2003) and the essay, 'Folly in the Garden', *The Hopkins Review* (new series), I/2 (2008).

2 For a complete review of these publications, see Eileen Harris, assisted by Nicholas Savage, *British Architectural Books and Writers, 1556–1785* (Cambridge, 1990). A chronology of titles (pp. 513ff) suggests the richness of the topic and its publications. Eileen Harris also introduced the splendid republication of *Arbours & Grottos* (London, 1979).

3 A remarkable republication of these *cahiers*, loosely bound for easy consultation, was issued by Editions Connaissance et Mémoires (Paris, 2004), from which I take my illustrations; an accompanying volume by Véronique Royet, *Georges Louis Le Rouge. Les jardins anglo-chinois*, was published by the Bibliothèque Nationale de France.

4 Michel Baridon, 'The Garden of the Perfectibilists: Méréville and the Désert de Retz', in *Tradition and Innovation in French Garden Art*, ed. John Dixon Hunt and Michel Conan, with the assistance of Claire Goldstein (Philadelphia, PA, 2002), and Yves Bonnefoy's essay, 'Le Désert de Retz et l'expérience du lieu', in his collection *Le Nuage rouge* (Paris, 1977).

5 See my *The Venetian City Garden* (Berlin, 2009), chapters 6 and 8, where some of the Picturesque gardens in the city were the product of theatre designers. For Villa Pallavicini, Fabio Calvi and Silvana Ghigino, *Villa Pallavicini a Pegli* (Genoa, 1998).

6 The first two are only briefly discussed here: more discussion is available in my *Nature Over Again: The Garden Art of Ian Hamilton Finlay* (London, 2008), and in Charles Jencks, *The Garden of Cosmic Speculation* (London, 2003); for Hobhouse, I am grateful for his personal communications and help.

15 The Invention of the Public Park

1 Anita Berrizbeitia, 'The Amsterdam Bos: The Modern Public Park and the Construction of Collective Experience', *Recovering Landscape*, ed. James Corner (New York, 1999), p. 187.

2 Quoted in Marc Treib and Dorothée Imbert, *Garrett Eckbo: Modern Landscapes for Living* (Berkeley, CA, 1997), p. 94.

3 Cited by Rachel Iannacone, 'Open Space for the Underclass: New York's Small Parks (1880–1915)', PhD dissertation, University of Pennsylvania, 2005. Here and elsewhere in this chapter I am indebted to this crucial enlargement of our understanding of American public parks.

4 I have discussed this in *The Venetian City Garden* (Basle, Boston and Berlin, 2009), chapter 6.

5 Martin Knuijt et al., eds, *Modern Park Design; Recent Trends* (Bussum, 1995), p. 38. For an excellent survey and description of various public parks see Alan Tate, *Great City Parks* (London, 2001).

6 Mohsen Mostafvi and Ciro Najle, eds, *Landscape Urbanism: A Manual for the Mechanic Landscape* (London, 2003), and Charles Waldheim, ed., *Landscape Urbanism Reader* (New York, 2006)

7 The Tuileries was surveyed by revolutionaries immediately after the fall of the Bastille and the elaborate surveys of it by Billiot l'Ainé d'Edine are preserved in the collections of the Oak Spring Garden Library, Upperville, Virginia.

8 I am indebted here for the French materials to Richard Cleary, 'Making Breathing Room: Public Gardens and City Planning in Eighteenth-century France', in *Tradition and Innovation in French Garden Art: Chapters of a New History*, ed. John Dixon Hunt and Michel Conan (Philadelphia, PA, 2002), pp. 68–81.

9 On 'leftovers' see Antoine Grumbach, 'The Promenades of Paris', trans. Marlène Barsoum and Hélène Lipstadt, *Oppositions*, 8 (Spring 1977), pp. 49–67, and for plans of these 'squares' see Jean Charles Adolphe Alphand, *Les Promenades de Paris* (Paris, 1869; republished 2003).

10 Frederick Law Olmsted, *Public Parks and the Enlargement of Towns* (Cambridge, MA, 1870).

11 There has been a huge increase in memorials, though the impact of their educational role is infinitely varied:

see Erika Doss, *Memorial Mania: Public Feeling in America* (Chicago, IL, 2010).

12 See the catalogue of the MOMA exhibition, *Groundswell: Constructing the Contemporary Landscape* (New York, 2005) and the collected entries for the Toronto park, *case: Downsview Park Toronto*, ed. Julia Czerniak (Munich, 2004), where many of these new parklands are discussed and illustrated.

13 See the collection of short pieces *Learning from Duisburg Nord* (Munich, 2010). Duisburg-Nord is justly famous, but it is not unique in the work of Peter Latz + Partner: Udo Weilacher, *Syntax of Landscape: The Landscape Architecture of Peter Latz and Partners* (Basle, Boston, Berlin, 2008) provides a rich analysis of the Duisburg project.

14 Cited by Heath Schenker, 'Parks and Politics during the Second Empire in Paris', *Landscape Journal*, 14 (1995), p. 205. To this, and to her new book, *Melodramatic Landscapes* (Charlottesville, VA, 2010), I am much indebted. Tuckerman's remark is from *Maga Papers about Paris* (New York, 1867), p. 157.

15 David Gray, *Thomas Hastings, Architect: Collected Writings. Together with a Memoir* (Boston, MA, 1933), p. 243.

16 Some attention to these sculpture gardens is given in Antonia Boström, ed., *The Fran and Ray Stark Collection of 20th-century Sculpture at the J. Paul Getty Museum* (Los Angeles, 2008). See also Philip Jodidio, *Tadao Ando at Naoshima: Art, Architecture, Nature* (New York, 2006). For more on the Getty campus design and the Irwin garden, see *Olin: Placemaking* (New York, 2008), pp. 64–87, Lawrence Weschler, *Robert Irwin Getty Garden* (Los Angeles, CA, 2002), and David R. Marshall, 'Gardens and the Death of Art: Robert Irwin's Getty Garden', *Studies in the History of Gardens and Designed Landscapes*, XXIV (2004), pp. 215–28.

16 National Parks and International Exhibition Gardens

1 See Paul Schullery, *Searching for Yellowstone: Ecology and Wonder in the Last Wilderness* (Boston, MA, and New York, 1997).

2 I have used the republication of this report by the Yosemite Association, with an introduction by Victoria Post Ranney (Yosemite National Park, CA, 1995); it is also published in *The Papers of Frederick Law Olmsted*, vol. 5 (Baltimore, MD, 2000). My in-text quotations are all taken from Olmsted's report.

3 I am particularly indebted in what follows to Hank Johnston, *The Yosemite Grant 1864–1906: A Pictorial History* (Yosemite, 1995).

4 A useful guide to the variety of French sites is provided by Catherine and Bernard Desjeux, *Les Parcs Naturels Régionaux de France* (Nonette, 1984).

5 David Matless, *Landscape and Englishness* (London, 1998), p. 249. I am indebted to this study in what follows.

6 Some years ago in Kirbymoorside, Yorkshire, a wonderful signpost announced 'to the surprise view', which must have spoilt the subsequent view over the moorland!

7 Robert Woof in *The Lake District: A Sort of National Property*, ed. John Murdoch (Cheltenham and London, 1986).

8 The 1748 text of the Stowe *Dialogue* is available in the Augustan Reprint Society, publication number 176 (1976); I quote here from the fourth edition (1800) of the Wye *Observations*.

9 Here I have used two books: Harold M. Abrahams, ed., *English National Parks* (London, 1959), where I have drawn upon Lord Birkett's lyrical and enthusiastic essay on the Lake District, and on Murdoch, ed., *The Lake District: A Sort of National Property*. On the Picturesque literature devoted to the Lakes, see the bibliographical study by Peter Bicknell, *The Picturesque Scenery of the Lake District 1752–1855* (Winchester, 1990).

10 Brian Redhead, *Inspiration of Landscape: Artists in National Parks* (Oxford, 1989). The exploration of French landscapes in the nineteenth century also opened them up to tourists – first by paintings displayed in the *salons*. See Nicholas Green, *The Spectacle of Nature: Landscape and Bourgeois Culture in Nineteenth-century France* (Manchester, 1990).

11 John E. Findling, ed., *Historical Dictionary of World's Fairs ad Expositions* (New York, 1990); Robert W. Rydell, John E. Findling and Kimberly D. Pelle, *Fair America: World's Fairs in the United States* (Washington, DC, and London, 2000); Cristina Della Coletta, *World's Fairs Italian Style: The Great Exhibitions in Turin and their Narratives, 1860–1915* (Toronto, 2006); Burton Benedict et al., *The Anthropology of World's Fairs: San Francisco's Panama Pacific International Exposition of 1915* (Berkeley, CA, 1983); Erik Mattie, *World's Fairs* (New York, 1998); Paul Young, *Globalization and the Great Exhibition: The Victorian New World Order* (Basingstoke, 2009); Matthew Rader, 'International Expositions Sites in the United States, 1850–1975: National Historic Landmarks Survey Theme Study', n.d., on file at the US National Parks Service.

12 A reproduction of contemporary newspaper articles reveals the scope of this material: *Le Livre des expositions universelles, 1851–1989* (Paris, 1983). And a trawl through collections of photographs in the United States Library of Congress, for example, reveals how rich this unexplored material is.

13 The bookseller Charles Wood issued one of his detailed and illustrated cataloues on national and international

Fairs and Expositions, no. 144 (2010), drawing also on Kenneth E. Carpenter's article, 'European Industrial Exhibitions before 1851 and their publications', *Technology and Culture* (1972).

14 I draw upon Hermione Hobhouse, *The Crystal Palace and the Great Exhibition* (London, 2002), Patrick Beaver, *The Crystal Palace, 1851–1936: A Portrait of Victorian Enterprise* (London, 1970), and for the later site, J. R. Piggott, *Palace of the People: The Crystal Palace at Sydenham, 1854–1936* (London, 2004), who does, appropriately, give the park a full chapter.

15 Illustrated by Hobhouse, *The Crystal Palace and the Great Exhibition*, plate XIV; see also Brent Elliott, *Victorian Gardens* (London, 1986), pp. 140–43.

16 Many of these are collected in the photographic album *The City of Palaces* (Chicago, IL, 1894). See also Rossiter Johnson, ed., *A History of the World's Columbian Exposition*, 4 vols (New York, 1897–8).

17 See Andrew Theokas, *Grounds for Review: The Garden Festival in Urban Planning and Design* (Chicago, IL, 2009).

17 Japanese Gardens and their Legacy to the West

1 Here alphabetically are the authorities I have consulted. Augustine Berque, *Japan: Nature, Artifice and Japanese Culture*, trans. from the French by Ross Schwartz (Northamptonshire, 1997); Mitchell Bring and Josse Wayembergh, *Japanese Gardens: Design and Meaning* (New York, 1981); Bruce A. Coats, 'In a Japanese Garden', *National Geographic*, CLXXVI/5 (1989); Loraine Kuck, *The World of the Japanese Garden: From Chinese Origins to Modern Landscape Art* (New York and Tokyo, 1968); Wybe Kuitert, *Themes, Scenes and Taste in the History of Japanese Garden Art* (Amsterdam, 1988); Sachimine Masui and Beatrice Testini, *San Sen Sou Moku. Il giardino giapponese nella tradizione e nel mondo contemporaneo* (Padua, 2007) – this is an excellent work by a former Penn student, not yet available in English; Gunter Nitschke, *Japanese Gardens* (Cologne, 1993); Jiro Takei and Marc P. Keane, *Sakuteiki. Visions of the Japanese Garden: A Modern Translation of Japan's Gardening Classic* (Boston, Rutland, VT, and Tokyo, 2001).

2 Harriet Hyman, 'Zen and the Art of Gardening', *Garden Design* (June/July 1994), pp. 53–61.

3 *The Japanese Garden: Islands of Serenity, Photographs by Haruzo Ohashi* (Tokyo, 1986). The photographs in Gunter Nitschke's volume are also immensely valuable.

4 I am quoting here from the translation of *Sakuteiki*, and its commentary by Marc Keane, to which I am much indebted. It contains a detailed glossary, and commentary on the various elements of the treatise. References in my text are to this edition. There are perhaps several authors involved, besides Tachibana no Toshitsuna, and occasionally contradictions occur and verbatim remarks get repeated (see p. 192n).

5 This is discussed and the sites illustrated by Wybe Kuitert, 'Two Early Japanese Gardens', in *The Authentic Garden*, ed. L. Tjon Sie Fat and E. de Jong (Leiden, 1991).

6 The best introduction to this complexity and its individual moments comes in Marc Treib and Ron Herman, *A Guide to the Gardens of Kyoto* (Japan, 5th printing, 1993), as much for the armchair traveller as the Japanese visitor. See also his 'Making the Edo Garden', *Landscape*, XXIV/1 (1980), pp. 24–9.

7 I rely here upon the work of Norris Brock Johnson, notably his 'Mountain, Temple, and the Design of Movement: 13th-century Japanese Zen Buddhist Landscapes', in *Landscape Design and the Experience of Motion*, ed. Michel Conan (Washington, DC, 2003).

8 I refer here to Robert Smithson's essay on Central Park in *The Collected Writings*, ed. Jack Flam (Berkeley, CA, 1996), p. 162. His own understanding of the famous 'Spiral Jetty' is itself premised on his discovery of crystalline forms in the adjacent topography of the lake. Another interesting approach to this issue is offered in the photographs and commentary of Pat Murphy and William Neill, *By Nature's Design* (San Francisco, CA, 1993).

9 Marc Treib, *Noguchi in Paris: The UNESCO Garden* (San Francisco and Paris, 2003).

10 Quoted in Udo Weilacher, *Between Landscape Architecture and Land Art* (Berlin and Boston, MA, 1999), p. 206. The range of readings at Ryoanji are from Bruce Coats's article; other tentative stabs at such meanings are essayed by Treib, *Noguchi in Paris*.

11 I am indebted to the collections of Newsom's Kyoto and Tokyo Study Books in the Graduate School of Design Special Collections, Harvard University, and to the brochure by Randall D. Bird for an exhibition there in 1996.

12 The remark is made by a Chinese character in Henning Mankell's *The Man from Beijing*, but the point seems useful here too: another character also advises that China is no more secretive than other countries, but it takes time and patience to master its culture.

18 Arts and Crafts Gardens: The Artist Back in the Garden

1 This is admirably described by David Ottewill, *The Edwardian Garden* (New Haven, CT, 1989), to which I am

indebted. See also Robert Williams, 'Edwin Lutyens and the Formal Garden in England', *Die Gartenkunst*, VII/2 (1995), pp. 201–9.

2 This is not the occasion for a review of the work of Lutyens or Jekyll, both have been well studied: see L. Weaver, *Houses and Gardens by E. L. Lutyens* (London, 1913), Jane Brown, *Gardens of a Golden Afternoon – The Story of a Partnership: Edwin Lutyens and Gertrude Jekyll* (New York, 1982), and Judith B. Tankard, *Gardens of the Arts and Crafts Movement* (New York, 2004). On Thomas and Peto there is nothing specifically on their work other than the good comments by Ottewill, *The Edwardian Garden*, and Williams, 'Edwin Lutyens and the Formal Garden in England'. A collection of *Country Life* photographs edited by Tim Richardson, *English Gardens in the Twentieth Century* (London, 2005) shows off these gardens to great advantage.

3 See what Judith Tankard calls 'halfmoon steps' at Folly Farm (*Gardens of the Arts and Crafts Movement*, p. 129), as well as other convex stairs that half-use or allude to Serlio's diagram of the Vatican Belvedere (see chapter 6, above). See also my illus. 218, where Thomas uses the same format. They were frequently used in early twentieth-century gardens around Philadelphia.

4 See Anne Helmreich, *The English Garden and National Identity: The Competing Styles of Garden Design, 1870–1914* (Cambridge, 2002), and David Matless, *Landscape and Englishness* (London, 1998).

5 I owe this moment of Englishness and its assimilation of old and new to Robert Williams, 'Edwardian Gardens, Old and New', *Journal of Garden History*, XIII (1993), pp. 90–103, not least for an understanding of the cultural and political climate of the years 1890–1910.

6 See Eric Hobsbawm and Terence Ranger, eds, *The Invention of Tradition* (Cambridge, 1983), especially the former's introduction, 'Inventing Tradition'.

7 See Benedetta Origo, Morna Livingston, Laurie Olin and John Dixon Hunt, *La Foce: A Garden and Landscape in Tuscany* (Philadelphia, PA, 2001). With further bibliography.

8 For Platt and his own book on *Italian Gardens* (New York, 1894), see *Shaping an American Landscape: The Art and Architecture of Charles A. Platt*, ed. Keith Morgan et al. (Hanover, NH, 1995). There have been several collections of writings of this phase of American gardening: see *The Once and Future Gardener*, ed. Virginia Tuttle Clayton (Boston, MA, 2000). Probably every local designer with some well-to-do clients wrote about foreign styles and forms. F.W.G. Peck, a Philadlephia designer (1910–1998), wrote pieces on French, English, Spanish and Italian gardens 'for American estates' in *Country Life*, vol. 65 and 66 (1934).

9 See Francis Duncan, 'The Gardens of Cornish', *The Century Magazine* (May 1906).

10 There have been several magazine pieces about this: the latest is by James Regina, 'Bunny Mellon's Secret Garden', *Vanity Fair* (August 2010), but two better pieces came earlier in the British weekly magazine *Country Life*: 'A Garden Furnished with Books', CXCI/51–2 (18/25 December 1997), and 'To Walk in a Vision of Economy and Grace', CXCII/1 (1 January 1998), both written by Christopher Ridgway.

11 Detailed discussion, plans and photographs are available in the report prepared by Emily T. Cooperman, for the Cultural Landscape Inventory (2000).

12 Quoted by Tankard, *Gardens of the Arts and Crafts Movement*, who devotes the last section of her book to American work.

13 See the collection of essays on different gardens in New England by Alan Emmet, *So Fine a Prospect* (Hanover, NH, 1996).

14 An extended analysis of their Culbertson garden in Pasadena gives a good sense of their range of reference and, craftsmanship, see Henry Hawley, 'An Italianate Garden by Greene and Greene', *Journal of Decorative and Propaganda Arts* (Summer/Fall 1986), pp. 32–45. See also Tankard's views of another great garden at Green Gables, Woodside, California.

19 The Prose and Poetry of Modern Landscape Architecture

1 Fletcher Steele, 'Public Splendour and Private Satisfaction', *Landscape Architecture* (January 1941), pp. 69–71. Steele himself was capable of both these prosaic as well as poetic gestures: his famous stairs at Naumkeag in Massachusetts were both very functional – allowing his patron to negotiate a steep slope to her vegetable garden – and a wonderful gesture in itself. That he combined recollections of Italian stairs and grottoes with modern materials – painted concrete and industrial railings – shows how well he understood that a site's 'poetry' can also respond to contemporary requirements. The following distinction between soul and body by Henry V. Hubbard comes from the same issue as Steele's essays.

2 'The Geometry of Landscape', *The Iconography of Landscape*, ed. Denis Cosgrove and Stephen Daniels (Cambridge, 1988), pp. 270–71.

3 *The Landscape Approach of Bernard Lassus* (Philadelphia, PA, 1998), p. 80. For more detailed discussion of these issues see Elizabeth K. Meyers, 'The Post-Earth Day Conundrum:

Translating Environmental Values into Landscape Design', in *Environmentalism in Landscape Architecture*, ed. Michel Conan (Washington, DC, 2000).

4 Raffaella Fabiano Giannetto, *Paolo Bürgi, Landscape Architect: Discovering the (Swiss) Horizon: Mountain, Lake and Forest*, Source Books in Landscape Architecture 5 (New York, 2009), p. 29; I have found this an excellent introduction to places that I know personally and for which Bürgi also provides commentary. See also various essays by Bürgi contributed to the magazine *Topos*, 19 (1997), 36 (2001), 50 (2005), and his essay 'Memory and Imagination: History as a Source of Inspiration', in *Historic Gardens Today*, ed. Michael Rohde and Rainer Schomann (Leipzig, 2004).

5 *The Landscape Approach of Bernard Lassus* contains a series of essays on different games and their relations to his landscape projects: see 'The Game of Red Dots' (pp. 16–20) and 'The Garden Game' (pp. 35–9), and also Lassus's book *Jeux* (Paris, 1977).

6 His project was not accepted, but was published in a pamphlet by the Coracle Press in London, *Le Jardin des Tuileries de Bernard Lassus* (1991), with short commentaries by eleven contributors. The site as it came to be laid out is a wholly uninteresting facsimile of tired French garden items.

7 Paolo Bürgi, *Feldstudien. Zur neuen Ästhetik urbaner Landwirtschaft / Field Studies: The New Aesthetics of Urban Agriculture* (Basel, 2010), especially pp. 96–101 for Bürgi's essays, to which I am indebted. The remark about the ethical nature of his work, quoted in my text, was a personal communication about this work in October 2010. Very interestingly, Bürgi also responds to a not dissimilar appeal from Bernard Lassus to make some new sense of the no-man's land between urban development and rural land (see *The Landscape Approach*, pp. 65ff.)

8 See the booklet by various authors and edited by Georges Descombes, *Voie suisse l'itinéraire genevois. De Morschan à Brunnen* (Canton of Geneva, 1991), and his own account of it, 'Shifting Sites: The Swiss Way, Geneva', in *Recovering Landscapes*, ed. James Corner (New York, 1999), pp. 81–5.

9 In his meditations *Between Landscape Architecture and Land Art* Udo Weilacher quotes Ian Hamilton Finlay's remark that landscape architecture is 'not at all lyrical' and this has nothing to do with inscriptions but with the relation of 'things to each other in a certain kind of way' (p. 102). Weilacher also comments upon the different kinds of 'poetry' that he finds, particularly in the designs of Dieter Kienst (see Weilacher 's pp. 150 and 152 for the Garden of the Mathematician).

10 The invigorating and innovative treatment of this topic is set out in Alan Berger, *Drosscape: Wasting Land in Urban America* (New York, 2006); the quotation in the next paragraph is by Lars Lerup, p. 242. Robert Smithson, *The Collected Writings*, ed. Jack Flam (Berkeley, CA, 1996), pp. 68–74 for the Passaic site.

11 The gardens were designed by Christine and Michel Péna: see their *Pour une Troisième Nature / For a Third Nature* (Paris, 2010).

12 See *Ken Smith: Landscape Architect*, with an introduction by John Beardsley (New York, 2009), and for the next paragraph, *The Vanguard Landscapes and Gardens of Martha Schwartz*, ed. Tim Richardson (London, 2004).

20 The Once and Future Garden

1 *The Landscape Approach of Bernard Lassus*, p. 50. I am returning here to some of the themes I explored in my book *Greater Perfection: The Practice of Garden Theory* (London and Philadelphia, PA, 2000).

2 See André Vera, *Le Nouveau Jardin* (Paris, 2009). For Repton's Red Books, see André Rogger, *Landscapes of Taste: The Art of Humphry Repton's Red Books* (London, 2007).

3 Quoted in Marc Treib and Dorothée Imbert, *Garrett Eckbo: Modern Landscape Architecture for Living* (Berkeley, CA, 1997), p. 94.

4 In *Learning from Duisburg Nord*, p. 66. In what follows I refer to the very influential essay by Rosalind Krauss, 'Sculpture in the Expanded Field', *October*, 8 (1979).

5 There is a more favourable piece on the Irwin garden by Lawrence Weschler, 'When Fountains Collide', in *The New Yorker* (8 December 1997) and in his book, *Robert Irwin Getty Garden* (Los Angeles, CA, 2002). There is also David Marshall, 'Gardens and the Death of Art: Robert Irwin's Getty Garden', *Studies in the History of Gardens and Designed Landscapes*, XXIV (2004), pp. 215-28.

6 See the interviewers conducted by Udo Weilacher in *Between Landscape Architecture and Land Art* (Basel, 1996).

7 Ingrid and Konrad Scheurmann, *Homage to Walter Benjamin: 'Passages', Places of Remembrance at Portbou* (Mainz, 1995); title and text also in German. On the mania for memorials more generally, see Erika Doss, *Memorial Mania: Public Feeling in America* (Chicago, IL, 2010).

8 The essays in Charles Waldheim, ed., *The Landscape Urbanism Reader* (New York, 2006) endeavour to explain this new disciplinary realignment, though its debts to earlier thinking seem sometimes to be occluded, presumably to bolster its own new and 'emerging' position. See Alan Berger, *Drosscape: Wasting Land in Urban America* (New York, 2006).

9 M. Jeffrey Hardwick, *Mall Maker: Victor Gruen, Architect of*

an American Dream (Philadelphia, PA, 2004). Others have also lent their skill to mall-making, including Lawrence Halprin and Martha Schwartz

10 Something very noticeable in England before the 'Beeching' cuts decimated the British railway system in the 1960s. See also Charlotte Lagerberg Fogelberg and Frederik Fogelberg, '100 Years of Gardening for the Public Service – The Horticultural Heritage of Swedish State Railways', *Studies in the History of Gardens and Designed Landscapes*, XXXI, forthcoming.

11 On Crazannes, see Michel Conan, *The Crazannes Quarries by Bernard Lassus: An Essay Analysing the Creation of a Landscape* (Washington, DC, 2004), and on the Nîmes rest area, my *Greater Perfections*, pp. 227–8. On the new interest in autoroutes there are the annual awards to different aspects of the *paysages routiers* by the Ministry of Transport (the Ruban d'Or, and other prizes). See also Christian Leyrit and Bernard Lassus, eds, *Autoroutes et Paysages* (Paris, 1994).

12 *Learning from Duisburg Nord*, p. 29. On the 'modernity' of the Picturesque see my essay, entitled 'John Ruskin, Claude Lorrain, Robert Smithson, Christopher Tunnard, Nikolaus Pevsner, and Yve-Alain Bois walked into a bar . . .', in *The Hopkins Review* (2011).

13 Sutherland Lyall, *Designing the New Landscape*, p. 25. This site is illustrated and discussed by Alison Hirsch, 'Lawrence Halprin's Public Spaces: Design, Experience and Recovery. Three Case Studies', *Studies in the History of Gardens and Designed Landscapes*, XXVI (2006).

14 Anita Berrizbeitia, ed., *Reconstructing Urban Landscapes: Michael Van Valkenburgh Associates* (New Haven, CT, 2009).

15 J. B. Jackson's essay, first published in *Landscape*, XXIII/1 (1987) was republished in *The Necessity for Ruins and other Topics* (Amherst, MA, 1980).

16 *Learning from Duisburg Nord*, p. 30.

17 See Hirsch, 'Lawrence Halprin's Public Spaces'.

18 I rely here upon the extraordinary essay by Yve-Alain Bois, 'A picturesque stroll around *Clara Clara*', *October*, 29 (1984), pp. 32–62.

19 *Recovering Landscape*, ed. James Corner (New York, 1999).

ACKNOWLEDGEMENTS & PHOTO ACKNOWLEDGEMENTS

This collection of essays would also not have been possible without, on the one hand, years of discussing with students and other interested audiences the why and the wherefore of garden-making, and also without the help and advice of friends: Patrick Bowe, Paolo Bürgi, Alberta Campitelli, Stanislaus Fung, Alison Hardie, Ron Henderson, Bernard Lassus, David Leatherbarrow, Sachimine Masui, Laurie Olin, Anatale Tchkine and James Wescoat.

The author and publishers wish to express their thanks to the below sources of illustrative material and/or permission to reproduce it. In some cases locations of items are also given.

Photos courtesy of the Academy of Natural Sciences, Philadelphia, PA: 201, 203; photo airunp 5; from Rito Akisato, *Miyako Rinsen meisho zue* ['Illustrated Manual of Celebrated Gardens in the Capital'] (Kyoto, 1799): 210; from Giovanni Battista Aleotti, *Gli artifitiosi et curiosi moti spiritali di Herrone* (Ferrara, 1589): 81; from Alphonse Alphand, *Les Promenades de Paris. Histoire – Description des embellissements – dépenses de création et d'entretiens des Bois de Boulogne et de Vincennes – Champs-Elysées – Parcs – Squares – Boulevards – Places plantées. Etude sur l'art des jardins et arboretum* (Paris, 1867–73): 181, 184; Art Institute of Chicago (gift of Mrs Richard E. Damielson and Mrs Chauncey McCormick): 58; photos by or courtesy of the author: 1, 2, 7, 19, 37, 38, 39, 66, 83, 84, 85, 90, 94, 107, 114, 129, 133, 144, 145, 172, 173, 174, 175, 188, 189, 191, 205, 206, 207, 216, 219, 227, 230, 231, 232, 234, 245, 250; postcard in the author's collection: 185; from Jacques Aymard, *Essai sur les chasses romaines, des origines à la fin du siècle des Antonins-Cynegetica* (Paris, 1951): 14; Bayerische Verwaltung der staatlichen Schlösser, Munich: 182; from Wilhelm Gottlieb Becker, *Das Seifersdorfer Thal* (Leipzig, 1792): 170; Biblioteca Reale, Turin (*Codice Torinese Saluzziano* 148): 17; Bibliothèque Municipale, Metz: 56 (MS 1486); Bibliothèque Nationale de France, Paris: 61 (MS 2810), 70 (MS Arsenal 5064); Bibliothèque Royale Albert 1er, Brussels: 67 (MS KBR 10308); from George Bickham, *The Beauties of Stow: or a Description of the Most Noble House, Gardens & Magnificent Buildings Therein . . .* (London, *c.* 1753): 169; drawing courtesy of Gregg Bleam: 138; Bodleian Library, Oxford 103 (Ashmole MS 1461), 104 (B.1.17 Med.), 116 (Gough Collection, MSGD a4), 118 (MS Gough drawings a.4), 121, 122, 123, 124 (all MS Top. Gen. D.14); British Library, London (photo © The British Library Board): 45 (Add. MS 18113), 50 (IOL 169B (3).), 57 (MS 18720), 59 (Add. MS 18855, folio 108), 60

(Add. MS 19720), 64 (Add. MS 19720), 65 (Harley MS 4425); British Museum, London (photos © The Trustees of the British Museum): 13, 196 (Department of Prints and Drawings); Buckinghamshire County Museum, Aylesbury: 141; from Robert Castell, *The Villas of the Ancients Illustrated* (London, 1728): 33; from Salomon de Caus, *Les Raisons des forces mouvantes, avec diverses machines tant utilles que plaisantes: aus quelles sont adioints plusieurs desseings de grotes et fontaines* . . . (Frankfurt, 1615): 115; Château d'Heudicourt, Eure, Normandy: 146; photo Laurent Châtel: 130; Trustees of the Chatsworth Settlement: 251, 252; thanks to Chen Congzhou: 155; from *The City of Palaces: Picturesque World's Fair* (Chicago, 1894): 202; Commanderie des Templiers de Coulommiers: 69; photos Emily T. Cooperman: 4, 101, 131, 132, 143, 149, 150, 157, 162, 168, 176, 177, 186, 190, 221, 222, 226, 233, 235, 236, 246, 247, 249; Country Life Archives: 53; from George Cumberland, *An Attempt to Describe Hafod, and the neighbouring scenes* . . . (London, 1796): 194; photo Jennifer Current: 248; from Marc'Antonio Dal Re, *Ville di Delizia o siano* PALAGI CAMPAREGGI NELLO STATO DI MILANO . . . (Milan, 1726): 109; Detroit Institute of Arts (photo The Bridgeman Art Library): 208; from Harold Donaldson Eberlein and Cartwright Van Dyke Hubbard, *The Practical Book of Garden Structure and Design* (Philadelphia, 1937): 229; photos Esto Photographics Ltd: 237, 238, 239; from John Evelyn, 'Elysium Britannicum, or the Royal Gardens' (formerly at Christ Church, Oxford, now at the British Library, London [Evelyn Papers MS 45]): 99; from Giovanni Battista Falda, *Le Fontane di Roma Nelle Piazze, e Luoghi Publici Della Citta, con il Loro Prospetti, come sono al Presente* (Rome, 1675(?)): 82; from Linda Farrar, *Ancient Roman Gardens* (Stroud, 1998): 29; The Fitzwilliam Museum, University of Cambridge: 23 (photo Fitzwilliam Museum, University of Cambridge / The Bridgeman Art Library); photo Martin Goalen: 31; after Marie-Luise Gothein, *Geschichte der Gartenkunst*, I: *Von Ägypten bis zur Renaissance in Italien, Spanien und Portugal* (Jena, 1914): 40; Harvard University Graduate School of Design, Cambridge, Mass.: 105, 209 (Frances Loeb Library, Special Collections); photo courtesy of David Grandorge: 179; Graphische Sammlung Albertina, Vienna: 75, 80, 86, 87, 89, 112; from Johann Gottfried Grohmann & Friedrich Gotthelf Baumgärtner, *Ideen-magazin für Liebhaber von Gärten, Englischen Anlagen und für Besitzer von Landgütern um Gärten und ländliche Gegenden* . . . (Leipzig, 1796–1806): 164; photo © Philip Halling: 140; photo courtesy of Ron Henderson: 161; from Giovanni Filippo Ingrassia, *Informazione del pestiforo e contaggioso morbo il quale affigge et have afflicto la citta di Palermo . . . nell' anno 1575 e 1576* (Palermo, 1576): 35; from 'J.W. Gent' [John Worlidge], *Systema Agriculturae; the mystery of husbandry discovered: treating of the several new and most advantagious ways of tilling, planting, sowing, manuring, ordering, improving of all sorts of gardens, orchards, meadows, pastures, corn-lands, woods & coppices . . .* (London, 1681): 126; renderings created by James Corner Field Operations (use courtesy of the City of New York): 137; photo Stanley Jashemski: 22; from Gertrude Jekyll and Lawrence Weaver, *Gardens for Small Country Houses* (London, 1912): 215; © Office of Dan Kiley: 138, 148; from Jan Kip and Leonard Knyff, *Britannia Illustrata or views of several of the Queen's Palaces as also of the principal Seats of the Nobility and Gentry of Great Britain, curiously engraven* . . . (London, 1707): 93, 119; Kupferstichkabinett, Berlin: 72; photos Marilena La China: 34, 163; from Comte Louis-Joseph-Alexandre de Laborde, *Description des nouveaux jardins de la France et de ses anciens chateaux, mêlée d'observations sur la vie de la campagne et la composition des jardins* (Paris, 1808): 167; from Comte Louis-Joseph-Alexandre de Laborde, *Voyage Pittoresque et historique de l'Espagne par le Comte de Laborde*, III (Paris, 1812): 36; photo Bernard Lassus: 95; from Pierre Le Lorrain, Abbe de Vallemont, *Curiositez de la nature et de l'art sur la végétation ou l'agriculture et le jardinage dans leur perfection* . . . (Paris, 1705): 92; photos Michael Leaman / Reaktion Books: 48, 49, 52, 55, 223; Leiden University Library: 100; Library of Congress, Washington, DC (Department of Prints and Photographs): 9, 21; Library of Congress, Washington, DC (General Collections): 106; from David Loggan, *Oxonia Illustrata* . . . (Oxford, 1675): 98; photo LuxTonnerre: 183; courtesy of the McLean Library, Pennsylvania Horticultural Society: 203; photo Marianne Majerus: 117; from Francesco di Giorgio Martini, *Architettura ingegneria e arte militare* . . . (*Codice Torinese Saluzziano* 148, Biblioteca Reale, Turin): 17; thanks to Sachimine Masui: 212; Musée des Beaux-Arts, Brussels: 78; Musée National du Château de Versailles et du Trianon, Versailles: 16; Museo di Roma: 74; Museo Storico Topografico, Florence (photos courtesy Musei Civici, Florence): 18, 88; Museum of Fine Arts, Boston: 8 (gift of Martha C. Karolik for the M. and M. Karolik Collection of American Paintings); National Galleries of Scotland, Edinburgh: 32; National Palace Museum, Taipei: 159, 160; from Timothy Nourse, *Campania Fœlix, or, A discourse of the benefits and improvements of husbandry: containing directions for all manner of tillage, pasturage, and plantation* . . . (London, 1700): 127; thanks to Haruzo Ohashi: 204; photo courtesy Olin Studio, Philadelphia: 147; from Frederic Law Olmsted, 'Landscape', in *American Florist* (15 January 1896): 203; Osaka Municipal Museum of Art (Abe Collection): 152; Parc de Bercy, Paris: 187; from John Parkinson, *Theatrum Botanicum: The Theater of Plants. Or, An Herball of a Large Extent* . . . (London, 1640): 104; Pennsylvania Horticultural Society, Philadelphia, PA: 201; from Victor Petit, *Habitations Champetres: Recueil de Maisons, Villas, Chalets, Pavillons, Kiosques, Parcs et Jardins, dessinées par Victor Petit* (Paris, n.d. [*c.* 1855]): 166; courtesy Philadelphia Print Shop: 200, 201;

from Girolamo Porro, *L'Horto dei semplici di Padova* (Venice, 1591): 96; reconstruction by E. G. Price from Eddie Price, *A Romano-British Settlement: Its Antecedents and Successors* (Gloucester and District Archaeological Research Group, Stonehouse, Gloucestershire, 2000), I: 24; private collections: 110, 113, 128 (reproduced with permission of Charles Cottrell-Dormer), 136 (photo courtesy of Hobhouse Ltd, London), 139; from J. B. Pyne, *Lake Scenery of England* (London, 1859): 195; Real Biblioteca del Monasteria de San Lorenzo del Escorial: 79; Record Office, Carlisle: 142; Chris Reed: 233; from *Roma antica e moderna* (Rome, 1750): 71; after D. Fairchild Ruggles, *Islamic Gardens and Landscapes*: 42; photo D. Fairchild Ruggles: 51; photo Nicolas Sapieha: 25; photo courtesy of Martha Schwartz Partners: 240; from Sebastiano Serlio, *Tutte l'Opere d'architettura e prospettiva . . .*, Book III [1540] (Venice, 1619): 77; from A. A. Cooper, 3rd Earl of Shaftesbury, *Characteristicks of Men, Manners, Opinions, Times* (London, 1714): 125; photos Leanda Shrimpton: 217, 220, 224; Osvald Sirén, *Gardens of China* (New York, 1950): 156; courtesy Ken Smith Landscape Architect: 237, 238, 239; photo Mark Edward Smith: 30; photo Lady Beatrix Stanley: 53; Stourhead House, Wiltshire: 12; Sutro Library, San Francisco: 102; from Stephen Switzer, *Ichnographia Rustica; or, The nobleman, gentleman, and gardener's recreation . . .* (London, 1718): 120; from Pompilio Totti, *Ritratto di Roma antica nel qvale sono figvrati i principali tempij, teatri, Anfiteatri, Cerchi, Naumachie, Archi Trionfali . . .* (Rome, 1627): 26; Tate, London (photo © Tate, London 2011): 11; from the *Third Annual Report of the Board of Commissioners of the Central Park* (New York, 1860): 180; from Gabriel Thouin, *Plans raisonnées de toutes espèces de jardins* (Paris, 1828): 165; photo Marc Treib: 211; Galleria degli Uffizi, Florence (Gabinetto Disegni e Stampe): 73, 111 (photo Scala, Florence/Art Resource); UNESCO Archives: 213, 214; University of California, Berkeley Environmental Design Archives: 225; University of Pennsylvania, Philadelphia, Architectural Archives: 3 (Lawrence Halprin Archives), 228; from Anna Hagopian van Buren, 'Reality and Literary Romance in the Park of Hesdin', in Elizabeth B. MacDougall, ed., *Mediaeval Gardens*, Dumbarton Oaks Colloquium on the History of Landscape IX (Washington, DC, 1986): 15; Victoria and Albert Museum, London: 20, 44, 46, 134, 135, 197, 198; courtesy Udo Weilacher: 166; courtesy Udo & Rita Weilacher: 178; photo James Wescoat Jr: 47; from J. D. Whitney, *The Yosemite Guide-Book: A Description of the Yosemite Valley and the Adjacent Region of the Sierra Nevada . . .* (Cambridge, MA, 1871): 192; photo Dianna Wojciechowski: 243; from Thomas Wright, *Six Original Designs of Arbors* (Book I of *Universal Architecture*) (London, 1755): 163; after Luigi Zangher, *Il Giardino Islamico* (Florence, 2006): 41.

INDEX